Lecture Notes in Control and Information Sciences

Edited by A. V. Balakrishnan and M. Thoma

For further listing of published volumes please turn over to inside of back cover.

Lecture Notes in Control and Information Sciences
Edited by A.V. Balakrishnan and M. Thoma

IFIP

25

Stochastic Differential Systems

Filtering and Control

Proceedings of the IFIP-WG 7/1 Working Conference
Vilnius, Lithuania, USSR, Aug. 28–Sept. 2, 1978

Edited by B. Grigelionis

Springer-Verlag
Berlin Heidelberg New York 1980

Editor
Bronius Grigelionis

Organized by:
Institute of Mathematics and Cybernetics
of the Lithuanian Academy of Sciences and
the Steklov Mathematical Institute of
the Academy of Sciences of the USSR

ISBN 978-3-540-10498-8 ISBN 978-3-540-38503-5 (eBook)
DOI 10.1007/978-3-540-38503-5

Originally published by Springer-Verlag Berlin Heidelberg New York in 1980.
2061/3020-543210

PREFACE

The Conference on Stochastic Differential Systems was held in Vilnius, August 28 - September 2, 1978. It was organized by the Institute of Mathematics and Cybernetics of the Lithuanian Academy of Sciences and the Steklov Mathematical Institute of the Academy of Sciences of the USSR sponsored by the International Federation for Information Processing, W.G.7.1.

A wide field of problems connected with Itô stochastic differential equations and its applications to control and filtering of stochastic differential systems was discussed.

The 103 participants of the Conference represented the USSR(87), GDR (3), USA (3), Japan (2), Hungary (2), Bulgaria (1), France (1), FRG (1), India (1), Romania (1) and United Kingdom (1). The Soviet participants were from Moscow (32), Vilnius (15), Kiev (10), Donetsk (6), Leningrad (5), Tbilisi (4) and other cities.

There were 10 sessions all in all in which 21 50-minute lectures and 24 25-minute talks were presented. This volume contains a major part of the texts of these lectures.

We should like to take this opportunity to thank other members of the Organizing Committee, lecturers and participants for their contributions to the success of the Conference.

Bronius Grigelionis

CONTRIBUTORS

S. V. Anulova
Moskvoskii
Ekonomiko-Statistitscheskii Institute
B. Savvinovski Per., 14
Moscow 119435 - USSR

B. I. Arkin
Moskva. ZEMI AN USSR
Leninskii Prospekt 62/I, Kw 433
Moscow 117296 - USSR

A. V. Balakrishnan
Sytems Science Department
University of Los Angeles
Los Angeles 90042 - USA

Ya. I. Belopolskaya
Kiev - USSR

C. Bromley
University of Minnesota
Minneapolis - USA

R. J. Cameron
Department of Computing and Control
Imperial College of Science and Technology
London SW 7 2BZ - UK

J. M. C. Clark
Department of Computing and Control
Imperial College of Science and Technology
London SW 7 2BZ - UK

M. Cranston
University of Minnesota
Minneapolis - USA

Yu. L. Dalecky

O. B. Enchev
Institute of Mathematics
Bulgarian Academy of Sciences
Sofia 1000 - Bulgarian

H. J. Engelbert
University of Jena
Jena - GDR

H. J. Fischer
Kiev State University
Kiev - USSR

L. I. Galtchouk
Department of Mathematic and Mechanic
Moscow State University
Moscow 117234 - USSR

V. L. Girko

O. A. Glonti
Institute of Economics and Law
Academy of Sciences of the Georgian SSR
Makharadze Street 14
Tbilisi 380 007 - USSR

B. Grigelionis
Institute of Mathematics and Cybernetics
Academy of Sciences of Lithuanian SSR
University of Vilnius
Vilnius - USSR

R. Z. Hasminskii,
Institute of Information Transmission
Steklov Institute of Mathematics
Leningrad Branch
Leningrad - USSR

J. Heß
University of Jena
Jena - GDR

I. A. Ibragimov
Institute of Information Transmission
Steklov Institute of Mathematics
Leningrad Branch
Leningrad - USSR

K. Itô
Research Institute for Mathematical Sciences
Kyoto University
Kyoto 606 - Japan

V. V. Jurinskii
Siberian Branch of the USSR Academy of Sciences
Institute of Mathematics
Novosibirsk 90 - USSR

Yu. M Kabanov
Moscow

G. Killianpur
University of Minnesota
Minneapolis - USA

G. L. Kulinič
Kiev State University
Vladimirskaja 64
Kiev 252017 - USSR

V. A. Lebedev
Lomonosov State University
Department of Mechanics and Mathematics
Moscow 117 234 - USSR

R. Sh. Liptser
Moscow

V. Mackevičius
Faculty of Mathematics
University of Vilnius
Vilnius 232006 - USSR

R. Mikulevicius
Institute of Mathematics and Cybernetics
Academy of Sciences of Lithuanian SSR
K. Pozelos 54
Vilnius 620024 - USSR

S. Ja. Mahno
Donetsk Institute of Applied Mathematics and Statistics
Academy of Sciences of Ukranian SSR
Universitetskaya Street 77
Donetsk 340048 - USSR

R. Morkvénas
Institute of Mathematics and Cybernetics
Academy of Sciences of the Lithuanian SSR
K. Požélos 54
Vilnius 620024 - USSR

A. A. Novikov
Steclov Mathematical Institute
Academy of USSR
Vavilov 42
Moscow 117966 - USSR

S. Orey
University of Minnesota
Minneapolis - USA

G. C. Papanicolaou
Courant Institute of Mathematical Sciences
New York University
New York 10006 - USA

E. Platen
Berlin - GDR

H. Pragarauskas
Institute of Mathematics and Cybernetics
Academy of Sciences of the Lithuanian SSR
K. Požėlos 54
Vilnius 620024 - USSR

U. Rösler
University of Minnesota
Minneapolis - USA

H. Rost
Institute for Applied Mathematics
Im Neuenheimer Feld 294
D 69 Heidelberg - FRG

B. L. Rozovsky
Institute Povyschenija Kvalifikazii MHP
Schtscherbakovskaja d. 3
Moscow 105318 - USSR

M. T. Saksonov
Moskva. ZEMI AN USSR
Belovezheskaja 57, Kw 23
Moscow 121353 - USSR

A. N. Shiryayev
Moscow

J. M. Stoyanov
Institute of Mathematics
Bulgarian Academy of Sciences
Sofia 1000 - Bulgarian

A. F. Taraskin
Aviation Institute
Molodogvardejskaja 151
Kuibyshev - USSR

S. R. S. Varadhan
Courant Institute of Mathematical Sciences
New York University
New York 10006 - USA

A. Yu. Veretennikov
Institute of Problems of Control
Moscow 117342 - USSR

D. Vermes

M. I. Višic
MGU, Mech.-Math.
Kafedra Differenzialnich Uravenii
Moscow B-234, 117234 - USSR

A. M. Yaglom
Institute of Atmospheric Physics
Academy of Sciences of the USSR
Moscow - USSR

CONTENTS

SOME ESTIMATION PROBLEMS FOR STOCHASTIC DIFFERENTIAL EQUATIONS

R. Z. Hasminskii, I. A. Ibragimov
Steklov Institute of Math.
Leningrad Branch, USSR

The well-known classification divides the parametric estimation problems of statistics from the nonparametric ones. But certainly any nonparametric statistical problem can be parametrized by introducing of a properly choosen parameter. In fact, an important feature of parametric problems consists in possibility to imbed the parameter set into finite dimensional Euclidean space and use the nice structure of this space. Suppose, one is given a nonparametric problem where the parameter set is a subset of some infinite dimensional metric space. One can find reasonable to use the structure of the space and to treat the problem as a parametric one but with infinite dimensional parameter.

Our main object here is to study in a such way one of the simplest nonparametric problem. Assume we are observing on the interval $0 \le t \le 1$ a random function $X_\varepsilon(t)$ where

$$dX_\varepsilon(t) = S(t)\,dt + \varepsilon\,dw(t) \tag{1.1}$$

Here an unknown parameter S belongs to a subset $\sum$ of the Hilbert space $L_2(0,1)$, w is the standard Wiener process, $\varepsilon > 0$ is a small parameter. Note that the set $\sum$ and ε are known to the statistician (the parameter ε can be estimated without error).

In this paper we consider the problem of estimating, the value $F(S)$ of a given functional $F : \Sigma \to R^1$ at an unknown point S (note that the problem of estimating S is considered in [2], [3]).

We shall denote by $P_S^{(\varepsilon)}(\cdot)$ the probability distribution in the space C(0,1) of continuous functions generated by $X_\varepsilon(t)$. The symbol $E_S^{(\varepsilon)}(\cdot)$ denotes the expectation with respect to $P_S^{(\varepsilon)}$. We agree to let $\|\cdot\|$ and $(\cdot,\cdot)$ designate the norm and the scalar product in the Hilbert space $L_2(0,1)$. The symbol $\|\cdot\|$ will also be used to designate the norm of linear functionals and linear operators in $L_2(0,1)$. Henceforth $U_\delta(S)$ will always denote the ball in $L_2(0,1)$ of radius δ with center in S.

The Kolmogorov's n -th diamater $d_n(\Sigma)$ of the set Σ is defined by

$$d_n(\Sigma) = \inf_{M_n} \sup_{x\in\Sigma} \inf_{y\in M_n} \|x-y\|$$

Where infimum is taken over all n -dimensional linear manifolds $M_n \subset L_2$ (see [5]).

2. We give at first a lower bound for the quality of estimation of differentiable functionals F(S) based on the observations (1.1). By a direction we mean any unit vector v in $L_2(0,1)$. We say that a direction v belongs to the set Σ at the point S if the vectors $S + vt \in \Sigma$ for all sufficiently small $|t|$. Let V(S) denote the set of all directions belonging to the Σ at the point S, and for any Fréchet-differentiable functional F let

$$\|F'(S)\|_V = \sup_{v\in V(S)} |(F'(S),v)| .$$

<u>Theorem 1</u>. Let F(S) be a Fréchet-differentiable functional on $\sum$. Then the inequality

$$\text{(2.1)}\quad \lim_{\delta\to 0}\ \overline{\lim_{\varepsilon\to 0}}\ \sup_{S\in\Sigma\cap U_\delta(S_o)} E_S^{(\varepsilon)}\ell\left(\frac{T_\varepsilon - F(S)}{\varepsilon\|F'(S_o)\|_V}\right) \geq \frac{1}{\sqrt{2\pi}}\int_{-\infty}^{\infty}\ell(x)e^{-\frac{x^2}{2}}\,dx\ .$$

holds for every even function ℓ monotonically nondecreasing on the positive half-line, every estimate $T_\varepsilon(X_\varepsilon) = T_\varepsilon$ of F(S) and all points $S_o \in \sum$.

<u>Proof</u>. Let $v \in V(S_o)$ and therefore $S = S_o + tv \in \sum \cap U\delta(S_o)$ for $|t| \leq \delta$ and all small δ. Let

$$\phi(t) = F(S_o+tv),\quad E^{(t)}(\cdot) = E^{(\varepsilon)}_{S_o+tv}(\cdot)\ .$$

Evidently

$$\sup_{s\in\sum\cap U_\delta(S_o)} E_S^{(\varepsilon)}\ell\left(\frac{T_\varepsilon - F(S)}{\varepsilon\|F'(S_o)\|_V}\right) \geq \sup_{|t|\leq\delta} E^{(t)}\ell\left(\frac{T_\varepsilon - \phi(t)}{\varepsilon|\phi'(0)|}\right).$$

Note that

$$\lim_{\delta\to 0}\ \overline{\lim_{\varepsilon\to 0}}\ \sup_{|t|\leq\delta} E^{(t)}\ell\left(\frac{T_\varepsilon - \phi(t)}{\varepsilon\phi'(0)}\right) \geq \frac{1}{\sqrt{2\pi}}\int_{-\infty}^{\infty}\ell(x)e^{-\frac{x^2}{2}}\,dx$$

and (2.1) is a self-evident consequence of (2.2). The proof of the "parametric" inequality (2.2) can be modeled upon that of Hajek's theorem from [4].

3. The following question arises in connection with Theorem 2.1: when do exist estimates T_ε of F(S) for which the equality sign in (2.1) is achieved. To make things simpler we shall consider as

a normalizing factor $\| F'(S) \|$ instead of $\| F'(S) \|_V$.

Definition. We call an estimate T_ε of the functional F(S) asymptotically efficient (with respect to a quadratic loss function) in Σ if

$$\lim_{\varepsilon \to 0} \sup_{S \in \Sigma} E_S^{(\varepsilon)} \frac{|T_\varepsilon - F(S)|^2}{\varepsilon^2 \| F'(S) \|^2} = 1.$$

Example 3.1. Let F be a bounded linear functional in L_2 such that

$$F(S) = \int_0^1 f(t) S(t)\, dt$$

and let $\Sigma = U_1(0)$. Then the estimate $F_\varepsilon = \int_0^1 f(t) dX_\varepsilon(t)$ is asymptotically efficient in Σ, and the random variables $\varepsilon^{-1}(F_\varepsilon - F(S))$ are normally distributed (with respect to $P_S^{(\varepsilon)}$) with mean 0 and variance $\| f \|^2 = \| F'(S) \|^2$.

Theorem 3.1. Suppose the Kolmogorov diameters $d_n(\Sigma)$ of the set Σ satisfy the condition $d_n(\Sigma) = 0(n^{-\beta})$, $\frac{1}{2k} < \beta \le \frac{1}{2(k-1)}$, $k \ge 1$, is an integer. Suppose the functional F(S) is k times Fréchet-differentiable, where k -th derivative $F^{(k)}(S)$ satisfies a Hölder condition with exponent $\gamma = \frac{1}{2}\beta + 1 - k$. Moreover, let the Hilbert-Schmidt norms of the operators $F^{(j)}(S)$, $j \le k$ be uniformly bounded in the ball $U_1(0)$. Then there exists an estimate F_ε of the functional F(S) which is asymptotically efficient in Σ, $E_S^{(\varepsilon)}(F_\varepsilon - F(S)) = 0(\varepsilon)$ and the difference $\varepsilon^{-1}(F_\varepsilon - F(S))$ is asymptotically normal with parameters 0, $\| F'(S) \|^2$.

The proof will be given only for the simplest case k = 1. In this case $\beta > \frac{1}{2}$ and the Hilbert-Schmidt norm of F'(S) coincides with $\|F'\|$. By Hölder condition there exists a constant B such that

$$\|F'(S_2) - F'(S_1)\| \leq B\|S_2-S_1\|^{\gamma} \tag{3.1}$$

Note that if T_ε is an estimate of S the estimate $F(T_\varepsilon)$ of F(S) will be very far from optimal one even in the case of linear bounded functionals F. In the last case we considered the estimate $F_\varepsilon = F(\dot{X}_\varepsilon)$ (Ex. 3.1) but the estimate of such type can not be defined for nonlinear functionals. To prove our theorem we act in the following way. At first we find a good estimate S_ε for S and then construct the estimate F_ε of F as

$$F_\varepsilon = F(S_\varepsilon) + (F'(S_\varepsilon), \dot{X}_\varepsilon - S_\varepsilon)$$

Of course, the problem remains to define $(F'(S_\varepsilon), \dot{X})$ but it is not so difficult because of linearity.

Suppose for the sake of simplicity that the set $\sum$ is centrally symmetric. In this case the n -th diameter

$$d_n(\textstyle\sum) = \inf_{L_n} \sup_{S\in\sum} \inf_{y\in L_n} \|S-y\|$$

where infimum is taken over all n-dimensional subspaces of the space L_2. Fix some subspace L_n for which

$$\sup_{\sum} \inf_{L_n} \|S-y\| \leq \sqrt{2}\, d_n(\textstyle\sum) ,$$

let $\phi_{1n},\ldots\phi_{nn}$ be an orthonormal basis in L_n and let $\phi_{n+1,n},\ldots$

denotes its orthocompliment. It follows that S can be written in the form

$$S = \sum_{1}^{n} a_{jn} \phi_{jn} + \sum_{n+1}^{\infty} a_{jn} \phi_{jn}, \quad a_{jn} = (S,\phi_{jn}) .$$

Evidently

$$(3.2) \qquad \| S - \sum_{1}^{n} a_{jn} \phi_{jn} \|^2 = \sum_{n+1}^{\infty} a_{jn}^2 \le 2d_n^2 .$$

Let us define the estimate S_n^* of S as

$$S_n^* = \sum_{1}^{n} (\phi_{jn}, \dot{X}_\varepsilon) \phi_{jn}$$

where by definition

$$(3.3) \quad (\phi_{jn}, \dot{X}_\varepsilon) = \int_0^1 \phi_{jn}(t) dX_\varepsilon(t) = a_{jn} + \varepsilon\xi_{jn}, \quad \xi_{jn} = \int_0^1 \phi_{jn} dw .$$

Note that ξ_{jn}, $j = 1,\dots n$ are iid normal variables with $E\xi_{jn} = 0$ $E\xi_{jn}^2 = 1$. We have by (3.2), (3.3) that

$$E_S^{(\varepsilon)} \| S - S_n^* \|^2 = \varepsilon^2 \sum_{1}^{n} E\, \xi_{jn}^2 + \sum_{n+1}^{\infty} a_{jn}^2 \le n\varepsilon^2 + 2d_n^2 .$$

Choose $n = n(\varepsilon)$ so that

$$2d_n^2 < n\varepsilon^2 \le 2d_{n+1}^2$$

and define

$$S_\varepsilon = S_{n(\varepsilon)}^* .$$

Note at first that $n(\varepsilon) = 0\left(\varepsilon^{-\frac{2}{2\beta+1}}\right)$ and

$$(3.4)\qquad E_S^{(\varepsilon)}\| S-S_\varepsilon\|^2 \le 2n\varepsilon^2 = 0\left(\varepsilon^{\frac{4\beta}{2\beta+1}}\right) .$$

Moreover, for any $p > 0$

$$(3.5)\qquad E_S^{(\varepsilon)}\| S-S_\varepsilon\|^p = 0\left(\varepsilon^{\frac{2p\beta}{2\beta+1}}\right) .$$

Indeed,

$$E_S^{(\varepsilon)}\| S-S_\varepsilon\|^p \le \left(E_S^{(\varepsilon)}\| S-S_\varepsilon\|^2\right)^{p/2} = 0\left(\varepsilon^{\frac{2p\beta}{2\beta+1}}\right),\ p \le 2$$

and

$$E_S^{(\varepsilon)}\| S-S_\varepsilon\|^p \le \left(\varepsilon^2 E^{\frac{2}{p}}\left(\sum_1^n \xi_{jn}^2\right)^{p/2} + 2d_n^2\right)^{p/2} = 0\left(\varepsilon^{\frac{2p\beta}{2\beta+1}}\right),\ p > 2.$$

Our next aim is to define some kind of stochastic integrals to give a sense to the expression $(F'(S_\varepsilon),\dot{X}_\varepsilon)$. For a nonrandom function $\phi \in L_2(0,1)$ the integrals $\int_0^1 \phi(t)dX_\varepsilon$ and $\int_0^1 \phi(t)dw$ have the usual meaning. For a random function $\phi \in L_2(0,1)$ we define integrals $\int_0^1 \phi(t)dX_\varepsilon$ and $\int_0^1 \phi(t)dw$ as

$$(\phi,\dot{w}) = \int_0^1 \phi(t)dw = \sum_1^\infty \int_0^1 \phi_{jn}(t)\phi(t)dt \int_0^1 \phi_{jn}(t)dw = \sum_1^\infty \xi_{jn}(\phi,\phi_{jn}) ,$$

$$(\phi,\dot{X}_\varepsilon) = \int_0^1 \phi(t)dX_\varepsilon(t) = \sum_1^\infty \int_0^1 \phi_{jn}(t)dX_\varepsilon \int_0^1 \phi(t)\phi_{jn}(t)dt$$

if these series converge. Certainly it is not the best of all possible definitions but it will prove to be reasonably good for our aims. There exists such element $\psi(S;\cdot) \in L_2$ that

$$(F'(S),\phi) = \int_0^1 \psi(S;t)\phi(t)\, dt$$

and we define

$$(F'(S_\varepsilon),\dot{X}_\varepsilon) = \int_0^1 \psi(S_\varepsilon;t)dX_\varepsilon .$$

<u>Lemma 3.1</u>. Denote $\mathfrak{A}_r$ the collection of all random functions $\phi \in L_2$ which are measurable with respect to σ-algebra generated by the random variables $\xi_{1n},\dots\xi_{2n}$. Let $\phi,\ \psi \in \mathfrak{A}_r$. Then

(α) Integrals $I = \int_0^1 \phi dw,\ I_\varepsilon = \int_0^1 \phi dX_\varepsilon$ are well defined and

$$\int_0^1 \phi dX_\varepsilon = \int_0^1 \phi(t)S(t)dt + \varepsilon \int_0^1 \phi(t)dw ;$$

(β) $$\int_0^1 (\phi+\psi)dX_\varepsilon = \int_0^1 \phi dX_\varepsilon + \int_0^1 \psi dX_\varepsilon ;$$

(3.6)(γ) $|EI| \le \sqrt{r}\, E^{\frac{1}{2}}\, \|\phi\|^2$;

(3.7)(δ) $EI^2 \le (2r+1)E^{\frac{1}{2}}\, \|\phi\|^4$.

<u>Proof</u>. Let $a_j = (\phi,\phi_{jn})$. Then $a_j = a_j(\xi_{1n}\dots\xi_{2n})$ and therefore the sequences $\{a_j\}$, $\{\xi_{jn}, j \ge r+1\}$ are independent. It follows that the series

$$I = \sum_j a_j\, \xi_{jn}$$

converges with probability I. This completes the proof of (α), (β).

Since $E\, a_j\, \xi_{jn} = 0,\ j \geq r+1$ it follows that

$$|EI| = |\sum_1^r a_j\, \xi_{jn}| \leq E^{\frac{1}{2}}|\sum_1^r a_j^2|E^{\frac{1}{2}}|\sum_1^r \xi_{jn}^2|^2 \leq \sqrt{r}\, E^{\frac{1}{2}}\|\phi\|^2.$$

The analogous computation shows that

$$EI^2 = E\left(\sum_1^r a_j\, \xi_{jn}\right)^2 + E\left(\sum_{r+1}^{\infty} a_j^2\right)^2 \leq (2r+1)E^{\frac{1}{2}}\|\phi\|^4 .$$

The proof of Lemma is completed.

Let us return to our estimate. We have that

$$F_\varepsilon = F(S_\varepsilon) + (F'(S_\varepsilon), \dot{X}_\varepsilon - S_\varepsilon)$$

by definition and

$$F(S_\varepsilon) = F(S) - (F'(\tilde{S}), S-S_\varepsilon)$$

by Lagrange formula. Here the random elements S_ε, $\tilde{S} \in L_2$ depends on $\xi_{1n}, \ldots \xi_{nn}$ only and $\|\tilde{S}-S\| \leq \|S_\varepsilon - S\|$.

It follows that

$$(3.8) \qquad F_\varepsilon - F(S) = \varepsilon(F'(S), \dot{w}) + R$$

where

$$R = (F'(S)-F'(S_\varepsilon), S-S_\varepsilon) + (F'(\tilde{S})-F'(S_\varepsilon), S-S_\varepsilon) + \varepsilon(F'(S_\varepsilon)-F'(S), \dot{w}) .$$

The first right hand term in (3.8) is a standard stochastic integral $\int_0^1 \psi(S;t)\, dw$ and $E_S^{(\varepsilon)}(F'(S),\dot{w}) = 0$, $E_S^{(\varepsilon)}(F'(S),\dot{w})^2 = \|F'(S)\|^2$.

It follows from (3.1) and (3.5) that

$$E_S^{(\varepsilon)}|(F'(S) - F'(S_\varepsilon),\ S-S_\varepsilon)| \le B\ E_S^{(\varepsilon)} \| S-S_\varepsilon \|^{1+\gamma} = o(\varepsilon),$$

(3.9)

$$E_S^{(\varepsilon)}|(F'(\tilde{S})-F'(S_\varepsilon),\ S-S_\varepsilon)| \le B\ E_S^{(\varepsilon)} \| S-S_\varepsilon \|^{1+\gamma} = o(\varepsilon)\ .$$

By Lemma 3.1

$$|E_S^{(\varepsilon)}(F'(S_\varepsilon) - F'(S),\dot{w})| \le \sqrt{n(\varepsilon)}\ E_S^{(\varepsilon)} \| F'(S_\varepsilon)-F'(S) \| \le$$

(3.10)

$$\le B\ \sqrt{n(\varepsilon)}\ E_S^{(\varepsilon)} \| S-S_\varepsilon \|^{\gamma} = O\left(\varepsilon^{\frac{2\beta\gamma}{2\beta+1}} . \varepsilon^{-\frac{1}{2\beta+1}}\right) = o(1)\ .$$

Hence $E_S^{(\varepsilon)}(F(S)-F_\varepsilon) = o(\varepsilon)$. Similarly, by (3.1), (3.5) and Lemma 3.1, (δ) ,

$$E_S^{(\varepsilon)}R^2 \le 6\ B^2 E_S^{(\varepsilon)} \| S-S_\varepsilon \|^{2+2\gamma} + \varepsilon^2 E_S^{(\varepsilon)}(F'(S_\varepsilon)-F'(S),\dot{w})^2 \le$$

$$\le \varepsilon^2(2n(\varepsilon) + 1)\ E^{\frac{1}{2}} \| F'(S_\varepsilon)-F'(S) \|^4 + O\left(\varepsilon^{\frac{4+4\gamma}{2\beta+1}}\right) = o\ (\varepsilon^2)\ .$$

Hence $E_S^{(\varepsilon)}|F_\varepsilon-F(S)|^2 = \varepsilon^2 \| F'(S) \|^2 + o(\varepsilon^2)$. This completes the proof of Theorem.

4. <u>An example</u>. Let the set $\Sigma = \Sigma\ (\beta,L)$ consists of all functions $S \in U_1(0$ such that

$$S(t) = a_o + \sum_1^\infty \sqrt{2}\ a_j \cos \pi jt\ , \quad \sum_j |j|^{2\beta}|a_j|^2 \le L\ .$$

Let the functional $F(S) = \| S \|^2$. It is easy to see that

$$d_n^2\ (\Sigma) = \sup_\Sigma \sum_n^\infty |a_j|^2 \le L\ n^{-2\beta}\ .$$

Theorem 3.1 asserts that an asymptotically efficient estimator of

F(S) exists if $\beta > \frac{1}{2}$. In fact, it exists if $\beta > \frac{1}{4}$ and does not exist if $\beta \le \frac{1}{4}$.

Consider the estimate

$$\hat{F}_\varepsilon = \sum_1^\infty \left[\sqrt{2} \int_0^1 \cos \pi jt \, dX_\varepsilon(t) - \varepsilon^2\right] + \left[\int_0^1 dX_\varepsilon(t)\right]^2 - \varepsilon^2$$

where $n = n(\varepsilon)$ will be choosen later. The simple calculations give that

$$F_\varepsilon = \sum_0^{n-1} a_j^2 + 2\varepsilon \sum_0^{n-1} a_j \xi_j + \varepsilon^2 \sum_0^{n-1} (\xi_j^2 - 1) ,$$

$$\xi_o = \int_0^1 dw , \quad \xi_j = \sqrt{2} \int_0^1 \cos \pi jt dw.$$

It follows that $\hat{F}_\varepsilon - F = 2 \sum_0^{n-1} a_j \xi_j + R_\varepsilon$ where

$$E R_\varepsilon^2 = \left(\sum_n^\infty a_j^2\right)^2 + 4 \varepsilon^2 \sum_n^\infty a_j^2 + 2 n\varepsilon^4 = 0\left(\max\left(\varepsilon^2, \varepsilon^{\frac{1\,6\beta}{4\beta+1}}\right)\right)$$

if we choose $n(\varepsilon) \sim \varepsilon^{-\frac{4}{4\beta+1}}$. Therefore for all $\beta > 0$

$$(4.1) \qquad E_S|\hat{F}_\varepsilon - F|^2 = 0\left(\max\left(\varepsilon^2, \varepsilon^{\frac{1\,6\beta}{4\beta+1}}\right)\right)$$

and for $\beta > \frac{1}{4}$ the estimator $\hat{F}_\varepsilon$ is asymptotically efficient and asymptotically normal with parameters F(S), $\varepsilon^2 \| F'(S) \|^2$.

It is possible to show that we cannot find estimators which are considerable better than $\hat{F}_\varepsilon$, namely,

$$\inf_{F_\varepsilon} \sup_{S \in \Sigma} E_S^{(\varepsilon)} |F_\varepsilon - F(S)|^2 \asymp \varepsilon^{\frac{1\,6\beta}{4\beta+1}} , \quad \beta \le \frac{1}{4}$$

where infimum is taken over all possible estimators F_ε of F(S) (see [4]).

REFERENCES

1. Hajek, J., Local aymptotic minimax and admissibility in estimation, Proc. 6-th Berkeley Symp.Math. Statist. Prob., v. I, 1972, p. 175-194.

2. Ibragimov, I. A., Hasminskii, R. Z., On the estimation of an infinite-dimensional parameter in Gaussian white noise, Dokl. Akad. Nauk SSSR, v. 236, No. 5 (1977). Engl. transl. in : Soviet Math. Dokl. v. 18, No. 5 (1977).

3. Ibragimov, I. A., Hasminskii, R. Z., Asymptotic theory of estimation, Moscow, "Nauka", 1978 (in Russian).

4. Ibragimov, I. A., Hasminskii, R. Z., A problem of statistical estimation in Gaussian white noise, Dokl. Akad. Nauk SSSR, v. 236, No. 6 (1977). Engl. transl. in: Soviet Math. Dokl., v. 18, No.5.

5. Tihomirov, V. M., Diameters of sets in functional spaces and approximation theory, Uspehi Mat. Nauk, v. 14, No. 3 (1960). Engl. transl. in: Russian Math. Surveys, 15 (1960).

APPLICATIONS OF STOCHASTIC DIFFERENTIAL EQUATIONS TO THE DESCRIPTION OF TURBULENT EQUATIONS

A. M. Yaglom
Institute of Atmospheric Physics
Acad. Sci. of the USSR
Moscow, USSR

Introduction. The Markov stochastic processes with continuous time appeared first at the beginning of the century in the works devoted to the theory of Brownian motion. Two different approaches were used by physicists rather early and till now they continue to be the main methods for the description of continuous Markov processes. As early as 1905 A. Einstein obtained a parabolic partial differential equation for probability density of the coordinate of a free Brownian particle and a little bit later this equation was generalized by A. D. Fokker and M. Planck to a wide class of Brownian motions. On the other hand P. Langevin developed in 1908 a theory of the Brownian motion, which did not use partial differential equations for probability density but which was based on the so-called Langevin equation, i.e. on the stochastic differential equation with 'random force' term for the Brownian trajectory. The two above mentioned approaches are, in fact, equivalent to each other, but each of them has specific merits and demerits; they have been developing for many years simultaneously, partially competing with each other and partially complementing one another.

A strict mathematical theory of various continuous Markov processes was developed by A. N. Kolmogorov in early thirties.

Kolmogorov based his theory on equations for probability densities (or probability distributions functions) which he obtained independently from physicists. The influence of Kolmogorov's work was so great that many years most of the mathematicians considered the probability equation approach only to be mathematically rigorous, while the equations of the Langevin type were considered as purely heuristic tools unsuitable for mathematical works. However in the thirties S.N. Bernstein made an earlier attempt to justify rigorously the construction of Markov processes as solutions of stochastic differential equations. This program was fully satisfactorily carried out independently by K. Ito and I. I. Gihman in the early fifties. Beginning from that time the study of Markov processes by means of stochastic differential equations became especially popular among mathematicians, while the Fokker-Planck-Kolmogorov equations began to be considered as a tool valuable only for physicists and engineers.

The main advantage of stochastic differential equations is the possibility to use them for a direct construction of a wide class of diffusions basing on the given Brownian motion. This circumstance permits to apply stochastic equations to the existence proof of solutions of parabolic or elliptic partial differential equations which can be considered as equations for probability densities. On the other hand a constructive nature of stochastic differential equations and their intuitive clearness make these equations very attractive for applied sciences. Therefore it is not surprising that the applications of stochastic differential equations to physical and engineering problems are now flourishing too. A very wide range of both purely mathematical and applied investigations based on stochastic equations had led to the publication of a number of monographs which are often very different in their style and which are intended for quite different readers (see, e.g., [1-4].

Let us mention that many physical and engineering works which are known to most mathematicians contain heuristic construction of some specific stochastic processes which often differ from pure diffusions considered in mathematical literature on stochastic equations. The applications of stochastic differential equations to the construction of statistical models of turbulent diffusion, which will be considered in the present paper, are typical examples of such works. (A more detailed exposition of some of the considered models and of several other models of turbulent diffusion can be found in [5].)

1. The Fickian diffusion. The most usual description of turbulent diffusion uses the so-called Fick law, i.e. the hypothesis about the linear dependence of the turbulent flux of the diffusing substance on the mean concentration gradient. The hypothesis implies a parabolic partial differential equation for the concentration of the diffusing substance (i.e. for the probability density of the diffusing particle coordinates; see, e.g., [6], section 10.3). In a one-dimensional case (i.e., when only concentration distribution p(x,t) along the fixed x-asis is considered) the simplest diffusion equation has the form

$$\frac{\partial p(x,t)}{\partial t} = \frac{\partial}{\partial x}\left(K \frac{\partial p(x,t}{\partial x} \right) \tag{1}$$

where K is the non-negative eddy diffusivity ('coefficient of turbulent diffusion'). It is considered most usually that diffusibity K may depend on coordinate x only; however, recently some models have been considered, which correspond to the assumption that diffusivity K is dependent on time t after the moment of the particle emission by a fixed admixture source (see, e.g. [7,8]). It is well known that the diffusion with probability density satisfying equation (1) may be constructed as a solution of the simplest stochastic differential equation of the form

$$(2) \qquad \frac{dx(t)}{dt} = \sqrt{2k}\, w'(t).$$

Here x(t) is the trajectory of the diffusing particle and w(t) is the generalized stochastic process ('random distribution') of 'white noise', i.e. the Gaussian stochastic process satisfying the conditions $E\, w'(t) = 0$, $E\, w'(t_1)w'(t_2) = \delta(t_2-t_1)$ where E symbolizes a mathematical expectation and $\delta(\zeta)$ is Dirac's delta function. (It is easy to give a strick meaning to equation (2) and formula for $E\, w'(t_1)w'(t_2)$; however, we shall not dwell upon such questions in this paper). Similar stochastic equations for the trajectories of diffusing particles can be written out when the diffusion takes place in the multi-dimensional diffusion in the presence of wind). Stochastic differential equations for the trajectories can be used for the simple numerical simulation of the trajectory ensemble and the averaging over this ensemble gives the possibility to estimate the mean concentration p(x,t) and the other statistical characteristics of the diffusion. Such an estimation of the concentration p(x,t) is, in fact, one of the methods of the numerical solution of parabolic partial differential equations by Monte-Carlo technique. This solution can be obtained easily with modern hybrid computers (see, e.g., [9,10]).

2. Diffusion with a finite velocity. The Fickian diffusion corresponds to the unlimited velocity of the admixture propagation, since in the case of instantaneous point source at the moment $t = 0$ the solution of parabolic diffusion equation p(x,t) will be positive at any distance from the source for any $t > 0$. Therefore it was suggested by many authors that parabolic diffusion equation (1) represents unjustified mathematical idealization and it must be replaced by another equation which takes into account that the actual velocity of admixture propagation cannot exceed some bounded limiting velocity U. (A number of references to the related works can be found in [6], section 10.6; see also [11]). Models of the diffusion with a finite velocity correspond to the assumption that stochastic process x(t) is not a pure diffusion, but is is a discontinuous stochastic process. The simplest model of the one-dimensional diffusion with a finite velocity in x-direction in the field of stationary homogeneous turbulence is based on the assumption that the diffusing particle is moving permanently with the same absolute velocity U, but the direction of the motion changes to the opposite one at random time moment $t_k, k=\ldots,-1,0,1,\ldots$, forming the Poisson point process (see [6], section 10.6). This model can be described by the stochastic differential equation of the form

(3) $$\frac{dx(t)}{dt} = v(t),$$

where x(t) is trajectory and v(t) is the so-called 'random telegraph signal', i.e. a purely discontinuous stochastic process taking only two values + U and -U and changing its values at the random Poisson sequence of points t_k. Process x(t) is here non-Markovian, but it s a component of two-dimensional Markov process (x(t),v(t)). Let us consider the distribution functions $F_1(x_1,t) = P\{x(t) < x_1, v(t) = U\}$, $F_2(x_1,t) = P\{x(t) < x_1, v(t) = -U\}$, where the symbol $P\{\cdot\}$ denotes the probability of occurence of the relations which appear between the barces. Then it is easy to show that the differential equations of Markov process (x(t),v(t) can be written out as the following system of two equations for the unknown probability densities

$p_1(x,t) = d\,F_1(x,t)/\,dx$ and $p_2(x,t) = d\,F_2(x,t)/\,dx$

(4) $$\frac{\partial p_1}{\partial t} = -\frac{\partial(Up_1)}{\partial x} + a(p_2-p_1)\ ,\ \frac{\partial p_2}{\partial t} = \frac{\partial(Up_2)}{\partial x} + a(p_1-p_2)$$

where a is the mean density of the Poisson point process t_k.

The system (4) implies a telegraph partial differential equation for the total probability density $p(x,t) = p_1 + p_2$ of the random variable x(t):

(5) $$\frac{\partial^2 p}{\partial t^2} + 2a\ \frac{\partial p}{\partial t} = U^2\ \frac{\partial^2 p}{\partial x^2}\ .$$

Stochastic equation (3) can be used for the numerical simulation of the trajectories x(t) and, consequently, for the numerical Monte-

Carlo solution of the hyperbolic partial differential equation (5).

The principal possibility of the numerical solution of some hyperbolic equations by the simulation of purely discontinuous stochastic processes was noted in [12]; however, I do not know whether such a method was used in practice or not. The problem of the description of all the hyperbolic equations and systems of equations accessible to such a method of solution is also unsolved till now. Only a few examples of purely discontinuous stochastic processes leading to hyperbolic equations for probabilities can be found in papers devoted to the theory of the diffusion with a finite velocity. For example, Monin [13, 14] (see also [6], section 10.6) considered the case, when the values $\pm U$ of the discontinuous process $v(t)$ and the density a of its discontinuity points t_k depend upon the value of $x(t)$. In this case a hyperbolic partial diferential equation can be derived for the distribution function $F(x,t) = F_1 + F_2$ of $x(t)$. A specific example $U = \text{const}$, $a(x) - cU/x$, $0 < x < \infty$, was considered by Monin as a model of the vertical diffusion in the atmospheric surface layer. In [5, 8] another model of the same vertical diffusion was considered: it was assumed here that $U = \text{const}$, and the discontinuity points t_k of purely discontimuous process $v(t)$ form a non-stationary Poisson point process with the density function $a(t) = c_1/t$, $0 < t < \infty$. In this case equation (5) for the probability density is valid, but the coefficient a = constant in it must be replaced by the function $a(t)$.

3. <u>Diffusion and phase space</u>. Obukhov [15] proposed to describe turbulent diffusion in three-dimensional space by the model of the

continuous Markov process $(\bar{x}(t),\bar{v}(t))$ in six-dimensional phase space of particle coordinates $\bar{x} = (x_1,x_2,x_3)$ and velocities $\bar{v} = (v_1, v_2, v_3)$. This model is equivalent to the systems of stochastic differential equations

$$(6) \qquad \frac{d\bar{x}(t)}{dt} = \bar{v}(t), \quad \frac{d\bar{v}(t)}{dt} = \bar{w}'(t)$$

where $\bar{w}' = (w_1', w_2', w_3')$ is three-dimensional white noise, so that $Ew_i'(t_1)w_j'(t_2) = B\,\delta_{ij}\,\delta(t_2-t_1)$, $B = \text{const}$. Obukhov also mentioned the possibility of taking into account the supplementary weak friction force acting on the particle. In such a case equations (6) must be replaced by the equations

$$(7) \qquad \frac{d\bar{x}(t)}{dt} = \bar{v}(t), \quad \frac{d\bar{v}(t)}{dt} = -\frac{\bar{v}(t)}{T} + \bar{w}'(t)$$

where T is rather large 'relaxation time' of the process v(t). The partial differential equation for the probability density $p(\bar{x},\bar{v},t)$ of the process $(\bar{x}(t),\bar{v}(t))$ satisfying (7) has the form

$$(8) \qquad \frac{\partial p}{\partial t} = -\sum_{j=1}^{3} v_j \frac{\partial p}{\partial x_j} + \frac{1}{T}\sum_{j=1}^{3} \frac{\partial (v_j p}{\partial v_j} + \frac{B}{2}\sum_{j=1}^{3} \frac{\partial^2 p}{\partial v_j^2} .$$

Obukhov [15] considered only the equation for the probability density $p(\bar{x},\bar{v},t)$. Stochastic differential equation (7) were studied, in particular, by Novikov [16] and Krasnoff and Peskin [17]. The latter authors assumed also that the 'random force' $\bar{w}'(t)$ can differ from 'white noise' and can be characterized by finite (but small) correlation time T_w. Then process $\bar{w}'(t)$ will be ordinary (not generalized) Gaussian stationary stochastic process, process $(\bar{x}(t),\bar{v}(t))$ will be non-Markovian and $\bar{x}(t)$ will be a linear transformation of the process $\bar{w}'(t)$.

4. Relative diffusion. Obukhov [15] applied the derived equation for the probability density $p(\bar{x},\bar{v},t)$ to the description of the relative turbulent diffusion, i.e. relative motion of two particles in a turbulent flow. It was shown in [15] (see also [16] and [17]), that equation (8) with $T = \infty$ implies the values of the second moments of the six-dimensional random vector $(\bar{x}(t),\bar{v}(t))$ which agree with the predictions of Kolmogorov's theory of locally isotropic turbulence (the theory is expounded, for example, in [6], chapter 8). However, this agreement is a restrictive one: it is related only to the so-called inertial subrange of intermediate (not very great and not very small) scales and to obtain the agreement it is necessary to suppose that $B = c\varepsilon$, where c is dimensionless constant and ε is mean rate of turbulent energy dissipation. The inclusion of the supplementary 'relaxation term' $-v/T$ in the right-hand side of (7) permits to obtain correct asymptotic behaviour of the above mentioned moments in the range of very large scales too. However the relative diffusion in the range of very small scales cannot be described by equations (6), (7) and (8), since the diffusion in the range is affected by the molecular viscosity ν which does not enter these equations.

A modified model stochastic differential equation for relative velocity $v(t) = v_2(t) - v_1(t)$ of two diffusing particles was proposed by Levin [18]. This equation permits to take into account the influence of the viscosity and it has the form

$$(9) \qquad dv(t) = -v(t)dt/T + d\zeta(t)$$

(only the simplest one-dimensional diffusion in the direction of the given axis is considered here). Process $\zeta(t)$ in the right-hand side of (9) is the generalized Poisson process which takes constant values Z_k on the intervals $t_k < t \le t_{k+1}$, where $\{t_k\}$ is the Poisson sequence of time points and the differences $V_k = Z_{k+1}-Z_k$ are mutually independent (and independent of random sequence t_k) identically distributed random variables having zero mean value, variance σ^2 and probability density $g(v)$. If we now assume that $E(t_{k+1}-t_k) = T_o = \lambda(\nu/\varepsilon)^{\frac{1}{2}}$ and $\sigma^2 = \mu(\nu\varepsilon)^{\frac{1}{2}}$ where λ and μ are dimensionless constants, then equation (9) implies the results which agree with the predictions of the Kolmogorov theory at all values of t. The solution $v(t)$ of equation (9) belongs to the class of the discontinuous Markov processes studies by Feller [19]. The direct evaluation of the probability density increment $p(x,t+\Delta t)-p(x,t)$, where Δt is a small time interval, leads to the following integro-differential equation for the probability density of $v(t)$:

$$(10) \qquad \frac{\partial p(v,t)}{\partial t} = \frac{1}{T}\,\frac{\partial[vp(v,t)]}{\partial v} - \frac{1}{T_o}\left[p(v,t) - \int_{-\infty}^{\infty} g(v-v')p(v',t)dv'\right].$$

Similarly, the probability density $p(x,v,t)$ of the two-dimensional Markov process $(x(t),v(t))$, where $x(t) = x_o + \int_0^t v(t')\,dt'$, satisfies the equation

$$(11)\quad \frac{\partial p(x,v,t)}{\partial t} = -v\frac{\partial p}{\partial x} + \frac{1}{T}\frac{\partial(vp)}{\partial v} - \frac{1}{T_o}\left[p-\int_{-\infty}^{\infty} g(v-v')p(x,v',t)dv'\right].$$

It is easy to see that equation (9) turns into the second equation (7) and equation (11) into equation (8), when $V \to 0$ (i.e., $T_o \to 0$, $\sigma^2 \to 0$, $\sigma^2/T_o \to B$ = const).

Equation (9) describes a direct construction of the discontinuous Markov process $v(t)$ basing on the given generalized Poisson process $\zeta(t)$. This construction raises an interesting question about the classification of the whole set of Markov processes that can be constructed basing on the given generalized Poisson process $\zeta(t)$ (or on the ordinary Poisson process $\zeta_o(t)$, or on $\zeta_o(t)$ and the Brownian motion $w(t)$).

5. <u>The turbulent diffusion with two different significant scales</u>. It is known that the graph of the spectral density of atmospheric turbulent fluctuations contains a wide gap in the frequency region with the center near the frequency n_o=1 hour^{-1}, i.e., the period τ_o=1 hour (see, e.g., [6], section 23.6). Therefore we can assume that the wind velocity fluctuations $v(t)$ are composed of the large scale meso-meteorological ('synoptic') fluctuations $v_1(t)$ with periods much greater than 1 hour and of small scale micro-meteorological fluctuations $v_2(t) = v(t)-v_1(t)$ which are superimposed on the synoptic fluctuations, but do not depend on them. It is naturally to expect that two quite different 'relaxation times' T_1 and $T_2 \ll T_1$ correspond to these two types of fluctuations. A statistical model of the diffusion in the field of such two-scale

turbulence needs, at least, two stochastic differential equations. In particular, Levin [20] suggested the following system of equations to describe atmospheric diffusion:

$$(12) \quad d\, v_1(t) = -T_1^{-1} v_1(t)dt + d\, \zeta_1(t), \; dv_2(t) = -T_2^{-1} v_2(t)dt + d\zeta_2(t).$$

According to Levin $\zeta_1(t)$ must be considered as generalized Poisson process, while $\zeta_2(t)$ is independent of $\zeta_1(t)$ and may be considered either as discontinuous generalized Poisson process or as continuous Brownian motion. In both cases the integro-differential equations can be obtained for probability densities of Markov processes $(v_1(t), v_2(t))$ and $(x(t), v_1(t), v_2(t))$, where $x(t) = x_o + \int_0^t [v_1(t') + v_2(t')]\, dt'$. The related statistical model was considered, in a slightly different aspect, also by Bywater and Chung [21] (see also [22]). These authors use a system of coupled stochastic equations

$$(13) \quad \frac{dv_1}{dt} = -\frac{v_1}{T_1} - \beta(v_1 - v_2) + \frac{dw_1}{dt}, \; \frac{dv_2}{dt} = -\frac{v_2}{T_2} + \beta(v_1 - v_2) + \frac{dw_2}{dt}$$

where both the processes $w_1(t)$ and $w_2(t)$ are considered to be independent Brownian motions. In this case stochastic processes $(v_1(t), v_2(t))$ and $(x(t), v_1(t), v_2(t))$ prove to be continuous Markov processes with probability densities $p(v_1, v_2)$ and $p(x, v_1, v_2)$ satisfying some Fokker-Planck-Kolmogorov partial differential equations.

REFERENCES

1. McKean, H. P. Stochastic Inegrals. New York, Academic Press, 1969.

2. Gihman, I. I., Skorohod, A.V. Stochastic Differential Equations, Berlin, Springer, 1972.

3. Friedman, A. Stochastic Differential Equations and Applications, 2 vols. New York, Academic Press, 1975, 1976.

4. Srinivasan, S. K., Vasudevan R. Introduction to Random Differential Equations and their Applications. Amsterdam, Elsevier, 1971.

5. Yaglom, A. M. Semi-empirical Equations of Turbulent Diffusion in Boundary Layers. Fluid Dynamics Trans. (Warsaw), 1976, 7, pt. II, 99-144.

6. Monin, A. S., Yaglom, A. M. Statistical Fluid Machanics, 2 vols. Cambridge, Massachusetts, MIT Press, 1971, 1975.

7. Deardorff, J. W. Difference between Eddy Coefficients for Instantaneous and Continuous Vertical Diffusion into the Neutral Surface Layer. Boundary-Layer Meteor., 1974, 5, No. 4, 451-457.

8. Yaglom, A. M. On Equations with Time Dependent Coefficients for Diffusion in a Stationary Atmospheric Surface Layer, Izv. Akad. Nauk SSSR,ser. Fiz. Atm. Okeana (Bull. Acad. Sci, USSR, ser. Atmospheric and Oceanic Phys.), 1975, 11, No. 11, 1120-1128.

9. Bullin, J. A., Dukler, A. E. Stochastic Modeling of Turbulent Diffusion with Hybrid Computer. Environmental Sci. Techn., 1974, 8, No. 2, 156-165.

10. Lee, H., Dukler, A. E. Lagrangian Simulation of Dispersion in Turbulent Shear Flow with Hybrid Computer. AIChE Journal, 1976, 22, No. 3, 449-455.

11. Swenson, R. J. Heat Conduction-Finite or Infinite Propagation. J. Non-Equilibr. Thermodyn., 1978, 3, No. 1, 39-47.

12. Gel'fand, I. M., Frolov, A. S., Chentsov, H. H. Evaluation of functional Integrals by Monte-Carlo Methods. Izv. VUZ'ov, ser, Matem. (Bull. Univ. and Colleges USSR, Ser. Math.), 1958, No. 5, 32-45.

13. Monin, A. S. Diffusion with Finite Velocity. Izv. Akad. Nauk SSSR, Ser. Geophys., 1955, No. 3, 234-248.

14. Monin, A. S. Turbulent Diffusion in the Atmospheric Surface Layer. Izv. Akad. Nauk SSSR, ser. Geophys., 1956, No. 12, 1461-1473.

15. Obukhov, A. M. Description of Turbulence in Terms of Lagrangian Variables. Adv. in Geophys., 1959, 6, 113-115.

16. Novikov, E. A. Random Force Method in the Theory of Turbulence. Zhurn. Exper. Teor. Fir. (J. Exper. Theor. Phys.), 1963, 44, No. 6, 2159-2168.

17. Krasnoff, E., Peskin, R. L. The Langevin Model for Turbulent Diffusion. Geophys. Fluid Dynamics, 1971, 2, 123-146.

18. Levin, A. V. Ramdom Processes and Richardson-Obukhov Law, Trudy Ukrainsk. Nauchno-Issled. Gidromet. Inst. (Works of Ukrainian Research Hydrometeor. Inst.), 1971, No. 106, 3-12.

19. Feller, W. On the Integro-Differential Equations of Purely Discontinuous Markov Processes. Trans. Amer. Math. So., 1940,

48, 488-515.

20. Levin, A. V. Statistical Model of Atmospheric Diffusion with Two Relaxation Times. Trudy Ukrainsk. Nauchno-Issled. Gidromet. Inst. (Works of Ukrainian Research Hydrometeor. Inst.), 1973, No. 125, 150-157.

21. Bywater, R. J., Chung, P. M. Turbulent Flow Fields with Two Dynamically Significant Scales. AIAA Papers, 1973, No. 73-646, 1-10.

22. Bywater, R. J. A Statistical Description of Turbulent Shear Flows Characterized by a Finite Number of Nonequilibrium Degrees of Freedom. Ph. D. Thesis, Univ. of Illinois, Chicago, Illinois, 1974.

ON SEMIMARTINGALES WITH VALUES IN EUCLIDEAN HALFSPACES

B. Grigelionis, R. Mikulevičius

University of Vilnius

0. Introduction. General and important theorems on the integral representation of functionals of semimartingales (see [1]), on the absolute continuity of probability measures, corresponding to semimartingales (see [2], [3]), and some others are proved under the assumption that a triplet of characteristics of the semimartingale and initial distribution determine finite dimensional distributions in a sense uniquely. For a wide class of semimartingales taking values in Euclidean halfspaces the triplet of the characteristics is expressed by means of local characteristics, determining their behaviour inside the domain and on its boundary as well as by the local time on the boundary. Since the local time is uniquely determined by local characteristics as a complicated functional of the process, the conditions of theorems of semimartingales can be considered as efficient when they are expressed in terms of local characteristics only. In that connection it is important to investigate the properties of the local time. In the paper the main attention is paid to the investigation of conditions when there exist exponential moments of the local time and to the derivation of the effective conditions of the absolute continuity of measures corresponding to the semimartingales with values in Euclidean halfspaces.

All proof will be sketched, the details of which can be found in papers [4] and [5].

1. Definitions. In the paper we consider a semimartingale $X = (X^1,\dots,X^m)$ taking values in $R^m_+ = \{x : x = (x_1,\dots, x_m), x_1 \geq 0\}$ defined on the probability space $(\Omega, \mathcal{F}, P)$ with an increasing right continuous family of σ-algebras $F = \{\mathcal{F}_t, t \geq 0\}$ and having canonical representation (see [2], [6]):

$$X_t = X_o + \alpha_t + X^c_t + \int_0^t \int_{|x|\leq 1} x\, q(ds,dx) + \int_0^t \int_{|x|>1} x\, p(ds,dx)t, \geq 0,$$

with characteristics (α, B, Π), where $p(ds,dx)$ is the measure of jumps of the process X, $q(ds,dx) = p(ds,dx) - \Pi(ds,dx)$. $\Pi(ds,dx)$ is the dual predictable projection of p with respect to F and P,

$\alpha = (\alpha_1,\dots, \alpha_m)$, α_j are predictable and have locally summable variation, X^{cj} are continuous local martingales, $j = 1,\dots,m$, $B(t) = ||\beta_{jk}(t)||^m_1$, $\beta_{jk}(t) = \langle X^{cj}, X^{ck}\rangle_t$.

We shall assume that predictable random functions $\gamma(t) \geq 0$, $\delta(t) \geq 0$, $\hat{\alpha}(t) = (\hat{\alpha}_1(t),\dots, \hat{\alpha}_m(t))$, $\tilde{\alpha}(t) = \left(\tilde{\alpha}_1(t),\dots,\tilde{\alpha}_m(t)\right)$, $\hat{B}(t) = ||\hat{\beta}_{jk}(t)||^m_1$, $\tilde{B}(t) = ||\tilde{\beta}_{jk}(t)||^m_2$, $\hat{\pi}(t,\Gamma)$, $\tilde{\pi}(t,\Gamma)t \geq 0$, are the local characteristics of the process X with respect to the family F and the measure P if its characteristics have the following structure:

$$(1) \quad \Pi([0,t]\times \Gamma) = \int_0^t \chi_G(X_{s-})\hat{\pi}(s,\Gamma)ds + \int_0^t \tilde{\pi}(s,\Gamma)d\phi_s,$$

$$(2) \quad \alpha_1(t) = \int_0^t \chi_G(X_{s-})\hat{\alpha}_1(s)ds + \int_0^t \gamma(s)d\phi_s + \int_0^t \int_{|x|\leq 1} x\, \tilde{\pi}(s,dx)d\phi_s,^{*)}$$

* χ_A denotes the indicator of the set A, $G = \{x: x \in R^m_+, x_1 > 0\}$, $\partial G = \{x : x \in R^m_+ , x_1 = 0\}$.

(3) $\alpha_j(t) = \int_0^t \chi_G(X_{s-})\hat{\alpha}_j(s)ds + \int_0^t \tilde{\alpha}_j(s)d\phi_s, j = 2,\ldots,m,$

$\beta_{1k}(t) = \int_0^t \chi_G(X_{s-})\hat{\beta}_{1k}(s)ds, \; k = 1,\ldots,m \; ,$

(4) $\beta_{jk}(t) = \int_0^t \chi_G(X_{s-})\hat{\beta}_{jk}(s)ds + \int_0^t \tilde{\beta}_{j.k}(s)d\phi_s, j,k = 2,\ldots,m,$

for some continuous increasing process ϕ such that for all $t \geq 0$ a.e.

(5) $\int_0^t \chi_{\partial G}(X_{s-})d\phi_s = \phi_t$ and $\int_0^t \chi_{\partial G}(X_{s-})ds = \int_0^t \delta(s)d\phi_s,$

moreover, the matrixes $\hat{B}(t)$ and $\tilde{B}(t)$ are nonnegative defined $\hat{\pi}(t,\cdot)$ are $\tilde{\pi}(t,\cdot)$ are measures on $E = R^m\backslash\{0\}$,

$\hat{\pi}(t,\{x : x_1 + X^1_{t-} < 0\}) \equiv 0, \; \tilde{\pi}(t,\Gamma) \equiv \tilde{\pi}(t,\Gamma \cap R^m_+\backslash\{0\} \; ,$

$\int_0^t \int_E \chi_G(X_{s-})(|x|^2 \wedge 1)\hat{\pi}(s,dx)ds < \infty, \; \int_0^t \int_E (|x|^2 \wedge 1 + x_1 \wedge 1)\tilde{\pi}(s,dx)d\phi_s < \infty.$

A process ϕ will be called the local time on the boundary.

2. Properties of the local time. The triplet (α, B, Π) and the local characteristics $\hat{\alpha}$, $\hat{B}$, $\hat{\pi}$ of the semimartingale X are defined uniquely up to the equivalence. The local characteristics γ, δ, α, $\tilde{B}$ and $\tilde{\pi}$ depend on the choice of the local time which can be nonunique. The following statement holds.

Lemma 1. Under fixed γ, δ, $\tilde{B}$ and $\tilde{\pi}$, such that for all $t \geq 0$ a.e.

(6) $\gamma(t) + \delta(t) + \mathrm{Sp}\,\tilde{B}(t) + \int_E |x|^2 \wedge 1 \; \tilde{\pi}(t,dx) > 0,$

the local time ϕ is defined uniquely.

Proof. Assume that there exists another continuous increasing process ϕ' satisfying equalities (1)-(5). Using the uniqueness of the triplet (α, B, Π) we have that for all $t \geq 0$ a.e.

$$\int_0^t [\delta(s) + \gamma(s) + \mathrm{Sp}\, \tilde{B}(s) + \int_E |x|^2 \wedge 1\, \tilde{\pi}(s, dx)] d(\phi_s - \phi'_s) = 0.$$

From here and (6) it follows that for all $t \geq 0$ a.e. $\phi_t = \phi'_t$.

Lemma 2. Let $\psi(t)$, $t \geq 0$, be a positive bounded continuously differentiable function such that

$$\int_0^\infty |\psi'(s)| ds < \infty \ , \ \int_0^\infty \psi(s)\, ds < \infty$$

and the integrals

$$\int_0^\infty \int_E \psi(s)(x_1^2 \wedge 1)\hat{\pi}(s,dx)\, ds \ , \ \int_0^\infty |\alpha_1(s)|\psi(s)\, ds, \text{ and } \int_0^\infty \hat{\beta}_{11}(s)\psi(s)\, ds$$

be bounded.

If for all $t \geq 0$ and some constant $c > 0$ a.e.

$$(7) \qquad \gamma(t) + \delta(t) \geq c > 0$$

then

$$E \exp \{ \int_0^\infty \psi(s)\, d\phi_s \} < \infty .$$

Proof. For every function $\Phi \in C_b^2(R^1)$, $\Phi(0) = 0$, $\Phi'(0) = 1$, $\Phi(x_1) \geq 0$, $x_1 > 0$, denote

$$V_t = \int_0^t \{\chi_G(X_{s-}) \tfrac{1}{2} [c^{-1}\psi(s)\Phi''(X_s^1) - (c^{-1}\psi(s)\Phi'(X_s^1))^2]\hat{\beta}_{11}(s) +$$

$$+ c^{-1}\psi'(s)\Phi(X_s^1) + c^{-1}\psi(s)\hat{\alpha}_1(s)\Phi'(X_s^1) -$$

$$-\int_E (\exp\{-c^{-1}\psi(s)[\Phi(X_s^1+x_1) - \Phi(X_s^1)]\} - 1 +$$

$$+ \chi_{\{|x|\le 1\}}(x)\psi(s)c^{-1}\,\Phi'(X_s^1)x_1)\hat{\pi}(s,dx)\}\,ds + \int_0^t [c^{-1}\psi(s)\gamma(s) -$$

$$- \int_G (\exp\{-c^{-1}\psi(s)\Phi(X_1)\}-1)\tilde{\pi}(s,dx)]\,d\phi_s,\ t \ge 0.$$

By means of Ito's formula we found that

$$U_t = \exp\{-c^{-1}\psi(t)\Phi(X_t^1) + V_t\},\ t \ge 0,$$

is a supermartingale. Therefore, from our assumptions and (5) it follows that

$$E\,U_o \ge E\,U_t = E\exp\ \{V_t + c^{-1}\int_0^t \delta(s)\psi(s)\,d\phi_s -$$

$$-c^{-1}\int_0^t \chi_{\partial G}(X_{s-})\psi(s)\,ds - c^{-1}\psi(t)\Phi(X_t^1)\} =$$

$$(8)\quad = E\exp\ \{c^{-1}\int_0^t (\gamma(s) + \delta(s)\,d\phi_s - Z_t\},$$

where the process $|Z_t|$, $t \ge 0$, is bounded by some constant K. From (7) and (8) we have that

$$E\exp\{\int_0^\infty \psi(s)\,d\phi_s\} \le e^K\,E\,U_o < \infty.$$

3. Conditions for absolute continuity of measures. Let the process X be a semimartingale with respect to the measures P and P' having the local characteristics $(\hat{\alpha}, \hat{B}, \hat{\pi}, \gamma, \delta, \tilde{\alpha}, \tilde{B}, \tilde{\pi})$ and $(\hat{\alpha}', \hat{B},$

$\hat{\pi}'$, γ, δ, $\tilde{\alpha}'$, $\tilde{B}$, $\tilde{\pi}'$) correspondingly, such that

$$\hat{\alpha}'(t) = \hat{\alpha}(t) + g(t)\,\hat{B}(t) + \int_{|x|\le 1} x\, u(t,x)\hat{\pi}(t,dx),$$

$$\tilde{\alpha}'(t) = \tilde{\alpha}(t) + h(t)\tilde{B}(t) + \int_{|x|\le 1} \tilde{x}\, v(t,x)\tilde{\pi}(t,dx),$$

$$\hat{\pi}'(t,\Gamma) = \int_{\Gamma} (u(t,x)+1)\hat{\pi}(t,dx)$$

and

$$\tilde{\pi}'(t,\Gamma) = \int_{\Gamma} (v(t,x) + 1)\tilde{\pi}(t,dx) ,$$

where

$$\tilde{x} = (x_2,\ldots,x_m) .$$

Assume, that there exist nonrandom functions $\psi_j(t)$, $t \ge 0$, such that

$$\int_0^\infty \psi_j(t)\, dt < \infty, \quad j = 1,\ldots,4 ,$$

$\psi(t) = \frac{1}{2}\psi_2(t) + \psi_4(t)$, $t \ge 0$, satisfies the conditions of Lemma 2 and for all $t \ge 0$ a.e.

(9) $(g(t), g(t)\hat{B}(t)) \le \psi_1(t)$, $\quad (h(t), h(t)\hat{B}(t)) \le \psi_2(t)$,

(10) $\int_E H(u(t,x))\hat{\pi}(t,dx) \le \psi_3(t)$, $\quad \int_E H(v(t,x))\tilde{\pi}(t,dx) \le \psi_4(t)$,

where

$$H(y) = \begin{cases} (1+y)\ln(1+y) - y & \text{for } y > 1, \\ 1 & \text{for } y = -1. \end{cases}$$

Denote

$$L_t = \sum_{j=1}^{m} \int_0^t \chi_G(X_{s-})g_j(s)dX_s^{cj} + \sum_{j=2}^{m} \int_0^t \chi_{\partial G}(X_{s-})h_j(s)dX_s^{cj} +$$

$$+ \int_0^t \int_E \chi_G(X_{s-})u(s,x)q(ds,dx) + \int_0^t \int_E \chi_{\partial G}(X_{s-})v(s,x)q(ds,dx),$$

$$\varepsilon_t(L) = \exp\{L_t - \frac{1}{2}\langle L^c, L^c\rangle_t\} \prod_{s\le t} (1+\Delta L_s)e^{-\Delta L_s}, \; t \ge 0.$$

Let P_t (corresp. P_t') be the restriction of the measure P corresp. P') on the σ -algebra $\mathcal{F}_t$, $t \ge 0$.

The following statements hold.

<u>Theorem 1</u>. Let $P_o' \ll P_o$, assumptions (7), (9) and (10) be fulfilled and the measure P' be uniquely determined by the local characteristics $(\hat{\alpha}', \hat{B}, \hat{\pi}', \gamma, \delta, \tilde{\alpha}', \tilde{B}, \tilde{\pi}')$ together with P_o' .

Then $P' \ll P$ and

(11) $$\frac{dP_t'}{dP_t} = \frac{dP_o'}{dP_o} \varepsilon_t(L), \; t \ge 0 .$$

<u>Theorem 2</u>. Let $P_o \sim P_o'$, assumptions (7), (9), (10) be fulfilled, $u(t,x) > -1$, $v(t,x) > -1$ P - a.e. for all $t \ge 0$, $x \in E$ and the measure P be uniquely determined by the local characteristics $(\hat{\alpha}, \hat{B}, \hat{\pi}, \gamma, \delta, \tilde{\alpha}, \tilde{B}, \tilde{\pi})$ together with P_o.

Then $P' \sim P$ and formula (11) is true.

Proof. Using assumptions (7), (9) and (10) from lemma 2 and the results of paper [7] it follows that a martingale $\varepsilon_t(L)$, $t \geq 0$, is uniformly integrable and $\int_\Omega \varepsilon_\infty(L)\, dP = 1$.

Define the measure P" by the following equality:

$$P''(A) = \int_A \frac{dP'_o}{dP_o} \varepsilon_\infty(L) dP, \quad A \in \mathcal{F}.$$

After standard calculations (see [4]) we find that the process X with respect to the measure P" and the family F is a semimartingale having local characteristics $(\hat{\alpha}', \hat{B}, \hat{\pi}', \gamma, \delta, \tilde{\alpha}', \tilde{B}, \tilde{\pi}',)$ and restriction P" on $\mathcal{F}_o$ coincides with P'_o. Hence statement of theorem 1 follows immediately.

Further from the assumptions of theorem 2 and the results of paper [7] we find that $\varepsilon_\infty(L) > 0$ P -a.e. and $\int_\Omega \varepsilon_\infty^{-1}(L) dP' = 1$.

Taking

$$P'''(A) = \int_A \frac{dP_o}{dP'_o} \varepsilon_\infty^{-1}(L) dP', \quad A \in \mathcal{F},$$

by means of Ito's formula it is not difficult to be convinced that the process X is a semimartingale with respect to the measure P''' and the family F having local characteristics $(\hat{\alpha}, \hat{B}, \hat{\pi}, \gamma, \delta, \tilde{\alpha}, \tilde{B}, \tilde{\pi})$ and the restriction P''' on $\mathcal{F}_o$ coincides with P_o. Hence follows the statement of theorem 1.

Remark. Conditions, when the finite dimensional distributions are uniquely determined by the local characteristics and the initial distribution of a semimartingale, are investigated in papers [8]-[11]. The problem of transformation of local characteristics by changing the family of σ -algebras and the measurability properties of the local time are discussed in detail in papers [4]-[5].

REFERENCES

1. Jacod, J, A general theorem of representation for martingales. Proceedings of Symp. in Pure Math., vol. 31, 1977, 37-53.
2. Jacod, J., Mémin, J., Charactéristiques locales et conditions de continuité absolue pour les semimartingales. Z. Wahrscheinlichkeitstheorie verw. Gebiete, 1976, B. 35, p. 1-37.
3. Kabanov, J.M., Liptser, R.S., Shiryayev, A.N., Absolute continuity and singularity of locally absolute continuous distributions, I, II, Math. Sb., 107, 3 (1978), 364-415; 108, 1(1979),32-61.
4. Grigelionis, B., On statistical problems of stochastic processes with boundary conditions, Liet. matem. rinkinys, 1976, t. XVI, Nr. 1, p. 63-87.
5. Grigelionis, B., Mikulevičius, R., On semimartingales with values in R_+^m, Liet. matem. rinkinys, 1979, t. XIX, Nr. 2.
6. Grigelionis, B., On the martingale characterization of random processes with independent increments, Liet. matem. rinkinys, 1977, t. XVII, Nr. 1, p. 75-86.
7. Lepingle, D., Mémin J., Sur l'intégrabilité uniforme des martingales exponentielles, Z. Wahrscheinlichkeitstheorie

verw. Gebiete, 1978, B. 42, p. 175-203.

8. Anderson, R. F., Diffusions with second order boundary conditions I, II, Indiana Univ. Math. J., 1976, v. 25, No. 4, p. 367-395, No. 5, p. 403-441.

9. Watanabe, S., Constructions of diffusion processes with Wentzell's boundary conditions by means of Poisson point processes, Proc. Semester on Probab., Banach Center, Warsaw (to appear).

10. Mikulevičius, R., On the existence of solutions of the martingale problem, Lit. matem. rinkinys, 1977, t. XVII, No. 4, p. 149-167.

11. Mikulevičius, R., On the uniqueness of solutions of the martingale problem, Liet. matem. rinkinys, 1978, t. XVIII, No. 2, p. 63-73.

MULTIPLICATIVE OPERATOR FUNCTIONAL OF MARKOV PROCESSES AND THEIR APPLICATIONS

Yu. L. Dalecky

1°. Multiplicative operator functionals (m.o.f.) of Markov processes are the natural extension of scalar ones. Their means over trajectory space of the process posess an evolution property. It comes to be useful to represent the solutions of some evolution equations in the form of those means if one is going to study them.

In fact m.o.f. had been used for this purpose already in [1]. Explicitly they were introduced independently in [2] and [3] and applied to represent the solutions of a Cauchy problem for parabolic systems. Some aspects and applications of m.o.f. theory are treated in [4-11]. Here we extend the results of these works to a case of differential operators with unbounded operator coefficients.

2°. Let $\xi(t)(t \geq t_0)$ be a Markov process defined on a probability space $(\Omega, \mathcal{A}, P)$ and valued in a metric space X. F_t a flow of algebras, agreed with it, M_t - a conditional expectation with respect to F_t.

Let K, K^* be a dual pair of Banach spaces with form $\langle \phi, f \rangle_k$ ($\phi \in K$, $f \in K^*$). In the all below $L(K,K_1)$ is a space of bounded linear operator from K to K_1, $L(K) = L(K,K)$, $L_2(K,K_1)$ a Hilbert-Schmidt operator space for Hilbert spaces K and K_1. All the space under consideration are real and separable.

Let $B = B(K)$ be a space of K -valued random variables defined on Ω with a norm $||\eta||_B^2 = M||\eta||_K^2$, $B_t \subset B$ is a subspace of its F_t -measurable elements.

<u>Definition</u>. A bounded linear operator $T(t,\tau) \in L(B_\tau, B_t)$ $(\tau \le t)$ is called a m.o.f. of $\xi(t)$, if the next is valid

(1) $M_\tau\{||T(t,\tau)\eta||_K^2\} \le \text{const.}\ ||\eta||_K^2 \quad (\eta \in B_\tau,\ \tau \le t),$

(2) $\qquad T(t,\tau) = T(t,s)T(s,\tau) \quad (\tau \le s \le t).$

<u>Remark</u>. According to this definition every uniformly bounded F_t -measurable random function $T(t,\tau)$ valued in $L(K)$ and possesing the property (2) a.e. is M.o.f. We shall call them uniform (u.m.o.f.).

Denote $B(X,K)$ a space of bounded measurable K -valued functions on X, $C(X,K)$ a subspace of uniformly continuous functions. A relation

(3) $\langle y, U(\tau,t)f(x)\rangle_K = M_{\xi(\tau)=x} \langle T(t,\tau)y, f(\xi(t))\rangle_K, (y \in K, f \in B(X,K^*); t \ge \tau)$

defines a family $U = U^{(T)}(\tau,t)$ $(t \ge \tau)$ of bounded operators in $B(X,K^*)$ possesing an evolution property

(4) $\qquad U(\tau,t) = U(\tau,s)\,U(s,t) \quad (\tau \le s \le t),\ U(t,t) = I.$

For a u.m.o.f. this relations may be written in a form

$$U^{(T)}(\tau,t)f(x) = M_{\xi(\tau)=x}\{T^*(t,\tau)f(\xi(t))\}.$$

A linear operator in $B(X,K^*)$ defined by a relation

$$\mathscr{A}_t^{(T)} f(x) = \lim_{\Delta t \to 0} \frac{1}{\Delta t} [U(t,t+\Delta t) - I] f(x)$$

on a linear set $D_t^{(T)} \subset B(X,K^*)$ of those elements for which the limit exist we call a generator of evolution family $U^{(T)}$. In particular, if T = I. $\mathscr{A}_t^{(I)}$ is a generator of Markov process $\xi(t)$. Simple calculations lead us to a representation

(5) $$\mathscr{A}_t^{(T)} f(x) = \mathscr{A}_t^{(I)} f(x) + V^*(t,x)f(x) + \gamma_t f(x), \quad (f \in \tilde{D}_t^{(T)}),$$

where $V(t,x), \gamma_t$ are operators in $B(X,K^*)$, defined by relations

$$V(t,x)y = \lim_{\Delta t \to 0} \frac{1}{\Delta t} M_{\xi(t)=x}\{T(t+\Delta t,t)y-y\},$$

$$\langle y, \gamma_t f(x)\rangle_K = \lim_{\Delta t \to 0} \frac{1}{\Delta t} M_{\xi(t)=x} \{\langle T(t+\Delta t,t)y-y, f(\xi(t+\Delta t))-f(x)\rangle_K\},$$

$\tilde{D}_t^{(T)} \subset D_t^{(T)}$ is a set of those $f(x)$ for which all summands in (5) are defined. For functions $f \in C(X,K^*)$ the conditions $M||T(t+\Delta t,t)y-y||^2 = 0(\Delta t^2)$ involves $\gamma_t f = 0$.

3°. Let $c(t,x)(t \geq t_o, x \in X)$ be a function valued in a space of linear (not necessary bounded) operators in K.

Consider a differential equation

(6) $$\frac{d\eta}{dt} = c_\xi(t)\eta \quad , \quad c_\xi(t) = c(t,\xi(t)).$$

If its Cauchy problem is correct then an evolution operator

(7) $$S(t,\tau) = \overleftarrow{\exp} \int_\tau^t c_\xi(\Theta)d\Theta$$

is a u.m.o.f. of ξ .

Below we shall use the following conditions

A_1) Domain $\delta = \delta_c$ ($\bar{\delta} = K$) of an operator $c(t,x)(t \geq t_o, x \in X)$ is constant and an operator $k(t,x) = c(t,x)c^{-1}(t_o,x_o)$ is uniformly bounded for $t \geq t_o; x, x_o \in X$.

A_2) A Holder conditions is valid

(8) $||k(t,x) - k(t_1,k_1)|| \le \text{const.} \{|t-t_1|^{\alpha} + [\rho(x,x_1)]^{\beta}\}, \alpha > 0, \beta > 0,$

(9) $A_3)$ $||[c(t,x)-\lambda I]^{-1}|| \le \text{const.} (1+|\lambda|)^{-1}$, $\text{Re } \lambda \ge 0$.

If trajectories of $\xi(t)$ possess a Holder property

(10) $$\rho(\xi(t), \xi(t_1)) \le \text{const.} |t-t_1|^{\gamma}, \gamma > 0 \quad \text{(a.e.)}$$

then a Cauchy problem (6) is uniformly correct (see [12]) and its evolution operator may be represented in a form

(11) $$S(t,\tau) = \lim_{n\to\infty} \prod_{K=1}^{n} \exp \{c_{\xi}(\tau_K)(\tau_K-\tau_{K-1})\}, \ \tau_K = \tau + K \frac{t-\tau'}{n} .$$

<u>Theorem 1.</u> Under conditions A_1, A_2, A_3 and (10) the operator (7) defines a u.m.o.f. of process $\xi(t)$. An evolutionary family

(12) $$U^{(s)}(t,\tau) = \lim_{n\to\infty} \prod_{K=1}^{n} U^{(I)}(t_{K-1},\tau_K) \exp [c^{*}_{\xi}(\tau_K)(\tau_K-\tau_{K-1})$$

has a generator, the restriction of which to $\tilde{D}_t^{(s)}$ may be represented in a form (5) with $\gamma_t = 0$ and $V = c$.

The proof of the theorem follows from the considerations mentioned above. The representation (12) is obtained by putting (11) in (3).

The theorem 1 involves the conditions of correctness of a Cauchy problem for an equation of the form

$$\frac{du}{dt} = \mathscr{A}_t^{(I)} u + c(t,x)u$$

(see [13] where an analogous problem was treated under the conditions that values of operator function c(t,x) had to commute for different t,x).

4^0. Let X be a smooth Riemannian manifold without a boundary

($\dim X \le \infty$) uniform in the next sense; there is $R > 0$ such that in a neighbourhood $\{y \in T_x X : ||y|| \le R, x \in X\}$ an exponential map is invertible and posseses uniformly bounded derivatives up to the third order.

Let H be a Hilbert space, w(t) - a standart Wiener process in an extension $H_- \supset H$ (see [16]), $a(t,x) \in T_x X$, $A(t,x) \in L_2(H, T_x X)$ are respectively vector and operator field on X. One may describe a diffusion process in (see [14] as a solution of invariant stochastic Ito equation (see [15, 6, 7])

$$(13) \qquad d\xi(t) = \exp_{\xi(t)}\{a(t,\xi(t))dt + A(t,\xi(t))dw(t)\}.$$

Under usual assumptions about Lipschitz property of coefficients the solution of this equation exists and associated generator has a form

$$(14) \qquad \mathscr{A}_t^{(I)} = \langle A(t,x)A^*(t,x)\nabla,\nabla\rangle_{T_x X} + \langle a(t,x),\nabla\rangle_{T_x X} ,$$

where ∇ is a symbol of covariant derivative on X.

To construct more general then (7) m.o.f. of $\xi(t)$ let us consider in a Hilbert space K a linear stochastic equation

$$(15) \qquad d\eta = c(t,\xi(t)\ dt + \alpha(t,\xi(t),\ dw(t),\eta),$$

where c(t,x) is an operator in K, $\alpha(t,x,\phi,y) = B(t,x,y)\phi$, $B(t,x,\cdot) \in L(K, L_2(H,K))$.

<u>Theorem 2</u>. Let A_1, A_3 is valid and moreover

$$(16) \qquad \langle c(t,x)\phi,\phi\rangle \le \text{const}\ ||\phi||^2 \quad (\phi \in \delta)$$

and operator functions $k(t,x)$ and $c(t_o,x_o)k(t,x)C^{-1}(t_o,x_o)$ are uniformly bounded along with their first order derivatives with respect to t and first, second order derivatives with respect to x. Let in addition an estimate

$$(17) \quad \sup_{t,x} ||C^j(t_o,x_o)B(t,x,\cdot)C^{-j}(t_o,x_o)||_{L(K,L_2(H,K))} < \infty, \quad (j=0,1,2)$$

is hold. Then a Cauchy problem for the equation (15) is correct and its solution defines a m.o.f. $T : \eta(t) = T(t,\tau)\eta(\tau)$. If coefficients a, A, B, k are smooth enough then evolution operator $U^{(T)}$ leaves a class of appropriately smooth functions from C(X,K) to be invariant. The restriction of the generator $\mathscr{A}^{(T)}$ to those function valued in $\delta(C^*)$ has a form (5) where $\mathscr{A}^{(I)}$ is an operator (14),

$V = c$, $\gamma_t f = \alpha^*(t,x,A^*(t,x)\nabla,f)$; $\quad \langle h,\alpha^*(z,f)\rangle = \langle \alpha(z,h),f\rangle$.

To prove the theorem we change the equation (15) to analogous equation with bounded operator $c_n = -nc(c-nI)^{-1}$(see [12] for deterministic equations). Properties of the solutions of those equations are known well enough ([5, 16]). We prove the convergence in a square mean of the sequence of solutions η_n and weak convergence of images $c\eta_n$. The limit process $\eta(t)$ comes to be an unknown solution. We prove that any solution $\eta(t)$ may be approximated by processes $\eta_n(t)$ that involves the uniqueness of solution as well as multiplicative property. To prove smooth property for solution we consider equations for derivatives of unknown solution obtained in a formal way.

Remark 1. A family of operator U, constructed above, is acting in a space C(X,K) of sections of a trivial vector bundle X x K.

Changing the construction one may consider an evolution family acting in a space of sections of nontrivial vector bundle $\pi : E \to X$. In this case m.o.f. is acting from a layer $K_{\xi(t)}$ to a layer $K_{\xi(t)}$ of the bundle π.

Remark 2. In theorem 1 and 2 it is enough to assure the conditions being valid for a subset $X_1 \subset X$ to which process $\xi(t)$ belongs with probability 1.

5°. The theorem 2 leads to state the conditions of correctness of Cauchy problem for a parabolic system

$$\frac{\partial u}{\partial t} + \mathscr{A}_t u \equiv \frac{\partial u}{\partial t} + [\tfrac{1}{2}\langle AA^*(t,x)\nabla,\nabla\rangle + \langle a(t,x)\nabla\rangle]u + \alpha^*(t,x,A^*(t,x)\nabla,u) + \tag{18}$$
$$+ c^*(t,x)u = 0, \quad (t \le \tau), u(\tau,x) = u_0(x).$$

The solution of the problem is a vector function valued in K and defined on X (a section of vector bundle π) that may be represent as a mean of a certain functional over the trajectory space of diffusion process. This comes to be very convenient for investigation and estimation of parabolic systems solutions (especially for infinite dimensional spaces X,K as far as traditional method being unacceptable in this situation).

Following to an idea [17] on this way we investigate a Cauchy problem for a quasilinear system of the type (18) (for bounded c see [8, 9]).

Let both random process and m.o.f. of it depend on a certain parameter

$$\xi_z = \xi(t,z) \quad , \quad T_z = T(\tau,t,z). \tag{19}$$

Then an evolution family

(20) $U_z = U^{(T_z)}(\Theta,t) : u(t,x) = U_z(t,\tau)f(x), \quad (\Theta \le t \le \tau)$

depend on it also. Putting

(21) $$z = u(t,x)$$

we shall have a closed system of equations that defines a Markov process $\tilde{\xi}(t)$, a m.o.f. of it $\tilde{T}(\tau,t)$ and a function $\tilde{u}(t,x) = U_{\tilde{u}}(t,\tau)f(x)$ which is natural to take for a generalised solution of a Cauchy problem for a quasilinear equation

(22) $$\frac{d\tilde{u}(t,x)}{dt} + \mathscr{A}_t^{(\tilde{T}_{\tilde{u}})}\tilde{u}(t,x) = 0;\ t \le \tau;\ \tilde{u}(\tau,x) = f(x) .$$

This solution becomes a classical one if it belongs to a domain of operator $\tilde{\mathscr{A}}_t$. We may construct the solution of (19)-(21) by means of successive approximations

(23) $$\xi_K = \xi(t,u_K), T_K = T(\xi_K,t,\tau,u_K), U_K(\tau,x) = f(x), U_{K+1}(t,x) = U_{(t,\tau)}^{(T_K)}f(x)$$
$$(t \le \tau;\ K = 0,1,2,\dots) ,\ u_o(x) = f(x) .$$

Consider the case described in theorem 2 assuming for the sake of simplicity that X is a Hilbert space. Let the next conditions be valid:

B_1) estimates for coefficient growth

$$||a|t,x,z)||^2 + \mathrm{Sp}\, A^*A(t,x,z) \le \text{const.}\ (1+||x||^2+||Z||^{2p}) ,$$

$$||B(t,x,z)||^2 + ||k(t,x,z)||^2 \le \text{const}\ (1+||Z||^{2p}) ,\ (p > 0) ,$$

B_2) Lipschitz condition for coefficients a,A,B,k in corresponding norms with respect to x on the whole X and with respect to z on any sphere $\{||z|| \leq R\}$,

B_3) dissipation condition

$$\langle c(t,x,z)\phi,\phi\rangle \quad \leq \quad ||\phi||^2(\rho + \text{const. } ||Z||^p) ,$$

B_4^K) K -order smooth conditions that is condition B_1 is valid in corresponding norms for all derivatives of coefficients with respect to t,x,z up to order K.

Under those conditions we succeed to obtain with a help of stochastic equations some uniform estimates of successive approximations (23) and their derivatives up to necessary order, that garantees the convergence of these approximations.

Theorem 3. Let in addition to assumptions of theorem 2 conditions B_1, B_2, B_3, B_4^K be valid. Then there exists an inverval $\Delta_K = [t,\tau]$, depending only on parameters in estimates of coefficients and initial function, such that the system (19)-(21) has exactly one solution of the class C^K on it.

This interval comes to be unbounded, if $\rho < 0$ and $|\rho|$ is large enough.

Remark. For dim X < ∞, dim K < ∞ successive approximations uniformly converge along with their derivatives up to order j=1,...,k for every finite part of Δ_k. For infinite dimensional case this statement is valid under some natural additional assumptions.

6^0. It is known that a contractive semigroup in a Hilbert space is generated by a dissipative operator. It is easy to verify that if a correspondent Cauchy problem is correct in some space which is a dense subset of the domain of this operator then its restriction is essentially maximal (in particular, selfadjoint if the considered operator is a symmetric one, see [10, 11]). On this way theorems 1 and 2 involve conditions of the maximality of differential operators for such nonstandard cases when argument x would be of infinite dimension and some operator coefficients would be unbounded. We shall state one of those results (in other conditions see [10, 11]).

Let $X \supset X_o$ be extended Hilbert space, μ -canonically connected with X_o a Gaussian measure in X. In a Hilbert space $L_2(X,K,\mu)$ consider a differential operator

$$(24) \quad \mathscr{A} = \mathscr{A}_o + \mathscr{A}_1; \; \mathscr{A}_o = \langle AA^*(x)\nabla,\nabla\rangle + \langle a(x),\nabla\rangle, \; \mathscr{A}_1 = \alpha(x)(\nabla,\cdot) + c(x).$$

It is followed from formulas of integrating by part (see [10,11] that if

$$(25) \qquad \langle a(x),\phi\rangle = \langle \nabla - x, AA^*(x)\phi\rangle \quad (\phi \in X) ,$$

then an operator $\mathscr{A}_o$ is symmetric and nonpositive, and if

$$\langle\psi,\alpha(x)(A^*(x)x,\psi)\rangle - \sum_k\langle[\alpha(x)(A^*(x)e_k,\psi)]'e_k, \; \psi\rangle + \langle[c(x)+C^*(x)]\psi,\psi\rangle \leq 0$$

$(\psi \in K, \; \{e_k\}$ - an orthonormal system in X),

then an operator $\mathscr{A}_1$ is dissipative. For the dense subset, which is invariant with respect to the action of semigroup, generated by the operator (24), we choose a space $\mathcal{C}_2(X,c^*)$ of functions, which are bounded and continuous along with their first and second derivatives and valued in δ_{c^*}.

<u>Theorem 4</u>. If in addition to conditions of theorem 2 the condi-

tions (25), (26) are valid, then the operator (24) in space $L_2(X,K,\mu)$ possesses a maximal dissipative closure on the set $C_2(X,C^*)$, which is selfadjoint, if $\alpha = 0$, $c = c^*$.

REFERENCES

1. Dalecky, Yu. L., Functional integrals connected with operator evolution equations, Russ. Math. Surv., 1962, 17, 5, 1-108.
2. Pinsky, M., Multiplicative operator functionals of a Markov processes, BAMS, 77, 1971, 377.
3. Dalecky, Yu. L., Teterina, N.I., Multiplicative stochastic integrals, Usp. Mat. Nauk, 1972, 27, 2, 167-168.
4. Dalecky, Yu. L., Diffusion processes in infinite-dimensional spaces and differential equations, Materjaly Vsesojusnoj Shkoly po Differencialnym Uravnenijam s Beskonecnym Cislom Nesavisimyh Peremennyh (Dilijan, 21. 5-3.6. 1973), AN Arm. SSR, 1974, 5-49.
5. Belopolskaya, Ya. I., Dalecky, Yu. L., Diffusion processes in smooth Banach spaces and manifolds, Trudy Mosk. Math. Ob., 1978, 37, 107-142.
6. Dalecky, Yu. L., Multiplicative operators of diffusion processes and differential equations in sections of vector bundles, Usp. Mat. Nauk, 1975, 30, 2, 209-210.
7. Belopolskaja, Ya. I., Dalecky, Yu. L., Probability methods in some problems of global analysis. The 2nd Vilnius Conf. on Prob. and Stat., Theses, Vilnius, 1977, v. 1, 27-30.
8. Belopolskaja, Ya. I., Dalecky, Yu. L., Probability methods of

investigation of quasilinear parabolic systems, Usp. Mat. Nauk., 1978, 33, 5, 212-213.

9. Belopolskaja, Ya. I., Dalecky, Yu. L., Investigation of Cauchy problem for quasilinear parabolic systems with Markov processes, Izvestija VUZ, Matematika, 1978, 12, 5-17.

10. Dalecky, Yu. L., On selfadjoitness and maximal dissipativeness of differential operators for functions of infinite dimension arguments, DAN SSSR, 1976, 227, 4, 784-787.

11. Dalecky, Yu, L., Probability methods in some problems of infinite dimension analysis, Predelnie teoremy dlja sluchajnyh processov, Kiev, 1977, 108-124.

12. Krein, S. G., Linear differential equation in a Banach space, Moscow, "Nauka", 1967.

13. Grady, M. D., Trans. Amer. Math. Soc., 1976, 223, 181-203.

14. Ito, K., Stochastic differential equations in a differentiable manifold, (2), Mem. of the College of Sciences, Univ. of Kyoto, S. A., 28, Math., I, 1953, 81-85.

15. Dalecky, Yu. L., Schnaiderman, Ya. I., Diffusion and quasi-invariant measures on infinite dimensionalen Lie group, Funkc. Anal., 1969, 2, 88-90.

16. Dalecky, Yu. L., Infinite-dimensional elliptic operators and parabolic equations connected with them, Russ. Math. Surv., 1967, 22, 4, 3-54.

17. Freidlin, M. I., On existence "as a whole" smooth solutions degenerate quasilinear equations, Mat. Sbornik, 1969, 78, 3, 332-348.

ON THE PREDICTABLE JUMPS OF MARTINGALES

L. I. Galtchouk
Laboratory of Mathematical Statistic
Department of Mathematic and Mechanic
Moscow State University
Moscow, 117234, USSR

0. We are given a complete probability space $(\Omega, \mathcal{F}, P)$ and a family of σ-sub-algebras $(\mathcal{F}_t)$, $t \in R_+$, satisfying the "usual" conditions: $\mathcal{F}_t = \bigcap_{s>t} \mathcal{F}_s$ and $\mathcal{F}_o$ is completed by all P-null sets. Let $X = (X_t)$, $t \in R_+$ be a local martingale with right-continuous paths having left-hand limits. Denote by (T_n), (S_n), $n \in |N|$, sequences of totally inaccessible and predictable stopping times (s.t.'s), respectively, which absorb all discontinuity times of X. The graphs of all S and T are disjoint.

It is well known (cf. [1]) that

$$\sum_n [(\Delta X_{S_n})^2 I_{S_n \le t} + (\Delta X_{T_n})^2 I_{T_n \le t}] < \infty$$

a.s. for any $t \in T_+$, where $\Delta X_t = X_t - X_{t-}$. For predictable jumps of X we have the following result.

<u>Lemma 1</u>. Let X be a local martingale and let (S_n), $n \in N$, be a sequence of predictable s.t.'s mentioned above. Then $Y_t = \sum_{S_n \le t} \Delta X_{S_n} < \infty$ a.s. for any $t \in R_+$ and the process $Y = (Y_t)$ is a local martingale.

Proof. 1°. $X \in \underline{\underline{M}}$, i.e. X is a square integrable martingale. For any n $(\Delta X_{S_n} I_{S_n \le t}) \in \underline{\underline{M}}$ and these processes are orthogonal for $n \neq k$ because $[\![S_n]\!] \cap [\![S_k]\!] = \emptyset$.
We have $E|\sum_n \Delta X_{S_n}|^2 = E \sum_n (\Delta X_{S_n})^2 < \infty$. Hence $\sum_n \Delta X_{S_n} < \infty$ a.s.

2°. $X \in \underline{\underline{W}}$, i.e. X is a uniformly integrable martingale with bounded variation. Then $\sum_n |\Delta X_{S_n}| < \infty$ a.s.

3°. X is a local martingale. There is a sequence of s.t.'s $(R_k), k \in |N|$ such that $R_k \uparrow \infty$ a.s. and for any k stopped process $X^{R_k} = m^{R_k} + V^{R_k}$ where $m \in \underline{\underline{M}}$, $V \in \underline{\underline{W}}$ (cf. [1]). As $\Delta X_{S_n} I_{S_n \le R_k} = (\Delta m_{S_n} + \Delta V_{S_n}) I_{S_n \le R_k}$ for any k from the mentioned above we have

$$\sum_n \Delta X_{S_n} I_{S_n \le R_k} = \sum_n (\Delta m_{S_n} + \Delta V_{S_n}) I_{S_n \le R_k} < \infty \text{ a.s.}$$

For $R_k \uparrow \infty$ a.s. we have the statement of lemma. □

We shall give three applications of this result.

1. Let $X = (X^1, \dots, X^n)$ be an n-dimensional semimartingale. On the product space $(R_+ \times E, \mathscr{B}(R_+) \times \mathscr{B}(E))$, $E = R^n \setminus \{0\}$ we define two random integer-valued measures

$$\mu(\Gamma) = \sum_n I_\Gamma(T_n, \Delta X_{T_n}) \ , \ p(\Gamma) = \sum_n I_\Gamma(S_n, \Delta X_{S_n}) \ ,$$

where Γ is a Borel set, $I_\Gamma(\cdot)$ is the indicator of Γ, (T_n), (S_n) are sequences of totally inaccessible and predictable s.t.'s absorbing the jumps of X. Denote by ν and λ dual predictable projections of μ and p, respectively (cf. [3]).

We denote $g*(\mu-\nu)_t = \int_0^t \int_E g(s,x)(\mu-\nu)(ds,dx)$, $h*p_t = \int_0^t \int_E h(s,x)p(ds,dx)$ when the stochastic integrals are defined.

Also for n-dimensional function $f = (f^1, \dots, f^n)$ and semimartingale $X = (X^1, \dots, X^n)$ we denote $f_o X_t = \int_0^t f_s dX_s = \int_0^t \sum_{i=1}^n f_s^i \, dX_s^i$.

Lemma 2. Let $X = (X^1, \dots, X^n)$ be an n-dimensional semimartingale.

Then for any $t \in R_+$

$$X_t = X_o + a_t + m_t + U *(\mu-\nu)_t + V * \mu_t + (U + V) * p_t$$

a.s., where a, m are continuous n-dimensional processes starting from zero, a is a predictable with loc. integrable variation, m is a local martingale, the functions $U(w,s,x) = x\, I_{|x|\leq 1}$, $V(w,s,x) = x\, I_{|x|>1}$. The last right-hand term is equal to $\sum_n \Delta X_{S_n} I_{S_n \leq t}$ and

$V * \mu_t = \sum_n \Delta X_{T_n} I_{|\Delta X_{T_n}|>1} I_{T_n \leq t}$.

Proof. By lemma 1 the process $Z = (\sum_n \Delta X_{S_n} I_{S_n \leq t})$ is a semimartingale. The process $\tilde{X} = X - Z$ is a semimartingale, $\tilde{X}$ has the jumps in totally inaccessible s.t.'s. For such process the result (without the last term) is well-known (see [4] and [5] for the one-dimensional case). □

The given representation of the semimartingale is completely analogous to the famous Levy-Khintchin decomposition for independent increaments processes.

2. Using lemma 2 one can give another form for Ito's formula.

Lemma 3. Let $X = (X^1,\dots,X^n)$ be a semimartingale, $F(x) = F(x_1,\dots,x_n)$ be a twice continuously differentiable function on R^n. Then the process $F(X) = (F(X_t))$ is a semimartingale and for any $t \in R_+$ the representation holds

$$F(X_t) = F(X_o) + \int_0^t \sum_{i=1}^n D^i F(X_{s-}) d(a^i+m^i)_s +$$

$$+ \frac{1}{2} \int_0^t \sum_{i,j=1}^n D^i D^j F(X_{s-}) d\langle m^i,m^j\rangle_s + \int_0^t \int_{|x|\leq 1} G_F(X_{s-},x)x$$

$$x(\mu-\nu)(ds,dx) + \int_0^t \int_{|x|\le 1} [G_F(X_{s-},x) - \sum_{i=1}^n x_i D^i F(X_{s-})]\nu(ds,dx) +$$

$$+ \int_0^t \int_{|x|>1} G_F(X_{s-},x)\mu(ds,dx) + \int_0^t \int_E G_F(X_{s-},x)p(ds,dx),$$

where $G_F(X_{s-},x) = F(X_{s-}+x) - F(X_{s-})$, D^i is the operator of differentiating in x_i.

Proof. One can transform the term $\sum_{s\le t} [\cdot]$ of Ito's formula given in [1] to our form. □

Remark. In the given Ito's formula we have the so-called "Levy system". Our formula coincides with that given in [6], when the jumps of X are totally inaccessible.

3. Let $X = (X^1,\dots,X^n)$ be a martingale and $X^i \in H^1$, $i = 1,\dots,n$. Let m, μ, p, ν be a martingale and measures from descomposition of X given in lemma 2.

For any predictable function $f = (f^1,\dots,f^n)$ satisfying $E[\sum_{i,j=1}^n f^i f^j \circ \langle m^i, m^j\rangle_\infty]^{1/2} < \infty$ one can define a stochastic integral $f \circ m \in H^1$. Let g, h be functions of the space $(\Omega \times R_+ \times E, \tilde{\mathscr{P}})$, $\tilde{\mathscr{P}} = P \times \mathscr{B}(E)$ satisfying $E[g^2 * \mu_\infty]^{1/2} < \infty$, $E[h^2 * p_\infty]^{1/2} < \infty$, respectively. Besides the function h satisfies $E[h(T,\Delta X_T)|\mathscr{F}_{T-}] = 0$ a.s. on $(T < \infty)$ for any predictable s.t. T with $[\![T]\!] \subseteq \bigcup_n [\![S_n]\!]$. Then one can define stochastic integrals $g * (\mu-\nu) \in H^1$, $h * p \in H^1$.

Denote by $\mathscr{L}(X)$ the family of martingales $Y \in H^1$ such that for any $t \in R_+$

$$Y_t = f \circ m_t + g * (\mu-\nu)_t + h * p_t ,$$

where the functions f, g, h satisfy the conditions given above.

<u>Theorem</u>. 1) $\mathscr{L}(X)$ is a closed subspace of H^1.

2) For any martingale $Y \in H^1$ the decomposition $Y = Y' + Y''$ holds, where $Y' \in \mathscr{L}(X)$, $Y'' \in H^1$ Y'' is orthogonal to any martingale $Z \in \mathscr{L}(X)$ with bounded jumps. The decomposition is unique.

We shall give two auxiliary results before proving the theorem.

Lemma 4. Let $X = (X^1,\dots,X^n)$ be a continuous martingale $X \in H^1$. Then any $Y \in H^1$ has unique decomposition $Y = f \circ X + Z$, where $Z \in H^1$, $Z \perp \mathscr{L}(X)$, predictable function $f = (f^1,\dots,f^n)$ satisfies

$$E[\sum_{i,j=1}^{n} f^i f^j \circ \langle X^i, X^j \rangle_\infty]^{\frac{1}{2}} < \infty.$$

The proof follows from decomposition $Y = Y^c + Y^d$, $Y^d \perp \mathscr{L}(X)$ and from the fact that $Y^c \in H^2_{loc}$. For $Y \in H^2_{loc}$ the decomposition is well known. We omit the details. □

Let Y be $\mathscr{F} \times \mathscr{B}(R_+) \times \mathscr{B}(E)$ -measurable function such that $E[|Y| * \mu_\infty] < \infty$. On the space $(\Omega \times R_+ \times E, \tilde{\mathscr{P}})$ we define the measure $M_{Y*\mu}$ by

$$M_{Y*\mu}(\phi) = E[\phi Y * \mu_\infty]$$

where ϕ is a positive $\tilde{\mathscr{P}}$-measurable function. The measures $M_{Y*\mu}$ and and $M_{1*\mu}$ are σ-finit on $\tilde{\mathscr{P}}$ and $M_{Y*\mu} \ll M_{1*\mu}$. Therefore one can define conditional expectation $E_\mu[Y|\tilde{\mathscr{P}}]$ as some modification of Radon-Nicodym derivative of the measure $M_{Y*\mu}$ with respect to $M_{1*\mu}$. On the same way one define $E_p[Y|\tilde{\mathscr{P}}]$.

Lemma 5. 1) Let Y be a square-integrable martingale ($Y \in H^2$). Denote

$$g = E_\mu[\Delta Y|\tilde{\mathscr{P}}], \quad h = E_p[\Delta Y|\tilde{\mathscr{P}}].$$

Then

$$g^2 * \mu_T \le \sum_{s\le T} (\Delta Y_s)^2 \ , \ h^2 * p_T = \sum_{s\le T} (\Delta Y_s)^2, \text{ a.s.}$$

for any s.t. T on $(T < \infty)$.

2) Let $Y \in H^1$. Then

$$E[g^2 * \mu_\infty]^{\frac{1}{2}} < \infty, \ E[h^2 * p_\infty]^{\frac{1}{2}} < \infty .$$

Proof. 1) At first one give the inequalities for expectations and then for paths by the optional section theorem. The proof of 2) follows from the fact that the space H^2 is dense in H^1. We omit the details. □

Proof of the theorem. 1) The closeness of $\mathscr{L}(X)$ follows from the completeness of the spaces of functions f, g, h satisfying the conditions given before the theorem.

2) We have $Y = Y^c + Y^d$, Y^c, $Y^d \in H^1$. By lemma 4 $Y^c = f \circ m + Z$, where f satisfies the necessary conditions, $Z \in H^1$, $Z \perp \mathscr{L}(m)$. For Z continuous martingale it is orthogonal to all martingales of the form $g * (\mu-\nu) + h * p$. Hence $Z \perp \mathscr{L}(X)$. Define $\tilde{Y} = Y^d - g * (\mu-\nu) - h*p$, where $g = E_\mu[\Delta Y|\tilde{\mathscr{P}}]$, $h = E_p[\Delta Y|\tilde{\mathscr{P}}]$. We shall prove that $\tilde{Y}$ is orthogonal to any martingale $N \in \mathscr{L}(X)$ with bounded jumps, i.e. $N = \phi*(\mu-\nu)+\psi*p$, where $|\phi| \le c$, $|\psi| \le c$. We have for any s.t. T

$$[\tilde{Y},N]_T = \sum_{T_i\le T} (\Delta Y_{T_i} - g(T_i,\Delta X_{T_i}))\phi(T_i,\Delta X_{T_i}) +$$

$$+ \sum_{S_i\le T} (\Delta Y_{S_i} - h(S_i,\Delta X_{S_i}))\psi(S_i,\Delta X_{S_i}) .$$

Therefore

$$E[\tilde{Y},N]_T = E[(\Delta Y-g)\phi*\mu_T] + E[(\Delta Y-h)\psi*p_T] .$$

By definition of g and h the terms to the right-hand are equal to zero. Hence the process $[\tilde{Y},N]$ is a local martingale, i.e. $\tilde{Y} \perp N$. Denote

$$Y' = f \circ m + g*(\mu-\nu) + h*p,$$

$$Y'' = (Y^c - f \circ m) + \tilde{Y} \ (=Y-Y') .$$

The proof of uniqueness is simple and we omit it. □

REFERENCES

[1] C. Doléans, P. A. Meyer, Integrales stochastiques par rapport aux martingales locales, Semin. Probab. IV, Univ. de Strasbourg, Lectures Notes in Math., 129, Springer-Verlag (1970), Berlin.

[2] C. Dellacherie, Capacitées et processus stochastiques, Ergebnisse der Math. 67, Springer, 1972.

[3] J. Jacod, Multivariate point processes: predictable projection. Radon-Nicodym derivatives, representation of martingales, Z. Wahrscheinlichkeitstheorie, verw. Geb., 31, 3(1975), 235-253.

[4] L. I. Galtchouk, On existence and uniqueness of the solution for the stochastic equation with respect to semimartingale, Theor. Probab and appl., XXIII, 4 (1978).

[5] J. Jacod, J. Mémin, Caractéristiques locales et conditions de continuité absolue pour les semimartingales, Z. Wahrscheinlichkeitstheorie, verw. Geb. 35 (1976), 1-37.

[6] M. Yor, Sur les integrales stochastiques optionnelles et une

suite remarquable de formules exponentielles, Semin. Probab. X, Univ. de Strasbourg, Lectures Notes in Math., 511, Springer-Verlag (1976), Berlin.

ON THE EXISTENCE OF A SOLUTION OF
THE STOCHASTIC EQUATION WITH RESPECT TO
A MARTINGALE AND A RANDOM MEASURE

V. A. Lebedev
Lomonosov State University
Department of Mechanics and Mathematics
Laboratory of Mathematical Statistics
117234, Moscow

Let be given a complete probability space $(\Omega, \mathcal{F}, P)$ with a non-decreasing right-continuous family $\underline{\mathcal{F}} = (\mathcal{F}_t)_{t\in R+}$ of σ-algebras $\mathcal{F}_t \subset \mathcal{F}$ where $\mathcal{F}_o$ contain all P-null sets from $\mathcal{F}$. Let us introduce the following denotations. For some metric space X let $\mathcal{B}(X)$ denote its Borel σ-algebra. Let also Π (respectively $\mathcal{O}$ or $\mathcal{P}$) be the progressive (respectively optional or predictable) σ-algebra of subsets of $\Omega x R_+$ that is generated by all $\underline{\mathcal{F}}$-progressively measurable (respectively $\underline{\mathcal{F}}$ -adapted right-continuous or $\mathcal{F}$-adapted continuous) real-valued processes. It is known from [1] that $\mathcal{P} \subset \mathcal{O} \subset \Pi \subset \mathcal{F} \times \mathcal{B}(R_+)$.

Let also be given: $a = (a^1,\ldots,a^\ell)$ as an ℓ-dimensional right-continuous predictable (that is $\mathcal{P}$-measurable) process with $a_o = 0$ having locally finite variation, $m = (m^1,\ldots,m^\ell)$ as an ℓ-dimensional continuous local $(\underline{\mathcal{F}}, P)$ -martingale with $m_o = 0$, μ as an integer-valued random measure on $\mathcal{B}(R_+) \times \mathcal{B}(E)$, where $E = R^\ell \setminus \{0\}$, generated by jumps of some ℓ-dimensional $\underline{\mathcal{F}}$-adapted right-continuous and left-limited process, ν as its dual predictable projection.

Notice that the measure μ can be reduced to a measure generated by jumps of some right-continuous $(\underline{\mathcal{F}}, P)$-semimartingale as it was assumed in [2] and such a semimartingale can have even locally finite variation. Indeed one can prove that there exists a strictly increasing function ϕ on R_+ such that $\phi(o) = 0$, $\phi(x) = x$ for $x \geq 1$ and the integral $\phi(|u|)*\mu_t$ converges for any $t \in R_+$ as also the integral $\ell_t = (\Phi(|u|)\wedge 1)*\nu_t$ and then $\frac{\Phi(|u|)}{|u|} u * \mu$ is a required process.

Denote the set $J = \{(\omega,t)\in\Omega x R_+ : \nu(\omega,\{t\}xE) > 0\}$, its section $J(\omega)$ for a given $\omega \in \Omega$, and define the measures μ^d, μ^c, ν^d, ν^c so that for $B \in \mathscr{B}(R_+)$ and $\Gamma \in \mathscr{B}(E)$

$$\mu^d(Bx\Gamma) = I_{(J(\omega)\cap B)x\Gamma} *\mu\infty,\ \mu^c(Bx\Gamma) = \mu(Bx\Gamma)-\mu^d(Bx\Gamma),$$

$$\nu^d(Bx\Gamma) = I_{(J(\omega)\cap B)x\Gamma} *\nu\infty,\ \nu^c(Bx\Gamma) = \nu(Bx\Gamma)-\nu^d(Bx\Gamma).$$

Since the set J is predictable then the measures ν^d and ν^c are dual predictable projections of the measures μ^d and μ^c respectively. Since $J = \{(\omega,t): \Delta\ell_t(\omega) > 0\}$ it is known from [1] that there exists a sequence (τ_k) of predictable stopping times such that $J = \bigcup_{k=1}^{\infty} [\tau_k]$ and $[\tau_i]\cap[\tau_k] = \emptyset$ for $i \neq k$.

It is convenient to define μ, ν and consequently μ^d, μ^c, ν^d, ν^c as measures on $\mathscr{B}(R_+)x\,\mathscr{B}(E)$ where $\bar{E} = R^\ell$. So define for $B \in \mathscr{B}(R_+)$

$$\mu(Bx\{0\}) = \sum_{k=1}^{\infty} I_{B\cap[\tau_k(\omega)]}(1-\mu(\{\tau_k\}xE))$$

and

$$\nu(Bx\{0\}) = \sum_{k=1}^{\infty} I_{B\cap[\tau_k(\omega)]}(1-\nu(\{\tau_k\}xE)) .$$

Then

$$\mu^d(Bx\{0\}) = \mu(Bx\{0\}),\ \nu^d(Bx\{0\}) = \nu(Bx\{0\}), \mu^c(Bx\{0\}) = \nu^c(Bx\{0\}) = 0$$

and for every $(\omega,t)\in \Omega x R_+$ $\mu(\omega,\{t\}x\bar{E}) = \mu^d(\omega,\{t\}x\bar{E}) = \nu(\omega,\{t\}x\bar{E}) =$ $= \nu^d(\omega,\{t\}x\bar{E}) = I_y(\omega,t)$.

Introduce also the following denotations: $|da| = \sum_{j=1}^{\ell} |da^j|$,

$\hat{a}_t = \int_0^t |da_s|$, $\hat{a}_t^d = \sum_{o<s\le t} \Delta\hat{a}_s$, $\hat{a}^c = \hat{a}-\hat{a}^d$, $\langle m\rangle = \sum_{j=1}^{\ell} \langle m^j,m^j\rangle$, $M = \{M^{ij}\}$

where $Mij = d\langle m^i,m^j\rangle / d\langle m\rangle$, for $\Gamma \in \mathcal{B}(E)$ and $t \in R_+$

$$\rho_t(\Gamma) = \frac{d(I_\Gamma * \nu)}{d\ell}\Big|_t\ ,\quad \rho_t^d(\Gamma) = \frac{d(I_\Gamma * \nu^d)}{d\ell}\Big|_t\ ,\quad \rho_t^c(\Gamma) = \frac{d(I_\Gamma * \nu^c)}{d\ell}\Big|_t .$$

Define $\mathcal{D}_{[0,T]}(R^d)$ as the space of R^d-valued right-continuous and left-limited functions on [0,T] with $\mathcal{J}_1$-topology of Skorohod. This space is separable and being metrized as in [3] becomes a complete metric space.

Now let be given d x ℓ-matrix $\mathcal{P} \times \mathcal{B}(R^d)$ -measurable functions f, g on $\Omega x R_+ x R^d$ and d-vector functions h, k respectively on $\Omega x R_+ x(\bar{E}\cap\{|u| \le 1\}) x R^d$ and $\Omega x R_+ x(\bar{E}\cap\{|u| > 1\})xR^d$ which are respectively $\mathcal{P} \times \mathcal{B}(\bar{E}\cap\{|u| \le 1\}) \times \mathcal{B}(R^d)$- and $\Pi \times \mathcal{B}(\bar{E}\cap\{|u|> 1\}) \times \mathcal{B}(R^d)$ -measurable. Consider the equation

$$(1)\quad Y = N + f(Y_-).a + g(Y_-).m + h(Y_-)I_{\{|u|\le 1\}}*(\mu-\nu)+k(Y_-)I_{\{|u|>1\}}*\mu\ ,$$

where N is a given d-dimensional left-limited $\mathcal{O}$-measurable process, and values of the process Y_- for every $(\omega,t) \in \Omega x R_+$ are substituted into the functions f, g, h, k to the place of $x \in R^d$. Define for the equation (1) the notion of its strong and weak solution which is more precise than in [2]. At first notice that the functions f, g, h, k can be changed on some properly measurable sets so that each integral in (1) is P-indistinguishable from the original one for every d-dimensional left-limited $\mathcal{O}$-measurable process Y. For every fixed

$t \in R_+$ denote by $\mathscr{H}_t$ the intersection of all sub-σ-algebras of $\mathscr{F}$ each of which is generated by all P-null sets from $\mathscr{F}$, by values of the processes N, a, m for $0 \le s \le t$ and of the measures μ and ν on sets from $\mathscr{B}([0,t]) \times \mathscr{B}(\bar{E})$ and $\mathscr{B}([0,t]) \times \mathscr{B}(\bar{E} \cap \{|u| \le 1\})$ respectively, and also by values of some above-mentioned modification of the system (f,g,h,k) for $0 \le s \le t$. Call a solution of the equation (1) strong if it is $(\mathscr{H}_t)$ -adapted and weak otherwise or in general. Then it is possible that a weak solution of the equation (1) does not exist on the original probability space and therefore we say that a weak solution exists if there are a probability space (Ω, $\mathscr{F}$, P) with a family $\underline{\mathscr{F}}$, which possess above-mentioned properties, and system of random elements (N, a, m, μ, ν, f, g, h, k) on that space adapted properly to the family $\mathscr{F}$ and having the given distribution on the corresponding space, and a d-dimensional left-limited $\mathscr{O}$-measurable process Y which satisfies the equation (1). In the work [4] there are given some sufficient conditions for existence and uniqueness of a strong solution of the equation (1) for d = 1.

The main result of this paper is the following theorem.

<u>Theorem</u>. Let hold following assumptions:

I. $\frac{|fda|}{|da|} \le c_1$, $Tr(gMg^*) \le c_2$,

$$\int_{|u|\le 1} |h|^2 \rho.(du) - \Big|\int_{|u|\le 1} h\rho^d_{\cdot}(du)\Big|^2 \Delta\ell \le c_3$$

for some $\mathscr{P}$-measurable functions c_1, c_2, c_3 on Ω x R_+ such that the process $c_1.\hat{a} + c_2.\langle m\rangle + c_3.\,\ell$ is P-a.s. finite for any $t \in R_+$;

II. $|k| \le c_4$ where c_4 is some $\Pi \times \mathscr{B}(\bar{E} \cap \{|u| > 1\})$ -measurable function on $\Omega \times R_+ \times (\bar{E} \cap \{|u| > 1\})$ which is P-a.s. $\mu(R_+ \times (\bar{E} \cap \{|u| > 1\}))$ -almost everywhere finite;

III. The functions f, g, h, k, $\int\limits_{|u| \le 1} |h|^2 \rho^c_\cdot(du)$, $\int\limits_{|u| \le 1} h\, \rho^d_\cdot(du)$

are continuous in x P-a.s. almost everywhere respectively in the measures $|da|$ and $d\langle m \rangle$ on R_+, ν on $R_+ \times (\bar{E} \cap \{|u| \le 1\})$, μ on $R_+(\bar{E} \cap \{|u| > 1\})$, $d\ell$ and also $d\ell$ on R_+.

Then there exists a weak, with respect to the system (N, a, m, μ, ν, f, g, h, k), solution of the equation (1) for $t \in R_+$.

The main auxiliary result used in the proof of the theorem is formulated as follows.

<u>Lemma 1</u>. Let $(f^{(\alpha)})$, $(g^{(\alpha)})$, be families of $d \times \ell$ -matrix -measurable functions on $\Omega \times R_+$ and $(h^{c(\alpha)})$, $(h^{d(\alpha)})$, $(k^{(\alpha)})$ be families of d-vector functions on $\Omega \times R_+ \times \bar{E}$ which are respectively $\mathscr{P} \times \mathscr{B}(\bar{E})$- , $\mathscr{P} \times \mathscr{B}(\bar{E})$- and $\Pi \times \mathscr{B}(\bar{E})$-measurable, all these functions depending on a parameter α from some set A, and they are such that for any $\alpha \in A$ P-a.s.

$$\frac{|f^{(\alpha)} da|}{|da|} \le b_1, \quad Tr(g^{(\alpha)} M g^{*(\alpha)}) \le b_2,$$

$$\int\limits_{\bar{E}} |h^{c(\alpha)}|^2 \rho^c_\cdot(du) \le b_3,$$

$$\int\limits_{\bar{E}} |h^{d(\alpha)}|^2 \rho^d_\cdot(du) - \Big|\int\limits_{\bar{E}} h^{d(\alpha)} \rho^d_\cdot(du)\Big|^2 \Delta\ell \le b_4 ,$$

$$|k^{(\alpha)}| \le b_5$$

for some $\mathscr{P}$-measurable functions b_1, b_2, b_3, b_4 on $\Omega \times R_+$ and some

$\Pi \times \mathcal{B}(\bar{E})$ -measurable function b_5 on $\Omega \times R_+ \times \bar{E}$ such that the process $b_1 . \hat{a} + b_2 . \langle m \rangle + (b_3 + b_4) . \ell + b_5 * \mu$ is P-a.s. finite for any $t \in R_+$.

Then each of the sets of distributions for the families of stochastic integrals $(f^{(\alpha)}.a)$, $(g^{(\alpha)}.m)$, $(h^{c(\alpha)} * (\mu^c - \nu^c))$, $(h^{d(\alpha)} * \mu^d - \nu^d))$, $(k^{(\alpha)} * \mu)$ on $D_{[0,T]}(R^d)$ is relatively compact for any $T \in R_+$.

The proof of the lemma 1 is given in [5]. It is based on the change of time connected with majorant processes and under that transformation on the application of results of the works [6] and [7] for stochastically continuous processes.

For the proof of the theorem the equation (1) can be reduced to a more convenient equivalent form. Denote $\hat{f} = \frac{fda}{|da|}$, $\hat{g} = gQ^*$ where Q is a $\mathcal{P}$ -measurable orthogonal $\ell \times \ell$ -matrix function on $\Omega \times R_+$ such that the process $\hat{m} = Q.m$ has orthogonal components, $\hat{h} = h\, I_{\{|u| \le 1\}} - \int_{|v| \le 1} h(t,v,x)\nu(\{t\}, dv)$ on $\Omega \times R_+ \times \bar{E} \times R^d$, $\hat{k}$ to be an $\mathcal{O} \times \mathcal{B}(\bar{E} \cap \{|u|>1\}) \times \mathcal{B}(R^d)$ -measurable function on $\Omega \times R_+ \times (\bar{E} \cap \{|u|>1\}) \times R^d$ which is equal to k for every such (ω,t) that $\mu(\omega)\{t\} \times (\bar{E} \cap \{|u|>1\})) = 1$. Then the equation (1) is equivalent to

$$Y = N + \hat{f}(Y_-).\hat{a} + \hat{g}(Y_-).\hat{m} + \hat{h}(Y_-)I_{\{|u| \le 1\}} * (\mu^c - \nu^c) + \tag{1'}$$
$$+ \hat{h}(Y_-) * (\mu^d - \nu^d) + \hat{k}(Y_-)I_{\{|u|>1\}} * \mu.$$

Notice that the proof can be reduced to the case when the probability measure P is separable as it was assumed in [2]. Namely $\mathcal{F}$ has a countably generated sub-σ-algebra $\tilde{\mathcal{A}}$ such that for any

$t \in R_+$ $\mathscr{H}_t \subset \mathscr{A} \cap \mathscr{F}_t$, where $\mathscr{A}$ is the completion of $\tilde{\mathscr{A}}$ by all P-null sets from $\mathscr{F}$, and the system $(N, \hat{a}, \hat{m}, \mu, \nu, \hat{f}, \hat{g}, \hat{h}, \hat{k})$ is adapted properly to the family $\underline{\mathscr{A}} = (\mathscr{A} \cap \mathscr{F}_t)$ $t \in R_+$. If the measure P is separable then as it follows from [8] there exists a random η with values in the Cantor discontinuum C such that for any random element ξ of some complete separable metric space X and some $\mathscr{B}(C)$ -measurable X -valued function ϕ on C P-a.s. $\xi = \phi(\eta)$. For a separable probability measure P we have the following results which are used in the proof of the theorem.

<u>Lemma 2.</u> Let a sequence of distributions of random elements $\phi_n(\eta)$ of some complete separable metric space X converge weakly to some limit distribution. Then on some complete probability space $(\tilde{\Omega}, \mathscr{G}, \tilde{P})$ as which we can take the Lebesgue interval there exists a sequence (η_n) of random variables distributed as η which converges $\tilde{P}$-a.s. to a random variable η_o and possesses the following properties:

1) the sequence $(\phi_n(\eta_n))$ converges $\tilde{P}$-a.s. to a random element ξ with the limit distribution;

2) for any complete separable metric space Y and any $\mathscr{B}(C)$ -measurable Y-valued function ψ on C the sequence $(\psi(\eta_n))$ converges in probability to $\psi(\eta_o)$.

For its proof notice that if the measure P is separable then the space of random elements of $C_{[0,1]}(R)$ on $(\Omega, \mathscr{F}, P)$ (each of which is determined uniquely up to P-null sets) with convergence in probability is also separable. Then we take a sequence $(\psi_k(\eta))$ of random elements of $C_{[0,1]}(R)$ which is dense in this space and apply the Skorohod theorem ([9], Theorem 3.1.1) to the sequence of random elements $(\eta, \phi_n(\eta), \psi_1(\eta),\ldots,\psi_k(\eta),\ldots)$ of the space $C \times X \times (C_{[0,1]}(R))^{\aleph_0}$. So we prove the lemma 2 for $Y = C_{[0,1]}(R)$ and in general we embed Y into $C_{[0,1]}(R)$ as its closed subset according to the Banach-Mazur theorem [10].

This lemma implies immediately the following result.

Corollary. Let under conditions of the lemma 2 a sequence $(\psi_n(\eta))$ converges to $\psi(\eta)$ in probability on $(\Omega, \mathscr{F}, P)$. Then the sequence $(\psi_n(\eta_n))$converges to $\psi(\eta_o)$in probability on $(\tilde{\Omega}, \mathscr{G}, \tilde{P})$.

Moreover if the measure P is separable then according to [11] there exists a C-valued right-continuous strictly increasing $\mathscr{F}$-adapted process 0 such that any $\mathscr{O}$- or $\mathscr{P}$-measurable process is P indistinguishable respectively from some $\tilde{\mathscr{O}}$- or $\tilde{\mathscr{P}}$-measurable process where $\tilde{\mathscr{O}}$ and $\tilde{\mathscr{P}}$ are σ-algebra of subsets of $\Omega \times R_+$ generated respectively by 0 an 0_-.

So it suffices to prove the theorem in the case when the measure P is separable and the measure μ is generated by jumps of some ℓ-dimensional right-continuous $\mathscr{F}$-adapted process with locally finite variation so that $\ell = (|u|\wedge 1)*\nu$. First of all the functions $\hat{f}$, $\hat{g}$, $\hat{h}$ can be approximated by a convergent sequence $(\hat{f}^{(n)}, \hat{g}^{(n)}, \hat{h}^{(n)})$of functions satisfying conditions of the theorem from [4] which permits extension to any finite dimension d. Notice also that in [4] the left-limited $\mathscr{O}$-measurable process N can be not necessarily right-continuous. By force of the lemma 1 under conditions I and II of the theorem we have the relative compactness of the sequence of distributions on $D_{[0,T]}(R^m)$ for a respective dimension m of the processes $\hat{f}^{(n)}(Y_-^{(n)}).\hat{a}^c$, $\hat{f}^{(n)}(Y_-^{(n)}).\hat{a}^d$, $\hat{g}^{(n)}(Y_-^{(n)}).\hat{m}$,

$(\hat{h}^{(n)}(Y_-^{(n)})I_{\{|u|\le 1\}}*\mu^c-\nu^c)$, $u\, I_{\{|u|\le 1\}}*\mu^c)$,

$(\hat{h}^{(n)}(Y_-^{(n)})*(\mu^d-\nu^d),\ u * \mu^d)$, and $(\hat{k}(Y_-^{(n)})I_{\{|u|>1\}}*\mu,\ u\, I_{\{|u|>1\}}*\mu)$, where $Y^{(n)}$ is the solution of the equation of the form (1') for $(\hat{f}^{(n)}, \hat{g}^{(n)}, \hat{h}^{(n)})$ instead of $(\hat{f}, \hat{g}, \hat{h})$. Then we take some countable dense subset I of R_+ such that $\Delta\hat{a}_t + \Delta\ell_t = 0$ P-a.s. for any $t \in I$ and a sequence of natural numbers (n_r) such that the sequence of joint distributions for $T \in I$ of these processes with $n = n_r$, stopped at the stopping times from some properly chosen sequence, converges weakly to some limit distribution and after that we apply the lemma 2. Let A^c, A^d, M^c, Q^c, Q^d, P^e be the processes on $(\tilde{\Omega}, \mathscr{G}, \tilde{P})$ with limit distributions respectively for $\hat{f}^{(n_r)}(Y_-^{(n_r)}).\hat{a}^c$, $\hat{f}^{(n_r)}(Y_-^{(n_r)}).\hat{a}^d$, $\hat{g}^{(n_r)}(Y_-^{(n_r)}).\hat{m}$, $\hat{h}^{(n_r)}(Y_-^{(n_r)})I_{\{|u|\le 1\}}*(\mu^c-\nu^c)$, $\hat{h}^{(n_r)}(Y_-^{(n_r)})*(\mu^d-\nu^d)$, $\hat{k}(Y_-^{(n_r)})I_{\{|u|> \}}*\mu$. Denote for simplicity $(0, N, \hat{a}, \hat{m}, \mu, \nu, \hat{f}, \hat{g}, \hat{h}, \hat{k})$ on $(\tilde{\Omega}, \mathscr{G}, \tilde{P})$ the same functional of η_0 as so denoted one of η on $(\Omega, \mathscr{F}, P)$ and $Y = N + A^c + A^d + M^c + Q^d + P^e$. Denote also the family $\underline{\mathscr{G}} = (\mathscr{G}_{t+})_{t \in R_+}$ where for every fixed $t \in R_+$ $\mathscr{G}_t$ is the sub-σ-algebra of $\mathscr{G}$ generated by values of the processes 0 and Y for $0 \le s \le t$ and also by all $\tilde{P}$-null sets from $\mathscr{G}$. Then the system $(N, \hat{a}, \hat{m}, \mu, \nu, \hat{f}, \hat{g}, \hat{h}, \hat{k})$ is adapted properly to the family $\underline{\mathscr{G}}$. Now we apply the corollary using condition III of the theorem and show that

$A^c = \hat{f}(Y_-).\hat{a}^c$, $A^d = \hat{f}(Y_-).\hat{a}^d$, $M^c = \hat{g}(Y_-).\hat{m}$, $Q^c = \hat{h}(Y_-)I_{\{|u|\le 1\}}*(\mu^c-\nu^c)$, $Q^d = \hat{h}(Y_-)*(\mu^d-\nu^d)$, and $P^e = \hat{k}(Y_-)I_{\{|u|>1\}}*\mu$, so that the process Y is a weak solution of the equation (1') or (1).

REFERENCES

1. Dellacherie, C., Capacités et processus stochastiques, Springer-Verlag, 1972.

2. Lebedev, V. A., On the existence of a solution of the stochastic equation with respect to a martingale and a stochastic measure, International Symposium on Stochastic Differential Equations (abstracts of communications), Vilnius, 1978, 65-69.

3. Billingsley, P., Convergence of probability measures, J. Wiley, 1968.

4. Galtchouk, L.I., On the existence and uniqueness of solutions for stochastic equations with respect to martingales and random measures (in Russian), Second Vilnius Conference on Probability Theory and Mathematical Statistics (abstracts of communications), V.1, Vilnius, 1977, 88-91.

5. Lebedev, V. A., On the relative compactness for families of distributions of stochastic integrals with respect to a martingale and a random measure (in Russian), Theor. Ver. i Primen. (to appear).

6. Grigelionis, B., On relative compactness of the sets of probability measures in $D_{[0,\infty]}(\mathscr{X})$ (in Russian), Liet. mat. rink., XIII (1973), 4, 84-96.

7. Mackevičius, V., On the weak compactness of stochastic processes on the space $D_{[0,\infty]}(\mathscr{X})$ (in Russian), Liet. mat. rink., XIV (1974), 4, 117-121.

8. Szpilrajn, E., The characteristic function of a sequence of

sets and some of its applications, Fund. Math, 31 (1938), 207-223.

9. Skorohod, A. V., Limit theorems for stochastic processes (in Russian), Theor. Ver. i Primen., I (1956), 3, 289-319.

10. Banach, S., Théorie des opérations linéaires, Monogr. Math., Warszawa-Lwów, 1932.

11. Dellacherie, C., Stricker, C., Changements de temps et intégrales stochastiques, Lect. Notes Math., V. 581, Springer-Verlag, 1977, 365-375.

ON BELLMAN EQUATION FOR CONTROLLED DEGENERATE GENERAL STOCHASTIC PROCESSES

H. Pragarauskas

Vilnius, Institute of Mathematics and
Cybernetics of the Academy of Sciences
of the Lithuanian SSR
R. Požėlos 54, 620024

In the paper an optimal control problem of solutions of stochastic equations with diffusion, drift and jump terms is considered. The main result is the following statement: the reward function is a solution of the Bellman equation, which is in this case a singular nonlinear integro-differential equation. This statement was proved in [4], [5] under additional assumption that diffusion terms of controlled processes are "weakly non-degenerated". Here we do not assume this condition, so diffusion terms of controlled processes can be zero identically.

This problem for controlled diffusion processes was considered by KRYLOV [2]. The methods of this paper are based on the methods of paper [2].

R^d is a d-dimensional Euclidean space $T \in (0,\infty)$, $H_T = [0,T) \times R^d$, $\bar{H}_T = [0,T] \times R^d$, A is a separable metric space, integer $d_1 \geq 1$.

For all $\alpha \in A$, $t \in [0,T]$, $x, z \in R^d$ are defined: $d \times d_1$ -matrix $\sigma(\alpha,t,x)$, d-vectors $b(\alpha,t,x)$, $c(\alpha,t,x,z)$ and real $r(\alpha,t,x) \geq 0$, $f(\alpha,t,x)$, $g(x)$.

We shall denote: $\Pi(dz) = dz/|z|^{d+1}$, $\|\sigma\|^2 = \operatorname{tr} \sigma\sigma^*$,

$u_{(\ell)}(x) = \frac{1}{|\ell|} \sum_{i=1}^{d} u_{x_i}(x)\ell_i$, $u_{(\ell)(\ell)}(x) = \frac{1}{|\ell|^2} \sum_{i,j=1}^{d} u_{x_i x_j}(x)\ell_i\ell_j$, where σ^* is a conjugate matrix, $\ell \in R^d\backslash\{0\}$.

Let us introduce the following conditions.

I. There exist continuous in (t,x) partial derivatives $\frac{\partial}{\partial t}\gamma, \gamma_{(\ell)}, \gamma_{(\ell)(\ell)}$ for all $\ell \in R^d\backslash\{0\}$, $\gamma = \sigma$, b, c, r, f, g, function c is measurable in z, functions σ, b, r, f are continuous in α and for all $\alpha \in A$, $(t,x) \in \bar{H}_T$

$$\lim_{\beta\to\alpha} \int |c(\beta,t,x,z) - c(\alpha,t,x,z)|^2 \Pi(dz) = 0.$$

II. There exist constants $K \geq 0$, $m \geq 0$ such that for all $\alpha \in A$, $x,y \in R^d$, $\ell \in R^d\backslash\{0\}$, $t \in [0,T]$

a) $\| \sigma(\alpha,t,x)\|^2 + |b(\alpha,t,x)|^2 + \int |c(\alpha,t,x,z)|^2 \Pi(dz) \leq K(1+|x|)^2$,

b) $\| \sigma(\alpha,t,x) - \sigma(\alpha,t,y)\|^2 + |b(\alpha,t,x) - b(\alpha,t,y)|^2 +$

$+ \int |c(\alpha,t,x,z) - c(\alpha,t,y,z)|^2 \Pi(dz) \leq K|x-y|^2$,

c) $\int |c(\alpha,t,x,z)|^p \Pi(dz) \leq K(1+|x|)^p$, $p = (6m)\vee(3m+3)$,

d) $|\gamma(\alpha,t,x)| \leq K$, $\gamma = \sigma^{ij}_{(\ell)(\ell)}, \sigma^{ij}_{(\ell)}, \frac{\partial}{\partial t}\sigma^{ij}, b^i_{(\ell)(\ell)}, b^i_{(\ell)},$

$\frac{\partial}{\partial t} b^i, \int [|c^i_{(\ell)(\ell)}|^2 + |c^i_{(\ell)}|^2 + |c^i_{(\ell)}|^4 + |\frac{\partial}{\partial t} c^i|^2]\Pi(dz),$

$i = 1,\ldots,d$, $j = 1,\ldots,d_1$,

e) $(|\gamma| + |\frac{\partial}{\partial t}\gamma| + |\gamma_{(\ell)}| + |\gamma_{(\ell)(\ell)}|)(\alpha,t,x) \leq K(1+|x|)^m$, $\gamma = f,r,g$.

III. For every $t \in [0,T]$ and $R \in (0,\infty)$

$$\lim_{\rho\to 0} \sup_{\alpha\in A} \sup_{|x|\leq R} \int_{|z|\leq\rho} |c(\alpha,t,x,z)|^2 \Pi(dz) = 0.$$

Remark. Using the Dini theorem it is easy to prove that condition III follows from conditions I, II if A is a compact set of a separable metric space.

Let $(\Omega, \mathscr{F}, P)$ is a complete probability space, w_t is a d_1-dimensional Wiener process, z_t is a d-dimensional Cauchy process independent of w_t, p(dtdz) is a Poisson random measure on Borel sets of $[0,\infty) \times R^d$ constructed from the jumps of z_t, q(dtdz) is a Poisson martingale measure, i.e.,

$$q(dtdz) = p(dtdz) - E\,p(dtdz)\ ,\quad E\,p(dtdz) = dt\,\Pi(dz)\ .$$

We shall denote by $\mathscr{A}$ the class of all stochastic processes α_t taking their values in A and progressively measurable with respect to $\{\mathscr{F}_t \equiv \sigma(w_s, z_s, s \le t)\}$.

For every $\alpha \in \mathscr{A}$ and $(s,x) \in \bar{H}_T$ corresponds the solution $x_t^{\alpha,s,x}$ of stochastic equation

$$x_t = x + \int_0^t \sigma(\alpha_u, s+u, x_u)dw_u + \int_0^t b(\alpha_u, s+u, x_u)du + \int_0^t \int c(\alpha_u, s+u, x_u, z)q\,(dudz).$$

The reward function v is defined by the formula

$$v(s,x) = \sup_{\alpha \in \mathscr{A}} E\left[\int_0^{T-s} e^{-\phi_t^{\alpha,s,x}} f(\alpha_t, s+t, x_t^{\alpha,s,x})dt + e^{-\phi_{T-s}^{\alpha,s,x}} g(x_{T-s}^{\alpha,s,x})\right],$$

where $\phi_t^{\alpha,s,x} = \int_0^t r(\alpha_u, s+u, x_u^{\alpha,s,x})\,du$.

We shall denote by $W_{loc,m}^{1b,+}$ the class of all functions $u = u(t,x)$ defined on $\bar{H}_T$, which are continuous in (t,x) convex in x, for which partial derivative $\frac{\partial u}{\partial t}$ exists in Sobolev sense and is bounded on every compact set from $\bar{H}_T$ and such that for some constant $\mathscr{N} \ge 0$ for all $(t,x) \in \bar{H}_T$

$$|u(t,x)| \le \mathcal{N}(1+|x|).^{3m+2}$$

$\tilde{W}^{1b,+}_{loc,m}$ is the class of all functions $u = u_1 - u_2$,

where $u_1 \in W^{1b,+}_{loc,m}$, $u_2(x) = \mathcal{N}(1+|x|^2)^{\frac{3m}{2}+1}$, $\mathcal{N} \ge 0$.

By the Buseman - Feller theorem (see [8]) every convex function u has usual partial derivatives u_{x_i}, $u_{x_i x_j}$ almost everywhere. Denote $a = \frac{1}{2}\sigma\sigma^*$ and define an operator F for proper functions from $\tilde{W}^{1b,+}_{loc,m}$ by the formula

$$F\,u(t,x) = \sup_{\alpha\in A}\Bigg[\frac{\partial u}{\partial t}(t,x) + \sum_{i,j=1}^{d} a_{ij}(\alpha,t,x)u_{x_i x_j}(t,x) +$$

$$+\sum_{i=1}^{d} b_i(\alpha,t,x)u_{x_i}(t,x) + \int[u(t,x+c(\alpha,t,x,z))-u(t,x) - \sum_{i=1}^{d} u_{x_i}(t,x)c_i(\alpha,t,x,z)]\Pi(dz) -$$

$$-r(\alpha,t,x)u(t,x) + f(\alpha,t,x)\Bigg].$$

By lemma IV.2.5 [1] partial derivatives $u_{x_i x_j}$ of function $u \in \tilde{W}^{1b,+}_{loc,m}$ are measures of bounded variation on bounded sets from $\bar{H}_T$ in sense of theory of generalized functions.

<u>Remark</u>. By lemma 1.1 [2] partial derivatives $u_{x_i x_j}$ defined by the Buseman - Feller theorem are Randon-Nicodym derivatives of the measures $u^a_{x_i x_j}$ with respect to the Lebesque measure dtdx, where $u^a_{x_i x_j}$ in an absolutely continuous part of the measure $u_{x_i x_j}$ with respect to the Lebesque measure dtdx.

For proper (see § 1 [6]) $u \in \tilde{W}^{1b,+}_{loc,m}$ define a measure $G^\alpha u$, $\alpha \in A$

by the formula

$$G^{\alpha}u(dtdx) = \sum_{i,j=1}^{d} a_{ij}(\alpha,t,x)u_{x_ix_j}(dtdx) + \Big[\frac{\partial u}{\partial t}(t,x) + \sum_{i=1}^{d} b_i(\alpha,t,x)u_{x_i}(t,x) +$$

$$+ \int \Big[u(t,x+c(\alpha,t,x,z)) - u(t,x) - \sum_{i=1}^{d} u_{x_i}(t,x)c_i(\alpha,t,x,z)\Big]\Pi(dz) -$$

$$- r(\alpha,t,x)u(t,x) + f(\alpha,t,x)\Big] dtdx .$$

If there exists a measure ν of bounded variation on bounded sets from $\bar{H}_T$ such that for every $\alpha \in A$ $G^{\alpha}u \le \nu$ then by the corallary III.7.6 [7] there exists a measure

$$G\,u = \overline{\sup_{\alpha\in A}}\, G^{\alpha}u$$

as an upper bound in a structure of measures.

The main result of the paper is the following

Theorem. Let I-III hold. Then

(i) $\boldsymbol{\alpha} \in \tilde{W}^{1b,+}_{loc,m}$,

(ii) $G\,\boldsymbol{\alpha} = 0$, $v(T,x) = g(x)$,

(iii) $F\,\boldsymbol{\alpha} = 0$ a.e. H_T ,

(vi) the measure $\sum_{i,j=1}^{d} a_{ij}(\alpha,t,x)\boldsymbol{\alpha}_{x_ix_j}(dtdx)$ is absolutely continuous with respect to the Lebesque measure dtdx for every $\alpha \in A$.

Proof of this theorem is given below. Now, we shall consider the following construction.

Let us define $x_{t}^{\alpha,s,x}(\rho)$, v_ρ, F_ρ, G_ρ, $\rho \in (0,1)$ by the same formulas as $x_t^{\alpha,s,x}$, v, F, G are defined writing function $c_\rho \equiv c.1_{|z|>\rho}$ istead of function c.

Let $\tilde{w}_t$ be a d-dimensional Wiener process independent of $(\mathscr{F}_t)$, $\tilde{\mathscr{F}}_t = \sigma(\mathscr{F}_t, \tilde{w}_s, s \le t)$. We shall denote by $\tilde{\mathscr{A}}$ the class of all stochastic processes $\tilde{\alpha}_t$ taking their values in A and progressively measurable with respect to $(\tilde{\mathscr{F}}_t)$.

Let us denote by $x_{t\,(\varepsilon,\rho)}^{\tilde{\alpha},s,x}$, $\varepsilon,\rho \in (0,1)$, $(s,x) \in \overline{H}_T$, $\tilde{\alpha} \in \tilde{\mathscr{A}}$ the solution of stochastic equation

$$x_t = x + \int_0^t \sigma(\tilde{\alpha}_u, s+u, x_u)dw_u + \varepsilon\tilde{w}_t + \int_0^t b(\tilde{\alpha}_u, s+u, x_u)du +$$

$$+ \int_0^t \int c\rho(\tilde{\alpha}_u, s+u, x_u, z)q(dudz)$$

and by $v_{\varepsilon,\rho}$ the reward function which is constructed in the same way as the reward function v, writing $x_{t\,(\varepsilon,\rho)}^{\tilde{\alpha},s,x}$ instead of $x_t^{\alpha,s,x}$ and $\tilde{\mathscr{A}}$ instead of $\mathscr{A}$.

<u>Lemma</u>. Let I-III hold. Then:

(i) there exists constant $\mathscr{N} > 0$ such that for all ε, $\rho \in (0,1)$ $(t,x) \in \overline{H}_T$

$$|v_{\varepsilon,\rho}(t,x)| + |v_\rho(t,x)| + |\alpha(t,x)| \le \mathscr{N}(1+|x|)^m ,$$

$$v_{\varepsilon,\rho} = \tilde{v}_{\varepsilon,\rho} - u_o, \quad v_\rho = \tilde{v}_\rho - u_o, \quad v = \tilde{v} - u_o ,$$

where $u_o = \mathscr{N}(1+|x|^2)^{\frac{3m}{2}+1}$, $\tilde{v}_{\varepsilon,\rho}$, $\tilde{v}_\rho$, $\tilde{v} \in W^{1b,+}_{loc,m}$, $\mathscr{N} = \text{const} > 0$,

(ii) there exist locally bounded derivatives $\frac{\partial}{\partial t} v_{\varepsilon,\rho}$, $\frac{\partial}{\partial x_i} v_{\varepsilon,\rho}$, $\frac{\partial^2}{\partial x_i \partial x_j} v_{\varepsilon,\rho}$ in Sobolev sense for all ε, $\rho \in (0,1)$. Moreover

$$F_\rho \, \boldsymbol{a}_{\varepsilon,\rho} + \frac{1}{2}\varepsilon^2 \Delta \boldsymbol{a}_{\varepsilon,\rho} = 0 \quad \text{a.e. } H_T ,$$

where $\Delta u(t,x) = \sum_{i=1}^{d} u_{x_i x_i}(t,x)$,

(iii) $v_{\varepsilon,\rho} \to v_\rho$, $\varepsilon \to 0$ and $v_\rho \to v$, $\rho \to 0$ uniformly on every compact set in $\overline{H}_T$.

Proof. (i) can be proved in the same way as the corresponding properties of the reward function from § 3 [4]. (ii)follows from theorem 1.4 [3]. (iii) The first statement follows from lemma 2.1 [5]. We shall prove the second. We have

$$|\boldsymbol{a}(s,x) - v_\rho(s,x)| = |\sup_{\alpha\in\mathcal{A}}\ldots - \sup_{\alpha\in\mathcal{A}}\ldots| \le$$

(1)
$$\le \sup_{\alpha\in\mathcal{A}} E\Big\{ \int_0^{T-s} |f(\alpha_t,s+t,x_t^{\alpha,s,x}) - f(\alpha_t,s+t,x_{t\,(\rho)}^{\alpha,s,x})\, dt +$$
$$+ \int_0^{T-s} |f(\alpha_t,s+t,x_{t\,(\rho)}^{\alpha,s,x})| \int_0^t |r(\alpha_u,s+u,x_u^{\alpha,s,x}) - r(\alpha_u,s+u,x_{u\,(\rho)}^{\alpha,s,x})|\,du\,dt +$$
$$+ |g(x_{T-s}^{\alpha,s,x}) - g(x_{T-s\,(\rho)}^{\alpha,s,x})| + |g(x_{T-s\,(\rho)}^{\alpha,s,x})| \int^{T-s} |r(\alpha_t,s+t,x_t^{\alpha,s,x}) - r(\alpha_t,s+t,x_{t\,(\rho)}^{\alpha,s,x})|\,dt \Big\} .$$

Let Q be a compact set in $\overline{H}_T$. Using well known estimates for solutions of stochastic equations from (1) it is easy to derive that for

$$\lim_{\rho\to 0} \sup_Q |\boldsymbol{a}(s,x) - v_\rho(s,x)| = 0$$

it is sufficient to prove for every $t \in [0,T]$

(2)
$$\lim_{\rho\to 0} \sup_Q \sup_{\alpha\in\mathcal{A}} E|x_t^{\alpha,s,x} - x_{t\,(\rho)}^{\alpha,s,x}|^2 = 0 .$$

From the definition $x_t^{\alpha,s,x}$, $x_{t\,(\rho)}^{\alpha,s,x}$ and condition II follows that for some constant $\mathcal{N}$ independent of $\rho \in (0,1)$, $(s,x) \in Q$, $\alpha \in \mathcal{A}$

(3)
$$E|x_t^{\alpha,s,x} - x_t^{\alpha,s,x}(\rho)|^2 \leq \mathcal{N} \int_0^t E|x_u^{\alpha,s,x} - x_u^{\alpha,s,x}(\rho)|^2 du + $$
$$+ E \int_0^t \int_{|z|\leq\rho} |c(\alpha_u, s+u, x_u^{\alpha,s,x}, z)|^2 \Pi(dz)du.$$

Using the condition II it is easy to derive

(4)
$$\sup_Q \sup_{\alpha\in\mathcal{A}} E\int_0^{T-s} \int_{|z|\leq\rho} |c(\alpha_t, s+t, x_t^{\alpha,s,x}, z)|^2 \Pi(dz)dt \leq$$
$$\leq \sup_Q \sup_{\alpha\in\mathcal{A}} E \int_0^{T-s} K(1+|x_t^{\alpha,s,x}|)^2 1_{|x_t^{\alpha,s,x}| > R} dt +$$
$$+ \int_0^T \sup_{|x|\leq R} \sup_{\alpha\in A} \int_{|z|\leq\rho} |c(\alpha,t,x,z)|^2 \Pi(dz) dt .$$

From the estimates (3), (4) using the condition III we derive, that (2) holds.

<u>Proof of theorem</u>. (i) follows from (i) lemma. (ii). $\alpha (T,x) = g(x)$ follows from the definition v. The inequality $G v \leq 0$ was proved in § 3 [4] in the case $r \equiv 0$. The proof in general case is analogous. So we have only to prove $G v \geq 0$.

Let us introduce the operators

$$\tilde{G} u = G(u-u_o) , \tilde{G}_\rho u = G_\rho(u-u_o) ,$$

where u_o is defined in (i) of lemma. $\tilde{G}$, $\tilde{G}_\rho$ are operators of the same type as G, G_ρ and $\tilde{G}\tilde{v} = Gv$, $\tilde{G}_\rho \tilde{v}_{\varepsilon,\rho} = G_\rho v_{\varepsilon,\rho}$, where $\tilde{v}$, $\tilde{v}_{\varepsilon\ \rho}$ are defined in (i) of lemma.

By (ii) of lemma for all ε, $\rho \in (0,1)$.

$$G_\rho v_{\varepsilon,\rho}(dtdx) + \frac{1}{2}\varepsilon^2 \Delta \mathfrak{a}_{\varepsilon,\rho}(dtdx) = (F_\rho v_{\varepsilon,\rho} + \frac{1}{2}\varepsilon^2 \Delta \mathfrak{a}_{\varepsilon,\rho})\, dtdx = 0$$

consequently

$$(5)\quad \tilde{G}_\rho \tilde{v}_{\varepsilon,\rho}(dtdx) + \frac{1}{2}\varepsilon^2 \Delta\, \tilde{\mathfrak{a}}_{\varepsilon,\rho}(dtdx) = \frac{1}{2}\varepsilon^2 \Delta u_o dtdx \geq 0,$$

because u_o is a convex function. From (5) we derive, that for every $\delta > 0$ there exist k_o, n_o such that if $k \geq k_o$, $n \geq n_o$ then

$$(6)\quad \tilde{G}_{\frac{1}{k}}\, \tilde{v}_{\frac{1}{n},\frac{1}{k}}(dtdx) + \delta\, \Delta\, \tilde{v}_{\frac{1}{n},\frac{1}{k}}(dtdx) \geq 0.$$

From (6) and lemma follows that assumptions of theorem 1.3 [6] are fulfilled. By this theorem $\tilde{G}\, \tilde{v} \geq 0$. This is equivalent to $G\, v \geq 0$.

(iii) It is easy to derive that the Lebesque representation of the measure Gv has the form

$$(7)\quad Gv\,(dtdx) = F\, v(t,x)dtdx + \overline{\sup_{\alpha\in A}} \sum_{i,j=1}^{d} a_{ij}(\alpha,t,x) v^s_{x_i x_j}(dtdx)\ ,$$

where $v^s_{x_i x_j}$ is a singular part of measure $v_{x_i x_j}$ with respect to Lebesque measure dtdx. By (ii) of theorem $Gv = 0$, consequently $F\, \mathfrak{a} = 0$ a.e. H_T.

(iv) Using (i) of lemma we obtain

$$\sum_{i,j=1}^{d} a_{ij}(\alpha,t,x) v_{x_i x_j}(dtdx) = \sum_{i,j=1}^{d} a_{ij}(\alpha,t,x) \tilde{v}_{x_i x_j}(dtdx) -$$

$$- \sum_{i,j=1}^{d} a_{ij}(\alpha,t,x) u_{ox_i x_j}(x)dtdx \geq -\sum_{i,j=1}^{d} a_{ij}(\alpha,t,x) u_{ox_i x_j}(x)dtdx\ .$$

On the other hand by (ii) of theorem and (7)

$$\overline{\sup_{\alpha\in A}} \sum_{i,j=1}^{d} a_{ij}(\alpha,t,x)v^{s}_{x_i x_j}(dtdx) = 0 ,$$

consequently,

$$\sum_{i,j=1}^{d} a_{ij}(\alpha,t,x)v_{x_i x_j}(dtdx) = \sum_{i,j=1}^{d} a_{ij}(\alpha,t,x)v^{a}_{x_i x_j}(dtdx) + v^{s}_{x_i x_j}(dtdx)) \leq$$

$$\leq \sum_{i,j=1}^{d} a_{ij}(\alpha,t,x)v^{a}_{x_i x_j}(dtdx) ,$$

where $v^{a}_{x_i x_j}$ is an absolutely continuous part of the measure $v_{x_i x_j}$ with respect to the Lebesque measure dtdx.

REFERENCES

1. Krylov, N.V., Controlled processes of diffusion type, M., "Nauka", 1977 (in Russian).
2. Krylov, N.V., Some new results from the theory of controlled diffusion processes, Matem sb. (in Russian, to appear).
3. Krylov, N.V., Pragarauskas, H., On Bellman equation for uniformly non-degenerated general stochastic processes, Lietuvos matem. rink. (in russian, to appear).
4. Pragarauskas, H., Control of solution of stochastic equation with discontinuous paths, Lietuvos matem. rink., 1978, XVIII, 1, 147-167 (in Russian).

5. Pragarauskas, H., On Bellman equation for weakly degenerated general stochastic processes, Lietuvos matem. rink. (in Russian, to appear).

6. Pragarauskas, H., On limit transition in general degenerated Bellman equations I, II, Lietuvos matem. rink. (in Russian, to appear).

7. Dunford, N., Schwartz, J. T., Linear operators, general theory, Interscience Publishers, New York, London, 1958.

8. Aleksandrov, A. D., The existence almost everywhere of the second order differential of a convex function and certain related properties of convex surfaces, Sci. Notes LSU, 1939, No 37, ser. math., 6, 3-35.

ON THE EXISTENCE OF THE OPTIMAL POLICY FOR A MULTIDIMENSIONAL QUASIDIFFUSION CONTROLLED PROCESS

A. Yu. Veretennikov
Moscow, Institute of Problems of Control
Moscow, 117342 USSR

Let $(\Omega, \mathscr{F}, P, (w_t, \mathscr{F}_t))$ be a complete probability space with d-dimensional Wiener process $(w_t, \mathscr{F}_t)$. We use all the notations and assumptions of [1]. For every (admissible) policy $u \in U(s,x)$ (see [1], [2]) $x^{u,x}_{s,t}$ is some (any) fixed solution of a stochastic equation

$$(1) \qquad x_t = x + \int_0^t b(u_r, s+r, x_r)dr + w_t$$

Remember that for $u \in U(s,x)$

$$v^u(s,x) = E\left[\int_0^{T-s} f^{u_t}(s+t, x^{u,x}_{s,t}) \exp(\phi^{u,x}_{s,t})\, dt + \right.$$

$$\left. + g(x^{u,x}_{s,T-s}) \exp(\phi^{u,x}_{s,T-s})\right],$$

$$\phi^{u,x}_{s,t} = -\int_0^t c^{u_r}(s+r, x^{u,x}_{s,r})dr; \qquad v(s,x) \equiv \sup_{u \in U(s,x)} v^u(s,x)$$

<u>Theorem 1</u>. The stochastic equation

$$(2) \qquad x_t = x + \int_0^t b(r, x_r)dr + w_t$$

has a strong solution; the pathwise uniqueness holds for (2).

It follows from theorem 1 that $U(s,x) \neq \emptyset$: for instance, if $u_t \equiv \alpha \in A$, then $u_t \in U(s,x)$.

Theorem 2. If $g \in C(E_d)$ *), then the Cauchy problem

$$(3) \qquad \begin{cases} (Lw + f)(t,x) = 0 & \text{a.e. } H_T , \\ w(T,x) = g(x), \ x \in E_d \end{cases}$$

has a unique solution w in the class of functions $C(\overline{H}_T) \cap \cap_{p>1} W^{1,2}_{p,loc}(H_T)$ **). For every $t \in (0,T)$ and every bounded domain $\mathcal{D} \subset E_d$

$$(4) \qquad ||w||_{W_p^{1,2}([0,t] \times \mathcal{D})} \leq C(||f||_B + ||g||_B) .$$

These are the theorems 1 and 2 from [1].

Theorem 1 has been formulated in [3]. Now we are going to prove theorems 3 and 4 [1]. Denote $F_\alpha[u](s,x) =$

$$= [\frac{\partial}{\partial s} + \Delta_x + (b(\alpha,s,x), \frac{\partial}{\partial x}) - c^\alpha(s,x)]u(s,x) + f^\alpha(s,x) ,$$

$$F[u](s,x) = \sup_{\alpha \in A} F_\alpha[u](s,x) ; \ G_\alpha[u](s,x) = [(b(\alpha,s,x),\frac{\partial}{\partial x}) -$$

$$- c^\alpha(s,x)]u(s,x) + f^\alpha(s,x), \ G[u](s,x) = \sup_{\alpha \in A} G_\alpha[u](s,x).$$

Theorem 3. Suppose $g \in C(E_d)$. Then the Bellman's differential equation

---

*) $C(E_d)$ is a class of all bounded continuous functions on E_d.

**) W are Sobolev's classes of functions.

$$(5)\quad \begin{cases} F[\tilde{v}](s,x) = 0 & \text{a.e. } H_T , \\ \tilde{v}(T,x) = g(x), & x \in E_d \end{cases}$$

has a solution $\tilde{v} \in C(\bar{H}_T) \cap \bigcap_{p>1} W^{1,2}_{p,loc}(H_T)$. This solution is unique in $C(\bar{H}_T) \cap W^{1,2}_{p,loc}(H_T)$ for every $p \geq d+1$.

Proof. We use the Bellman-Howard's method in a like way as Fleming did [4]; see also theorem 1.1.5[2].

Let $\alpha_o(s,x),(s,x) \in H$, be any Borel function with values in A. The stochastic equation

$$x_t = x + \int_0^t b(\alpha_o(s+r,x_r),s+r,x_r)dr + w_t$$

has a strong solution $x^o_{s,t}$(theorem 1). Set $v_o(s,x) = v^{\alpha_o}(s,x)$ - the reward function which corresponds the Markovian policy $\alpha_o(\cdot)$. Due to theorem 2 the Cauchy problem

$$\begin{cases} F_{\alpha_o}[w](s,x) = 0 & \text{a.e. } H_T, \\ w(T,x) = g(x), & x \in E_d \end{cases}$$

has a unique solution w in $C(\bar{H}_T) \cap \bigcap_{p>1} W^{1,2}_{p,loc}(H_T)$. Let $\phi^{\alpha_o}_t = -\int_0^t c^{\alpha_o}(s+r,x^o_{s,r})\, dr$. Applying the Itô's formula ([2], ch. 2) to the expression $\exp(\phi^{\alpha_o}_t)w(s+t,x^o_{s,t})$, we obtain $w = v_o$. So $v_o \in C(\bar{H}_T) \cap \bigcap_{p>1} W^{1,2}_{p,loc}(H_T)$, $F_{\alpha_o}[v_o] = 0$ (a.e. in H_T), $v_o(T,x) = g(x)$, $x \in E_d$.

If the Markovian policies $\alpha_o, \alpha_1, \ldots, \alpha_n$ and reward functions $v_o, v_1, \ldots, v_n$ are defined, let α_{n+1} be a Borel function with values in A such that $F_{\alpha_{n+1}}[v_n] = F[v_n]$ (a.e.); such α_{n+1} exists, -see [5], Addition III, Theorem IX.

The Markovian policy α_{n+1} is admissible, for the equation

$$x_t = x + \int_0^t b(\alpha_{n+1}(s+r,x_r),\ s+r,x_r)\ dr + w_t$$

has a strong solution $x^{n+1}_{s,t}$ (theorem 1).

Prove that the consequence $\{v_n(s,x)\}$ tends to a limit and the limit function is a solution of (5). Let $\phi^n_t = -\int_0^t c^{\alpha_n}(s+r,x^n_{s,r})dr$.

The same arguments as for v_o prove that

$v_n \in C(\overline{H}_T) \cap \bigcap_{p>1} W^{1,2}_{p,loc}(H_T)$, $F_{\alpha_n}[v_n] = 0$ (a.e.),

$v_n(T,x) = g(x)$, $x \in E_d$. So $F_{\alpha_{n+1}}[v_n] \geq 0$ (a.e.).

Applying the Itô's formula to the expression $\exp(\phi^{n+1}_t).v_n(s+t,x^{n+1}_{s,t})$, we obtain

$$E \exp(\phi^{n+1}_{T-s})\ g\ (x^{n+1}_{s,T-s}) - v_n(s,x) =$$

$$= E\int_0^{T-s} \exp(\phi^{n+1}_r)\ L^{\alpha_{n+1}} v_n(s+r,x^{n+1}_{s,r})\ dr \geq$$

$$\geq -E\int_0^{T-s} \exp(\phi^{n+1}_r)\ f^{\alpha_{n+1}}(s+r,x^{n+1}_{s,r})\ dr =$$

$$= E\int_0^{T-s} \exp(\phi^{n+1}_r)\ L^{\alpha_{n+1}} v_{n+1}(s+r,x^{n+1}_{s,r})\ dr =$$

$$= E \exp(\phi^{n+1}_{T-s})\ g\ (x^{n+1}_{s,T-s}) - v_{n+1}(s,x)$$

due to the estimates ch. 2 [2]. So $v_n(s,x) \le v_{n+1}(s,x)$, $n = 0,1,\ldots$ For f and g are bounded, $\{v_n\}$ is bounded as well, and so there exists a limit $\tilde{v}(s,x) = \lim_{n\to\infty} v_n(s,x)$. We'll prove that $\tilde{v}$ is a solution of (5) in $C(\bar{H}_T) \cap \bigcap_{p>1} W^{1,2}_{p,loc}(H_T)$. Using the notion of the weak convergence and the estimate (4) one can easily see that $\tilde{v} \in W^{1,2}_{p,loc}(H_T)$, $p > 1$.

It follows from (4) and lemma 2.3.3 [6] that for every $t \in (0,T)$ and every bounded domain $\mathscr{D} \subset E_d$

(6) $$\sup_n ||v_{n,x}||_{H_\gamma([0,t]\times\mathscr{D})} < \infty \;{}^{*)}$$

for some $\gamma > 0$. By Arzela-Ascoli theorem the consequence $\{v_n\}$ is compact in the sense of the uniform convergence on $[0,t]\times\mathscr{D}$. So $\tilde{v} \in C(\bar{H}_T) \cap \bigcap_{p>1} W^{1,2}_{p,loc}(H_T)$.

Obviously $\tilde{v}(T,x) = g(x)$, $x \in E_d$. Check that $F[\tilde{v}] = 0$ (a.e.). It is sufficient to prove the equality

(7) $$\int_{H_T} F[\tilde{v}]\phi \, dsdx = 0$$

for every function $\phi \in C_0^\infty(H_T)$ such that $\operatorname{supp}\phi \subset [0,t]\times\mathscr{D}$, $t \in (0,T)$, $\mathscr{D} \subset E_d$ -bounded domain.

It is easily seen that $\{v_n\}$ tends to $\tilde{v}$ weakly in $W^{1,2}_p[0,t]\times\mathscr{D}$ for every $p > 1$, so

---

*) the definition of Hölder classes H_γ see in [6].

$$\int_{H_T} (v_{n,s}-\tilde{v}_s)\phi \, dsdx \to 0,$$

$$\int_{H_T} a(v_{n,xx}-\tilde{v}_{xx})\phi \, dsdx \to 0, \; n \to \infty .$$

It follows from Arzela-Ascoli theorem and (6) that $\{v_{n,x}\}$ is compact in the sense of the uniform convergence on $[0,t]\times\mathscr{D}$. Let $\{v_{n_k},x\}$ converge uniformly on $[0,t]\times\mathscr{D}$, $\tilde{v}^1$ is it's limit (on $[0,t]\times\mathscr{D}$). Obviously, v^1 is continuous. Set $\ell = \frac{x-y}{|x-y|}$, $x \neq y \in E_d$. When $k \to \infty$ we obtain making use of the Adamar's formula

$$v_{n_k}(s,x) - v_{n_k}(s,y) = \int_0^1 (v_{n_k},x(s,rx+(1-r)y),\ell)dr$$

the following equality

$$\tilde{v}(s,x)-\tilde{v}(s,y) = \int_0^1 (\tilde{v}^1(s,rx+(1-r)y),\ell) \, dr.$$

For $\tilde{v} \in W^{1,2}_{p,loc}(H_T)$, $p > 1$, it follows from this equality that $\tilde{v}^1 = \tilde{v}_x$ and $v_{n,x} \to \tilde{v}_x$ uniformly on $[0,t]\times\mathscr{D}$.

Item,

$$|G[\tilde{v}] - G[v_n]|(s,x) \leq$$

$$\leq \sup_{\alpha\in A} |b(\alpha,s,x)(\tilde{v}_x-v_{n,x})(s,x) - C^{\alpha}(s,x)(\tilde{v}-v_n)(s,x)| \leq$$

$$\leq C.|(\tilde{v}_x-v_{n,x}|+|\tilde{v}-v_n|)(s,x) \to 0, \; n \to \infty,$$

uniformly on $[0,t]\times\mathscr{D}$.

$$F[v_n] \;=\; F_{\alpha_{n+1}}[v_n] - F_{\alpha_{n+1}}[v_{n+1}] \;=\; (v_{n,s}-v_{n+1,s}) \;+$$

$$+ a(v_{n,xx}-v_{n+1,xx}) + G_{\alpha_{n+1}}[v_n] - G_{\alpha_{n+1}}[v_{n+1}], \text{ so}$$

$$\left|\int_{H_T} F[v_n]\phi\, dsdx\right| = \left|\int_{[0,t]x\mathcal{D}} \{(v_{n,s}-v_{n+1,s})\phi + \right.$$

$$\left. + a(v_{n,xx}-v_{n+1,xx})\phi + (G_{\alpha_{n+1}}[v_n] - G_{\alpha_{n+1}}[v_{n+1}])\phi\}dsdx\right| \le$$

$$\le \left|\int_{[0,t]x\mathcal{D}} (v_{n,s}-v_{n+1,s})\phi\, dsdx\right| + \left|\int_{[0,t]x\mathcal{D}} a(v_{n,xx}-v_{n+1,xx})\phi dsdx\right| +$$

$$+ C\int_{[0,t]x\mathcal{D}} (|v_{n,x}-v_{n+1,x}|+|v_n-v_{n+1}|)|\phi|dsdx \to 0,\ n \to \infty,$$

consequently,

$$\int_{H_T} F[\tilde{v}]\phi dsdx = \int_{H_T} F[v_n]\phi\, dsdx +$$

$$+ \int_{H_T} \{F[\tilde{v}] - F[v_n]\}\phi\, dsdx = \int_{H_T} F[v_n]\phi\, dsdx +$$

$$+ \int_{H_T} \{(\tilde{v}_s-v_{n,s}) + a(\tilde{v}_{xx}-v_{n,xx})\}\phi\, dsdx +$$

$$+ \int_{H_T} \{G[\tilde{v}] - G[v_n]\}\phi\, dsdx \to 0,\ n \to \infty .$$

This proves (7) and so $\tilde{v}$ is a solution of (5).

The statement of the uniqueness is a simple corollary of arguments in theorem 4 (the proof of theorem 4 uses the existence only). Thus the theorem is proved.

Theorem 4. Suppose $g \in C(E_d)$. Then $v(s,x) = \tilde{v}(s,x),(s,x)\in \bar{H}_T$;

the optimal Markovian policy does exist.

Proof. Let $u \in C(\overline{H}_T) \cap W^{1,2}_{p,loc}(H_T)$, $p \geq d+1$, be a solution of (4). There exists a Borel function $\overline{\alpha}(s,x),(s,x) \in H$, with values in A, such that $F[u](s,x) = F_{\overline{\alpha}(s,x)}[u](s,x)$(a.e.), - see [5], Addition III, Theorem IX. Show that $\overline{\alpha}$ is the optimal Markovian policy. The stochastic equation

$$x_t = x + \int_0^t b(\overline{\alpha}(s+r,x_r), s+r,x_r)dr + w_t$$

has a strong solution $x^{\overline{\alpha},x}_{s,t}$ due to theorem 1. Using the Itô's formula ([2], ch. 2), we obtain

$$u(s,x) = E\{\int_0^{T-s} f^{\overline{\alpha}}(s+r, x^{\overline{\alpha},x}_{s,r}) \exp(\phi^{\overline{\alpha},x}_{s,r})\, dr +$$

$$+ g(x^{\overline{\alpha},x}_{s,T-s}) \exp(\phi^{\overline{\alpha},x}_{s,T-s})\} = v^{\overline{\alpha}}(s,x) \leq v(s,x).$$

Conversely, for every $u \in U(s,x)$ by the Itô's formula again

$$u(s,x) = E\{\int_0^{T-s} f^{u_r}(s+r,x^{u,x}_{s,r}) \exp(\phi^{u,x}_{s,r})dr +$$

$$+ g(x^{u,x}_{s,T-s}) \exp(\phi^{u,x}_{s,T-s}) +$$

$$+ \int_0^{T-s} (-L^{u_r}u-f^{u_r})(s+r,x^{u,x}_{s,r}) \exp(\phi^{u,x}_{s,r})dr\} =$$

$$= v^u(s,x) - E\int_0^{T-s} (L^{u_r}u+f^{u_r})(s+r,x^{u,x}_{s,r}) \exp(\phi^{u,x}_{s,r})dr \geq v^u(s,x)$$

due to the estimates of ch. 2 [2] and the Bellman's equation (5). So $u(s,x) \geq v(s,x)$, and finally

$$u(s,x) = v(s,x) . \tag{8}$$

It follows from (8) that the solution u of (5) is unique in $C(\bar{H}_T) \cap W^{1,2}_{p,loc}(H_T)$, $p \geq d+1$. Moreover, $v \in C(\bar{H}_T) \cap \bigcap_{p>1} W^{1,2}_{p,loc}(H_T)$, and $\bar{\alpha}$ is the optimal Markovian policy, for $v^{\bar{\alpha}}(s,x) = u(s,x) = v(s,x)$. The theorem is proved.

Comment. All the statements of theorems 1-4 hold true in a more general case when $(w_t, \mathscr{F}_t)$ is a d_1 -dimensional Wiener process $(d_1 \geq d)$ and our controlled process has the following diffusion $d \times d_1$ -matrix $\sigma(t,x)$, $t \geq 0$, $x \in E_d$:

1^0. there exists a constant $\nu > 0$ such that for $t \geq 0, \lambda, x \in E_d$

$$\lambda^* a(t,x) \lambda \geq \nu |\lambda|^2 ,$$

where $a(t,x) \equiv \frac{1}{2}\sigma\sigma^*(t,x)$ is uniformly continuous in (t,x).

2^0. σ admits a representation

$$(9) \quad \sigma(s,x) = \sigma_L(s,x,\sigma^d(x), \sigma^{d+1}(s,x)),$$

where σ^d and σ^{d+1} are n-dimensional Borel vector-functions,

$$\sigma^d \in W^1_{2d,\ loc}(E_d),$$

$$\sigma^{d+1} \in W^{0,1}_{2d+2,\ loc}([0,\infty) \times E_d),$$

σ_L is a Borel function of 2n+d+1 variables,

$$\sigma_L(t,x,z,v) = \sigma_L(t,x^1,\ldots,x^d,z^1,\ldots,z^n,v^1,\ldots,v^n)$$

is Lipschitz continuous in (z,v) uniformly in (t,x) and for every t,z,v,x,y

$$|\sigma_L(t,x,z,v) - \sigma_L(t,y,z,v)| \leq C_t|x-y|, \int_0^t C_s^2 ds < \infty .$$

3^0. for every $t > 0$ and every bounded domain $\mathcal{D} \subset E_d$ there exists such $\varepsilon > 0$, that

$$\inf_{0\le\alpha\le 1} \inf_{\substack{x,y\in\mathcal{D} \\ |x-y|<\varepsilon}} \inf_{s\in[0,t]} \inf_{0\ne\lambda\in E_d} \frac{\lambda^*}{|\lambda|^2} [\alpha\sigma(t,x) +$$

$$+ (1-\alpha)\sigma(t,y)] [\alpha\sigma(t,x) + (1-\alpha)\, \sigma(t,y)]^* \lambda > 0$$

(this condition missed in [1]).

Note that (3) holds true, for instance, if $\sigma(s,\cdot)$ is continuous uniformly in $s \ge 0$; another example -if $d_1 = d$ and σ satisfies condition (§ 6 ch. 2 [2])

$$\inf_{s,x} \inf_{0\ne\lambda\in E_d} \frac{\lambda^*}{|\lambda|^2} \sigma(s,x)\, \lambda > 0 .$$

In the case $d_1 = d = 1$ similar results have been proved by A. K. Zvonkin [7].

The author expresses his deep gratitude to N. V. Krylov for help and consideration.

REFERENCES

1. A. Yu. Veretennikov, On the existence of the optimal strategy in a diffusion process control problem, International Symposium on Stochastic Diff. Equations, Vilnius, 1978, Abstr. of Comm., 174-177.

2. N. V. Krylov, The Controlled processes of diffusion type, Moscow, "Nauka", 1977 (Russian).

3. A. Yu. Veretennikov, On strong solutions of some stochastic equations, Usp. Mat. Nauk, 1978, 33,5,173-174 (Russian).

4. W. H. Fleming, Some Markovian optimization problems, J. Math. and Mech., 1963, 12, 1, 131-140.

5. M. A. Neumark, Normed rings, Moscow, "Nauka", 1968 (Russian).

6. O. A. Ladyzenskaja, V. A. Solonnikov, N. N. Ural'ceva, Linear and quasilinear equations of parabolic type, Moscow, "Nauka", 1967 (Russian).

7. A. K. Zvonkin, The transformation of the state space eliminating the drift, Matem. zborn., 1974, 93, 1, 129-149 (Russian) .

ON THE SEMIGROUP THEORY OF STOCHASTIC CONTROL

D. Vermes

1. Introduction

The aim of the present paper is to establish some regularity properties of the optimal expense function in a broad class of Markovian control problems with continuous time.

The underlying controllable objects are described by Markov processes, their state evolution can (but need not) include diffusion, drift and jump components, given by the corresponding coefficients and measures. In order to make semigroup-perturbation methods applicable [4], we assume that the highest order coefficients do not depend on control.

In the present paper we consider only piecewise continuous Markov strategies and time optimality. In other words the value of the control depends on the completely observable state only, and we want to minimize the expected hitting time of a fixed target set.

In the semigroup approach the Markovian control problems it is shown, that the optimal expense function belongs to the domain of the infinitesimal generator of the process corresponding to the optimal strategy. Moreover if the optimal expense function belongs to the intersection of the domains of generators, corresponding to all continuous strategies (e.g. if all domain coincide), then the abstract version of the Bellman equation is a necessary and sufficient condition of optimality [7]. In most of known concrete Markovian control problems the optimal expense belongs in fact to the intersection of the domains, though the domains do not coincide. The only known counterexamples are deterministic control problems and diffusions with degeneration or with controlled diffusion coefficient.

The aim of the paper is to point out a property of the infinitesimal generator, which is responsible for the nice behaviour of the "really" stochastic control problems compared with the mentioned excess classes. Loosely speaking this property is that the generator can be decomposed into the sum of a control dependent and a control independent part, where the latter is in some sense of higher order than the first one.

Under this assumption we show not only, that the optimal strategy and the expense function together suffice the Bellman equation, but also that the optimal expense is much more regular than a general element from the domain of the optimal generator. In fact it belongs also to the domains corresponding to all continuous strategies. This is an essential gain of information if the optimal strategy is discontinuous. In a forthcoming publication [8] we show that the Bellman equation together with the just mentioned regularity of the expense imply some continuity and extremality properties of the optimal strategy. In particular we point out in [8], that for important classes of problems the optimal strategy is discontinuous.

Not to get lost in a jungle of complicated definitions and notations, in most of the paper we treat a special class of problems including diffusions, Markovian and semi-Markovian jump processes, the piecewise monoton processes of queuing and storage theory e.t.c. Processes with possibly infinitely many jumps in finite time intervals will be considered in the last chapter.

We remark, that similar results were proved for more speficic classes of problems by Krylov and Pragarauskas. Their method is different from ours and it works also without the assumption of the existence of the optimal strategy [5], [6].

2. Statement of the Problem

As state space we regard a Borel set of the n-dimensional Euclidean space $E \subset R^n$ and denote by E_Δ its one-point-compactification, by Δ the point of infinity. Let Y denote the action space, which is a compact subset of R^m. Measurable mappings $u: E \to Y$ will be called feed-back (or pure Markov) strategies. In the present paper we regard only the case, where the set U of admissible strategies consist of all piecewise continuous mappings. In other words for every $u \in U$ and for a.e. $x \in E$ there exists a surrounding $\Gamma_x(u)$ of $x \in E$ such that u is continuous in $\Gamma_x(u)$. All results would remain valid if U contained only the piecewise constant strategies. We denote by $\mathcal{B}$ the space of all bounded measurable functions on E with the sup-norm, by $\mathcal{C}$ the space of continuous functions.

With each strategy $u \in U$ we associate a homogeneous continuous-time Markov process x_t^u with transition function $P^u(x,t,\Gamma)$ and semi-group $P_t^u f := \int P^u(x,t,dy)f(y)$. The $\mathcal{C}$-infinitesimal operator is defined by $L^u f := \lim_{t \to 0} (P_t^u f - f)/t$ for all $f \in \mathcal{D}(L^u) \subset \mathcal{C}$ for which the limes exists in $\mathcal{C}$. We denote the characteristic operator of x_t^u by $\mathcal{L}_u$, its domain at $x \in E$ by $\mathcal{D}_x(\mathcal{L}^u)$.

A strategy $v \in U$ is called to be optimal if for any starting point $x \in E$ the expected first exit time $\tau^v(\omega) := \inf\{t: x_t^u(\omega) = \Delta\}$ is not larger than that of the processes governed by other admissible strategies $u \in U$, ie. $E_t^v \tau^v \leq \inf_{u \in U} E_x^u \tau^u$ for all $x \in E$. (Here E_x^u denotes the expectation corresponding to $P^u(x,\cdot,\cdot)$). We assume the existence of an optimal $v \in U$ throughout the whole paper.

In the third chapter we shall consider processes x_t^u which are killed at the first exit τ^u from E, and whose characteristic operator is of the form

$$\mathscr{L}^u f(x) = \sum_{i,j=n_1+1}^{n_1+n_2} a_{ij}(x) f_{x_i x_j}(x) + \sum_{i=1}^{n_1+n_2} b_i(x,u(x)) f_{x_i}(x) + \tag{1}$$

$$+ \int_E Q_x^{u(x)}(dz)[f(z)-f(x)]$$

for any $f \in \mathscr{C}^\infty$ and for a.e. $x \in E$. Here $0 \leq n_1+n_2 \leq n$, $0 \leq n_1 \leq 1$ and a_{ij}, b_i are real-valued functions while Q_x^y finite measures. Regarding the objects defining our processes we make the following assumptions.

(i) The state space is $E = E_1 \times E_2 \times E_3$ with $E_1 \subset R^{n_1}$, $E_2 \subset R^{n_2}$, $E_1 \times E_2$ is a bounded Lipschitzian domain with its usual Euclidean topology, while E_3 is an arbitrary Borel subset of $R^{n-n_1-n_2}$ endowed with the finest topology. In the subsequent all continuity properties are ment in the product topology on E. Sometimes we write $x = (x',x'',x''')$ for $x \in E$ meaning $x' \in E_1$, $x'' \in E_2$, $x''' \in E_3$.

(ii) The coefficients $a_{ij}(x)$, $b_i(x,y)$ are bounded Lipschitzian functions. If $n_1 \neq 0$ when $b_1(x) \geq \delta > 0$ and do not depend on $y \in Y$. If $n_2 = 0$ then the matrix $a_{ij}(x)$ is uniformly positive definite, i.e. $\sum a_{ij}(x)\xi_i\xi_j \geq \delta|\xi|$ holds for all $\xi \in R^{n_2}$, $x \in E$ and for some $\delta > 0$.

(iii) The measures Q_x^y are uniformly bounded:
$\sup_{x,y} Q_x^y(E) < \infty$, and if
$\mathscr{H}^\alpha := \{f: |f(x',x'',x''') - f(y',y'',y''')| \le C(|x'-y'|) + |x''-y''|)^\alpha\}$ denotes
$\mathscr{H}^\alpha$ with $0 < \alpha < \infty$ then $Q^y_\cdot(f) \in \mathscr{H}^\alpha$ for any $f \in \mathscr{H}^\alpha$, $0 < \alpha < \infty$ and
$y \in Y$.

These assumptions ensure the existence of a unique Feller process corresponding to each admissible strategy. This process can be choosen to have right-continuous paths with left limits.

We call a set $\Omega \subset E$ a cylinder if $\Omega = \Omega_1 \times \Omega_2 \times E_3$ where $\Omega_1 = (a,b)$ an open interval from E_1 and Ω_2 a domain in E_2. For measurable functions defined on cylinder Ω we introduce the following L_p and W_p norms.

$$\|f\|_{\Omega,p} := \sup_{x'''\in E_3} \left(\int_{\Omega_1}\int_{\Omega_2} |f(x',x'',x''')|^p dx''dx'\right)^{1/p}; \quad |||f|||_{\Omega,p} := \sum_{i,j=n_1+1}^{n_1+n_2} \|f_{x_i x_j}\|_{\Omega,p} + \sum_{i=1}^{n_1+n_2} \|f_{x_i}\|_{\Omega,p} + \|f\|_{\Omega,p}$$

(If E_1 or E_2 are degenerated to one point, then the corresponding integral is deleted). By $\mathscr{C}_0^\infty(\Omega)$ we denote the infinitely often differentiable functions tending to zero on $\partial\Omega\setminus(\{a\} \times \Omega_2 \times E_3)$. The space $L_p(\Omega)$ consists of all f with $\|f\|_{\Omega,p} < \infty$, while the Sobolev space $W_p(\Omega)$ is the closure of $\mathscr{C}_0^\infty(\Omega)$ in the $|||\ |||_{\Omega,p}$ norm. We call f and g to be equal a.e. on Ω if $\|\chi_{\{f\neq g\}}\|_{\Omega,p} = 0$. A function $f \in L_p$ is called continuous if there is a continuous g with f = g a.e..

The essential property of $\mathscr{L}^y$ is that by (i)-(iii) it can be decomposed: $\mathscr{L}^y = \mathscr{A}^0 + \mathscr{A}^1 + \mathscr{B}^y$ with

(2)

$$\mathscr{A}^k f(x) = \sum_{i,j=n_1+1}^{n_1+n_2} a^k_{ij}(x) f_{x_i x_j}(x) + b^k_1(x) f_{x_i}(x) \text{ and}$$

$$\mathscr{B}^y f(x) = \sum_{i=n_1+1}^{n_1+n_2} b_i(x,y) f_{x_i}(x) + Q^y_x(f-f(x))$$

and on any fixed cylinder Ω operators $\mathscr{A}^o, \mathscr{A}^1, \mathscr{B}^u$ can be choosen such that

(a) $\mathscr{A}^o, \mathscr{A}^1$ do not depend on the control variable $y \in Y$.

(b) $\mathscr{A}^o$ do not depend on the state variables, the closure A^o of $\mathscr{A}^o | \mathscr{C}_0^\infty(\Omega)$ generates a Feller process, in fact a part of a process with independent increments. Consequently the resolvent $R_\lambda = (\lambda - A^o)^{-1}$ is bounded by $1/\lambda$ in the L_p- norm ($1 \leq p \leq \infty$).

(c) $\mathscr{A}^1$ and $\mathscr{B}^u$ are relatively bounded [4] in the L_p-norm w.r.t. $\mathscr{A}^o$. More precisely:

(3) $$\|\mathscr{A}^1 f\|_{\Omega,p} \leq K_1 \|f\|_{\Omega,p} + K_2 \|\mathscr{A}^o f\|_{\Omega,p}$$

with some constants K_1, K depending on Ω and p. If Ω is small enough, $\mathscr{A}^o$ and $\mathscr{A}^1$ can be choosen such that $K_2(\Omega,p) < 1/4$. The relative $\mathscr{A}^o$ -bound of $\mathscr{B}^u$ is zero, i.e.

(4) $$\|\mathscr{B}^u f\|_{\Omega,p} \leq K(\varepsilon) \|f\|_{\Omega,p} + \varepsilon \|\mathscr{A}^o f\|_{\Omega,p}$$

for any $\varepsilon > 0$, $u \in U$ and $f \in \mathscr{C}_0^\infty$.

These are the only properties of $\mathscr{L}^u$ which will be essentially used in the proofs. For purposes of § 3 a decomposition $\mathscr{A} + \mathscr{B}^u$ with $\mathscr{A} = \mathscr{A}^o + \mathscr{A}^1$ would suffice. Using the fact that the resolvent correspondint to $\mathscr{A}$ is L_p-bounded [5], the proof of Lemma 2 would became even somewhat simpler. But keeping more general processes in mind we do not use this relatively deep property of diffusions.

3. The Bellman Equation

The first lemma allows to compare the effectivity of two strategies differing on an open set only. Its proof, which we omit here, is straightforward but uses a deep result of Pittenger and Shih.

Lemma 1. Let u and v are two admissible strategies coinciding outside an open set G, and such that for each $x \in G$

$$(5) \quad E_x^v\,(\sigma^v + E^u_{x(\sigma^v)}\,\tau^u) \le E_x^u(\sigma^u + E^u_{x(\sigma^u)}\tau^u)$$

holds true where σ denotes the first exit time from G. Then $E_x^v\tau^v \le E_x^u\tau^u$ is valid for every $x \in E$. If there is an $x_o \in G$ with strict inequality in (5) then $E^v_{x_o}\,\tau^v < E^u_{x_o}\,\tau^u$.

Since A^o is a generator of a part of a Feller process on Ω, R_λ maps $\mathscr{C}(\Omega)$ into $\mathscr{D}(A^o) \subset \mathscr{C}_o(\Omega)$. By $\| R_\lambda \|_p \le 1/\lambda$ R_λ can be extended mapping L_p onto $W_\rho(\Omega)$. The operator A^o, consequently also $\mathscr{A}^1$ and $\mathscr{B}^u$ can be extended to $W_p(\Omega)$ we denote these extended operator by $\tilde{R}_\lambda$, $\tilde{A}^o$, $\tilde{A}^1$, $\tilde{B}^u$ resp. or if we want to emphasize the dependence on Ω, then by $\tilde{R}_{\Omega,\lambda}$, $\tilde{A}^o_\Omega$ etc.

Lemma 2. For arbitrary $v \in U$ and continuous $u \in U$ operator $\tilde{B}^u_E$ maps $\mathscr{D}(L^v)$ into $\mathscr{C}$, if $u \equiv y \in Y$ then into $\mathscr{H}^\alpha$ with some $0 < \alpha < \infty$.

Proof. If $\mathscr{A}^o + \mathscr{A}^1 \equiv 0$ then $L^v = B^v$, $E = E_3$ and the statement is trivial. We assume $\mathscr{A}_o \not\equiv 0$. It is enough to show $\mathscr{D}(L^v) \subset W_p(E)$ since $\tilde{B}^u, \tilde{B}^y$ maps W_p into $\mathscr{C}, \mathscr{H}^\alpha$ resp. if p is large and u is continuous. Let $Z(r,x_o):=\{x=(x',x'',x''')\in E: |x'-x_o'|<r, |x''-x_o''|<r\}$. By the Lipschitz continuity of a_{ij}, b_i there exists a $\rho > 0$ such that finitely many cylinders of radius ρ cover E and on every such cylinder the $\tilde{A}_z^o$ bound of $\tilde{A}_z^1$ is less than 1/4.

We fix any such cylinder Z and build the part of x_t^v on Z. Denote R_λ the resolvent of $A^o = A_z^o$, $\tilde{R}_\lambda$ its extension. We seek the resolvent $\tilde{G}_\lambda^v = (\lambda - \tilde{A}^o - \tilde{B}^v)^{-1}$ in the form $\tilde{G}_\lambda^v = \tilde{R}_\lambda\,(J-\tilde{B}^v\,\tilde{R}_\lambda)^{-1}$. By (4) we have $\|\tilde{B}^v\tilde{R}_\lambda\|_p \le K(\varepsilon)/\lambda + 2\varepsilon$. Hence if ε is small and λ large enough $\|\tilde{B}^v\tilde{R}_\lambda\|_p \le 1/2$ the Neuman series $(J-\tilde{B}^v\tilde{R}_\lambda)^{-1} = \sum_{n=0}^{\infty} (\tilde{B}^v\tilde{R}_\lambda)^n$ converges in L_p-norm, $\|(J-\tilde{B}^v\tilde{R}_\lambda)^{-1}\|_p \le 2$. Consequently if λ is large enough, $\|\tilde{G}_\lambda^v\| \le 2/\lambda$, and $\tilde{G}_\lambda^v$ maps L_p onto W_p. Since the $\tilde{A}_o$ -bound of $\tilde{B}^u$ is zero, (3) can be rewritten in the form $\|\tilde{A}^1 f\|_p \le K_1'\|f\|_p + K_2'\|(\tilde{A}^o+B^v)f\|_p$ with $K_2' < 1/4$. Repeating the above construction one can show that for large λ the resolvent $(\lambda - L_z^v)^{-1}$ maps $L_p(Z)$ onto $W_p(Z)$. We have shown $\mathcal{D}(L_z^v) \subset \mathcal{D}(\tilde{L}_z^v) = W_p(Z)$ for any cylinder $Z(\rho,x)$ with radius ρ fixed above.

Now remember, that $\mathcal{D}(L^v) = \bigcap_{x\in E} \mathcal{D}_x(\mathcal{L}^v) \cap \{f: \mathcal{L}f \in \mathcal{C}\}$. Take a sequence of neighbourhoods $\Gamma_n \downarrow x$ with $\Gamma_n \subset Z(\rho,x_o)$ for $|x-x_o| < \delta$. Then with abbreviation $x_n = x^v(\sigma(\Gamma_n))$ we have

$$\mathcal{L}^v f(x) = \lim_{n\to\infty} \frac{E_x^v f(x_n) - f(x)}{E_x^v \sigma(\Gamma_n)} = \lim_{n\to\infty} \frac{E_x^v\{f(x_n); x_n \in Z\} - f(x)}{E_x^v \sigma(\Gamma_n)} + \tag{6}$$

$$+ \lim_{n\to\infty} \frac{E_x^v\{f(x_n); x_n \notin Z\}}{E_x^v \sigma(\Gamma_n)} .$$

The second term converges for any $f \in \mathcal{L}$ to bounded limes function, while the first term tends to $\mathcal{L}^v f$ if $f \in \mathcal{D}_x(\mathcal{L}^v) x \in Z$. Consequently if

$f \in \mathcal{D}(L_E^v)$, then necessarily $f \in \cap_{x\in Z}\mathcal{D}_x(\mathcal{L}^v)$ and $\mathcal{L}_z^v f \in \mathcal{B}(Z)$ for any cylinder of radius ρ. But it implies that for any $\rho' < \rho$ and large enough λ there exists a $g \in L_p(Z(\rho,x_o))$ such that $f(x) = (\lambda-\tilde{L}_z^v)^{-1}g(x)$ for a.e. $x \in Z' = Z(\rho',x_o)$. In other words there exists a $\phi \in W_p(Z)$ such that it coincides with f on Z'. Since E can be covered by a finite number of cylinders of radius ρ' as well, it follows $f\in W_p(E)$. We have shown $\mathcal{D}(L^v) \subset W_p(E)$. Q.e.d.

Corollary: Decomposition $L^v f(x) = \tilde{A}f(x) + \tilde{B}^{v(x)}f(x)$ holds true for all $f \in \mathcal{D}(L^v)$ and for a.e. $x \in E$.

The following theorem shows that the optimal expense function $\psi(x) = E_x^v\tau^v = \inf_{u\in U} E_x^u\tau^u$ is much more regular than a general element of $\mathcal{D}(L^v)$. Moreover ψ is the solution of the Bellman equation. Denote v the optimal strategy and A the restriction of $\tilde{A}^o + \tilde{A}^1$ to $\{f \in W_p : (\tilde{A}^o + \tilde{A}^1)f \in \mathcal{C}\}$.

Theorem 1. (a) $\psi \in \mathcal{D}(L^v) \cap \mathcal{D}(A) \cap \cap_{u\in U}$, continuous $\mathcal{D}(L^u)$

(b) $L^v\psi(x)+1 = \min_{u(x)\in y} L^{u(x)}\psi(x)+1=0$ for a.e. $x \in E$.

Proof. $L^v\psi + 1 = 0$ is known. In order to show (b) assume that there exists a continuity point x_o of v and $y \in Y$ such that $\tilde{L}^y\psi(x_o) < -1$. By the continuity of v and Lemma 2 there exists a neighbourhood Γ of x_o such that $\tilde{A}\psi(x) = -1 - \tilde{B}^{v(x)}\psi(x)$ is continuous in Γ. Consequently $\tilde{L}^y\psi(x) = L^y\psi(x) < -1$ in a $\Gamma_1 \subset \Gamma$. Let u be the strategy which coincides with y inside Γ_1 and with v outside of it, and denote σ the first exit time from Γ_1. Since $L^u\psi(x) = L^y\psi(x) < -1$ inside Γ_1 and $E_x^u\sigma > 0$ we obtain for each $x\in\Gamma_1$.

$$E_x^u E_{x(\sigma)}^v \tau = E_x^u\psi(x_\sigma) = \psi(x) - E_x^u \int_0^\sigma L^y\psi(x_t)dt < E_x^v\tau - E_x^u\sigma .$$

This means (5) is fulfilled with strict inequality and so by Lemma 1 $E_x^u\tau < E_x^v\tau$ despite the assumed optimality of v. Hence (b) is proved for continuity points of v, i.e. for a.e. $x\in E$.

To show (a) observe that by Lemma 2 $\tilde{B}^y\psi\in\mathscr{H}^\alpha$ for any constant strategy $u\equiv y\in Y$. Consequently $\tilde{A}\psi = -1-\min \tilde{B}^y\psi \in \mathscr{C}$ proving $\psi\in\mathscr{D}(A)$. Together with $B^u f\in\mathscr{C}$ for continuous $u \in U$ we have also $\psi\in\mathscr{D}(L^u)$. Q.e.d.

4. Processes With Infinitely Many Jumps

Dr. H. Pragarauskas called the author's attention to the following straightforward generalization.

Let the state space as in § 2, but the operator $\mathscr{L}^u$ of the form

$$\begin{aligned}\mathscr{L}^u f(x) &= \sum a_{ij}(x) f_{x_i x_j}(x) + \sum b_i(x,u(x)) f_{x_i}(x) + \\ &+ \int Q_x^{u(x)}(dz)[f(z)-f(x)] + \\ &+ \int_{|z|<1} [f(x+z)-f(x) - \sum z_i f_{x_i}(x)]\Pi_x^{u(x)}(dz) + \\ &+ \int_{|z|>1} [f(x+z)-f(x)]\, \Pi_x^{u(x)}(dz).\end{aligned} \tag{7}$$

Here Π_x^y is a measure on E_2 and one of the following conditions hold

1) $a_{ij}(x)$ is uniformly non-degenerate on E_2 and there exist $K, \delta > 0$ and $0 < \beta < 2$ such that $\Pi_x^y(dz) \le K\, dz/|z|^{n_2+\beta}$ if $dz\subset\{z\in E_2: |z|< \delta\}$.

2) $a_{ij}\equiv 0$ and Π_x^y can be decomposed into two measures

$$\Pi_x^y(dz) = C(x,z)dz/|z|^{n_2+\alpha} + \tilde{\Pi}_x^y(dz) \tag{8}$$

Here $0 < \alpha < 2$ is called the characteristic exponent of Π. Function $c(x,z)$ is Lipschitzian on $E \times E_2$ and $0 < \gamma_1 \le c(x,z) \le \gamma_2 < \infty$ holds with some constants γ_1, γ_2. There exist $K, \delta > 0$ and $0 < \beta < \alpha < 2$ such that $\tilde{\Pi}^y_x(dz) \le K\, dz/|z|^{n_2+\beta}$ if $dz \subset \{z \in E_2 : |z| < \delta\}$. Moreover $\Pi^y_\cdot(f)$ is r-Hölderian on E for any $f \in \mathscr{H}^r$ such that $\int |f|dz/|z|^{n_2+\beta}$. If $\alpha \le 1$ then $b_i(x,y) \equiv 0$ for $i \ge 2$. Otherwise $a_{ij}(x)$, $b_i(x,y)$ and Q^y_x are as in § 2.

Operator $\mathscr{L}^u$ allows a decomposition similar to that of § 2, and all the results of § 3 remain valid for $\mathscr{L}^u$. If $\alpha < 2$ then in the proofs the Sobolev-space W_p is to be substituted by a suitable modification of the α-order Besov space [1].

The same method applies if in (8) the first term is an arbitrary measure $\Pi^o(dz)$ on E_2, not depending on x and such that $\int |z|^2(1-|x|^2)^{-1}\ \Pi^o(dz) < \infty$ and $\Pi^0(dz) \ge K^o dz/|z|^{n_2+\alpha}$ for $dz \subset \{z \in E_2;\ |z| < \delta\}$.

REFERENCES

1. Adams, R. A., Sobolev spaces, New York, San Francisco, London 1975.
2. Dynkin, E. B., Markov processes, Berlin, Heidelberg, 1965.
3. Fleming, W. H. and Richel, R. W., Deterministic and Stochastic Control, New York, Heidelberg, Berlin, 1975.
4. Kato, T., Perturbation theory for linear operator, Berlin, Heidelberg, New York, 1966.

5. Krylov, N. V., Controlled processes of diffusion type, Moscow, 1977 (Russian).

6. Pragarauskas, H., On the optimal control of discontinuous random processes (Russian), Trudi Skoli-seminara po Teorii Sluchaynih Processov, Vilnius, 1975.

7. Vermes, D. A., A necessary and sufficient condition of optimality for Markovian control problem, Acta. Sci. Math. 34 (1973), 401-413.

8. Vermes, D., Extremality properties of the optimal strategy in Markovian control problems. To appear in "Analysis and Optimization of Stochastic Systems", London, 1979.

STATIONARY SOLUTIONS OF THE STOCHASTIC NAVIER-STOKES EQUATIONS

M. I. Višic, A. I. Komech

1. Let us consider the stochastic Navier-Stokes equations

$$\dot{u}(t,x) + \sum_{i=1}^{n} \frac{\partial(u^i u)}{\partial x^i} = -\nabla p(t,x) + \nu\nabla u + \dot{w}(t,x),$$

(1.1)

$$(\nabla, u(t,x)) = 0, \; t > 0, \; x \in \Omega,$$

in the bounded domain $\Omega \subset \mathbb{R}^n$ with boundary condition $u(t,x) = 0$, $t > 0$, $x \in \partial\Omega$, where $u = u(t,x)$ is velocity, p -pressure, $\nu > 0$ viscosity, $w(t,x)$ - the Wiener process in $H^0 \equiv [L_2(\Omega)]^n$, hence a.s. $w \in \mathscr{C} \equiv C(\mathbb{R}^+; H^0)$; $\partial\Omega$ is supposed to be smooth. Denote by $Q: H^0 \to H^0$ the correlation operator of the process w: $Q \geq 0$, $\bar{S} \equiv SpQ < +\infty$ and denote by λ the distribution of w, i.e. the Borelian measure on $\mathscr{C}$ with the characteristic functional $\tilde{\lambda}(\varphi) \equiv \exp(-\frac{1}{2}B(\varphi,\varphi))$,
$B(\varphi,\varphi) = \int_0^\infty \int_0^\infty t \wedge s \langle Q\varphi(t,\cdot), \varphi(s,\cdot)\rangle dtds$, $\forall \varphi \in [C_0^\infty(\mathbb{R}^+ \times \Omega)]^n \equiv \mathscr{D}$.

Let us define now the functional spaces we need. Denote by $\langle \cdot,\cdot \rangle$ the duality between $[D'(\Omega)]^n$ and $[D(\Omega)]^n$ and also the scalar product in the spaces H^0 and $\mathscr{H}^0 \equiv \{u \in H^0: (\nabla, u(x)) = 0, x \in \Omega\}$.
Then $\mathscr{H}^0 \subset [D'(\Omega)]^n$. Let $e_j(x), j \in \mathbb{N}$, be the eigen-functions of the operator Δ in $\overset{\circ}{\mathscr{H}}{}^1 \equiv [\overset{\circ}{H}{}^1(\Omega)]^n \cap \mathscr{H}^0$: $\Delta e_j(x) = \lambda_j e_j(x)$; $0 < \lambda_j \to \infty$ as $j \to \infty$; $\langle e_j(x), e_\ell(x) \rangle = \delta_{j\ell}$. For $s \in \mathbb{N}$ and $u(x) = \sum_1^\infty u^j e_j(x) \in \mathscr{H}^0$ define the norm $\|u\|_s^2 \equiv \sum_1^\infty \lambda_j^s |u^j|^2$ and the space $\mathscr{H}^s \equiv \{u \in \mathscr{H}^0: \|u\|_s < +\infty\}$. Then $\mathscr{H}^s$ is the Hilbert space with the

norm $\|\cdot\|_s$ equivalent to the norm in $[H^s(\Omega)]^n$. For $s \in \mathbb{Z}$, $s < 0$ let $\mathcal{H}^s \equiv (\mathcal{H}^{-s})' \subset [D'(\Omega)]^n$. Define the spaces $\mathcal{L}_2 \equiv L_2^{loc}(\mathbb{R}^+;\mathcal{H}^o)$ and $\mathcal{L}_2^1 \equiv L_2^{loc}(\mathbb{R}^+;\overset{o}{\mathcal{H}}{}^1)$ with the corresponding seminorms

$$\|u\|^2_{\mathcal{L}_2;T} = \int_0^T \|u(t)\|_0^2\, dt < +\infty, \quad \|u\|^2_{\mathcal{L}_2^1;T} \equiv \int_0^T \|u(t)\|_1^2\, dt < +\infty, \forall T > 0.$$

It is evident, that $\mathcal{L}_2$ and $\mathcal{L}_2^1$ are Frechet spaces. Let $b \in \mathbb{N}$ and $p \geq 1$ are fixed. For mappings $u: \mathbb{R}^+ \to \mathcal{H}^{-b}$ and for $t \geq \tau \geq 0$ define

$$\operatorname*{osc}_{[\tau,t]} u \equiv \operatorname*{vrai\ sup}_{(\tau',t') \in [\tau,t]x[\tau,t]} \|u(t') - u(\tau')\|_{-b} .$$

<u>Definition</u> 1.1. The space BV_q^{-b} consists of functions of "bounded variations of degree q" which define mappings $\mathbb{R}^+$ into $\mathcal{H}^{-b}$ with finite seminorms: $\forall T \in \mathbb{N}$

$$(1.2) \quad \|u\|_{BV_q^{-b};T} \equiv \Big(\sup_{\mathcal{N}>0} \sup_{\{t_j\}} \sum_{j=1}^{\mathcal{N}} |\operatorname*{osc}_{\Delta t_j} u|^q\Big)^{1/q} + \operatorname*{vrai\ sup}_{[0,T]} \|u(t,\cdot)\|_{-b}$$

here $\sup\limits_{\{t_j\}}$ denotes the supremum over the set of all divisions $t_o = 0 < t_1 < \ldots < t_{\mathcal{N}} = T$ of $[0,T]$; $\Delta t_j \equiv [t_{j-1},t_j]$.

Similarly to the classical theory of functions of bounded variations the following lemma holds.

Lemma 1.1. Let $u \in BV_q^{-b}$. Then after changing of $u(t)$ on a subset of the interval $[0,T]$ of Lebesgue's measure zero, the mapping $t \to u(t)$ is continuous from $[0,T]$ into $\mathcal{H}^{-b}$ for $\forall t \in [0,T]$ except for a countable set of points of discontinuity of the first kind.

<u>Corollary</u> 1.1. We can assume, that all functions $u \in BV_q^{-b}$ are right continuous.

<u>Definition</u> 1.2. U is the space $\mathcal{L}_2^1 \cap BV_q^{-b}$ with the seminorms

$$\|u\|_{U;T} \equiv \|u\|_{\mathscr{L}_2^1;T} + \|u\|_{BV_q^{-b};T} < +\infty, \quad \forall T \in \mathbb{N}.$$

The solutions u(t,x) of (1.1) belongs to U. Let us exclude the pressure p from (1.1) and formulate the main result. For $u \in \mathscr{L}_2^1$ let $\mathscr{A}u \equiv \dot{u} + \sum_{i=1}^{n} \frac{\partial(u^i u)}{\partial x^i} - \nu\Delta u \in [D'(\mathbb{R}^+ \times \Omega)]^n$. Let $\mathscr{V}^* \equiv [H^{-b}_{loc}(\mathbb{R}^+ \times \Omega)]^n$, $\mathscr{V} \equiv (\mathscr{V}^*)' = \bigcup_{T>0} [H^b([0,T] \times \Omega)]^n$. Then $\mathscr{A}: \mathscr{L}_2 \to \mathscr{V}^*$ is a continuous mapping, if $b > n+1/2+2$ (see [1]). Denote $\mathscr{G} \equiv \{h \in \mathscr{V}^* : \exists p(t,x) \in D'(\mathbb{R} \times \Omega), h = \nabla p\}$, $g: \mathscr{V}^* \to \overline{\mathscr{V}^*} \equiv \mathscr{V}^*/\mathscr{G}$ the mapping $f \to gf \equiv f \bmod \mathscr{G}$. Let $G: \mathscr{C} \to \overline{\mathscr{V}^*}$ be the map $w \to Gw \equiv g\dot{w}$ and $\beta \equiv G^*\lambda$ is the measure on $\overline{\mathscr{V}^*}$ which is image of the measure λ with respect to the mapping G.

All the measures in the present paper are Borelian. Let $\mathscr{B}(X)$ denotes Borelian σ-algebra of a topological space X.

<u>Definition</u> 1.3. The measure P on $\mathscr{L}_2$ is the weak solution of (1.1), if $(g\mathscr{A})^*P = \beta$ that is $P((g\mathscr{A})^{-1}B) = \beta(B)$ for $\forall B \in \mathscr{B}(\overline{\mathscr{V}^*})$.

Let us denote $\mathscr{V}^0 = \{v \in \mathscr{V} : (\nabla, v(t,x)) = 0,\ t \ge 0,\ x \in \Omega\}$. Then $\mathscr{V}^0$ is isomorphic to $(\overline{\mathscr{V}^*})'$ and therefore $(g\mathscr{A})^*P = \beta$ is equivalent to

$$(1.3) \quad \int \exp(i\{\mathscr{A}u, v\}) P(u) = \tilde{\lambda}(-\dot{v}) \equiv \exp\{-\int_0^\infty \langle Qv(t,\cdot), v(t,\cdot)\rangle\, dt,\ \forall v \in \mathscr{V}^0;$$

here $\{\cdot,\cdot\}$ is the duality between $\mathscr{V}^*$ and $\mathscr{V}$ and also the scalar product in $[L_2(\mathbb{R}^+ \times \Omega)]^n$. For $\tau \ge 0$ let $\hat{\tau}$ be the operator in $\mathscr{L}_2$: $\hat{\tau}u(t) = u(\tau+t)$, $t \ge 0$.

<u>Definition</u> 1.4. We shall say that measure P on $\mathscr{L}_2$ is stationary if $\hat{\tau}^*P = P$ for $\forall \tau \ge 0$.

For $u \in \mathscr{L}_2$ and $t, \varepsilon > 0$ put $\gamma_t^\varepsilon u \equiv \frac{1}{\varepsilon}\int_0^\varepsilon u(t+\tau)\, d\tau \in \mathscr{H}^{-b}$. Then the operator $u \to \gamma_t u \equiv u(t) \equiv \lim_{\varepsilon\to 0+} \gamma_t^\varepsilon u$ is defined for $u \in U$ by the corollary 1.1. It follows from (1.2) that $U \in \mathscr{B}(\mathscr{L}_2)$. Introduce in U the $\mathscr{L}_2$-topology. Then $\mathscr{B}(U) = U \cap \mathscr{B}(\mathscr{L}_2)$ and $\gamma_t: U \to \mathscr{H}^{-b}$ is Borelian mapping as the limit of continuous ones $\gamma_t^\varepsilon : \mathscr{L}_2 \to \mathscr{H}^{-b}$. Hence for every measure P on $\mathscr{L}_2$, concentrated on U the restrictions $\mu(t,\cdot) \equiv \gamma_t^* P$ can be defined as follows: $\mu(t,B) \equiv P(\gamma_t^{-1} B)$ for $\forall B \in \mathscr{B}(\mathscr{H}^{-b})$.

Denote $\mathscr{A}(u,v) \equiv i \langle -(u,\nabla)u + \nu\Delta u, v \rangle - \frac{1}{2} \langle Qv, v \rangle$ for $u \in \overset{\circ}{\mathscr{H}}{}^1$ and $v \in V_\infty \equiv \overset{\infty}{\underset{1}{\cup}} V_m$ where $V_m = (e_1(x),\ldots,e_m(x))$. The main results of this work are formulated in the theorems 1.1 and 1.2.

Theorem 1.1. For every $\nu > 0$ there exists a stationary weak solution $\bar{P} = \bar{P}^\nu$ of the stochastic system (1.1). $\bar{P}$ is a measure, concentrated on $U(p > 2, b > \frac{n+1}{2} + 2)$. The measure $\bar{\mu} \equiv \gamma_t^* \bar{P}$ does not depend on t, it is concentrated on $\overset{\circ}{\mathscr{H}}{}^1$ and the following estimates holds true

$$(1.4) \quad \int \|\nabla u\|_0^2 \|u\|_0^{2k} \bar{\mu}(du) \le \frac{\bar{C}_k(\Omega,Q)}{\nu^{k+1}} < +\infty, \quad k \in \mathbb{Z}_+ \equiv \mathbb{N} \cup \{0\}$$

where $\bar{C}_0(\Omega,Q) = \frac{1}{2}\bar{S}$; $\bar{\mu}$ satisfies the stationary direct Kolmogoroff equation [2,3]:

$$(1.5) \quad 0 = \int \exp(i\langle u,v\rangle) \mathscr{A}(u,v)\bar{\mu}(du), \ \forall v \in V_\infty .$$

Theorem 1.2. Let n = 2. Then for every $\nu > 0$ the corresponding measure $\bar{P}^\nu$ constructed in theorem 1.1, satisfies the identity:

$$(1.6) \quad \nu \int \Big(\int_\Omega |\nabla u(t,x)|^2 dx \Big) \bar{P}^\nu(du) = \frac{1}{2}\, \mathrm{Sp} Q, \ \forall t \ge 0.$$

<u>Remark</u> 1.1. The integral in (1.6) is the mean velocity of energy dissipation at the moment t. It does not depend on t, because $\bar{P}^\nu$ is stationary. From (1.6) it follows, that for fixed Q the mean velocity of the energy dissipation does not depend on $\nu > 0$. The result corresponds to the Kolmogoroff guess he has formulated in his

report [4].

2. Proof of the theorem 1.1. We approximate the system (1.1) by the following Itô system in V_m:

$$(2.1)\quad \dot{u}_m(t,x) + \Pi_m(u_m,\nabla)u_m = \nu\Pi_m\Delta u_m + \dot{w}_m(t,x),\ t \geq 0,$$

where $w_m \equiv \Pi_m w$, Π_m is operator of orthogonal projection H^o on V_m. We are going to construct the measure $\bar{P}$ as the limit of measures $\bar{P}_m$, which are weak stationary solution of (2.1)

Lemma 2.1. For $\forall \nu > 0$, $\forall m \in \mathbb{N}$ there exists a stationary measure $\bar{P}_m$ on $\mathscr{L}_2$ satisfying (1.3) for $\mathfrak{v} \in \mathscr{V}_m^o \equiv \Pi_m \mathscr{V}^o$:

$$(2.2)\quad \int \exp\,(i\{\mathscr{A}u,\mathfrak{v}\})\bar{P}_m(du) = \tilde{\lambda}\,(-\dot{\mathfrak{v}}),\ \forall \mathfrak{v} \in \mathscr{V}_m^o;$$

the measures $\bar{P}_m$ are concentrated on U and $\forall m \in \mathbb{N}$

$$(2.3)\quad \int \|\nabla u(t)\|_0^2 \|u(t)\|_0^{2k}\ \bar{P}_m(du) \leq \frac{\bar{C}_k(\Omega,Q)}{\nu^{k+1}} < +\infty,\ \forall t > 0,$$

$$(2.4)\quad \int \|u\|_{BV_q^{-b};T}\ \bar{P}_m(du) \leq \bar{C}_T < +\infty,\ \forall T > 0;$$

$\bar{C}_o(\Omega,Q) = \frac{1}{2}\bar{S}$. The measure $\bar{\mu}_m \equiv \gamma_t^* \bar{P}_m$ is concentrated on V_m and satisfies (1.5) for $v \in V_m$:

$$(2.5)\quad 0 = \int \exp(i\langle u,v\rangle)\mathscr{A}(u,v)\bar{\mu}_m(du),\ \forall v \in V_m\ .$$

The existence of a stationary process $u_m(t) \in V_m$ satisfying (2.1), follows from the results of [5]. However for the proof of the estimates (2.3) - (2.4) we need to construct the stationary process from the nonstationary one by Bogoliubov method, similarly to [5]. Namely consider the Cauchy problem for (2.1) with the initial condition $u_m(0,x) = 0$. The existence of the solution $u_m(t,x)$ of this Cauchy problem follows from [5] in view of existence of "Liapunoff's function".

Denote by $\mathscr{E}$ the expectation operator on the probability space, where are defined the processes $w(t,x)$ and $w_m(t,x) \equiv \Pi_m w(t,x)$. For simplicity let us define $w(t,x)$ as the process in natural representation on $(\mathscr{C}, \mathscr{B}(\mathscr{C}), \lambda): w(t,x;\omega) \equiv w(t,x)$ for $\forall\, \omega \in \mathscr{C}$. Now we shall prove the estimate: for $\forall k \in \mathbb{Z}_+, t \geq 0$ and $m \in \mathbb{N}$

$$(2.6) \quad \mathscr{E}\left(\| u_m(t) \|_0^{2+2k} + (2+2k)\nu \int_0^t \| \nabla u_m(\tau) \|_0^2 \| u_m(\tau) \|_0^{2k} dt \right) \leq \frac{t C_k(\Omega, Q)}{\nu^k} < +\infty ,$$

where $C_o(\Omega, Q) = \bar{S}$. By the Itô formula

$$\begin{aligned} d\| u_m(t) \|_0^{2+2k} &= (2+2k) \| u_m(t) \|_0^{2k} \langle u_m(t), du_m(t) \rangle + \\ (2.7) \qquad & + \frac{1}{2}(2+2k)\, [2k \| u_m(t) \|_0^{2k-2} \langle Q u_m(t), u_m(t) \rangle + \| u_m(t) \|_0^{2k} \operatorname{Sp} \Pi_m Q \Pi_m]\, dt. \end{aligned}$$

From (2.1) $\langle u_m(t), du_m(t) \rangle = -\nu \langle \nabla u_m(t), \nabla u_m(t) \rangle dt + \langle u_m(t), dw_m(t) \rangle$. Hence by integrating (2.7) with respect to t and taking expectation $\mathscr{E}$, we get by using the initial condition $u_m(0,x) = 0$: $\mathscr{E}\,(\| u_m(t) \|_0^{2+2k} + (2+2k)\nu \int_0^t \| u_m(\tau) \|_0^{2k} \| \nabla u_m(\tau) \|_0^2 \, d\tau) \leq$

$$\leq (2+2k)\, \frac{1}{2}\, (2k+1)\, \bar{S}\, \mathscr{E} \int_0^t \| u_m(\tau) \|_0^{2k} \, d\tau .$$

Hence, by induction in $k = 0, 1, 2, \ldots$ we get (2.6).

Now we shall construct the measures $\bar{P}_m$. For $\forall\, \Theta > 0$ the process $u_{m,\Theta}$ is defined on the space $(\mathscr{C} \times [0,\Theta], \mathscr{B}(\mathscr{C} \times [0,\Theta]), \lambda \times \frac{d\tau}{\Theta})$ by the formula $u_{m,\Theta}(t,x;\omega) = u_m(\tau+t,x)$ for $\forall\, \omega = (w,\tau) \in \mathscr{C} \times [0,\Theta]$, where u_m

is the solution of (2.1), corresponding to $w_m \equiv \Pi_m w$. Denote by $\mathscr{E}_\Theta$ the corresponding expectation operator and by $P_{m,\Theta}$ the measure on $\mathscr{L}_2$ the distribution of $u_{m,\Theta}$.

Lemma 2.2. The set of measures $\{P_{m,\Theta}\}$, $m \geq 1$, $\Theta > 1$ is weakly compact on $\mathscr{L}_2$.

Proof. By using (2.6) we can estimate "the derivatives with respect to x" in the following way:

for $\forall T > 0$, $0 < t < T$, $m \in \mathbb{N}$, $\Theta > 1$

$$\mathscr{E}_\Theta \| \nabla u_{m,\Theta}(t) \|_0^2 \| u_{m,\Theta} \|_0^{2k} \leq \mathscr{E} \frac{1}{\Theta} \int_0^{\Theta+T} \| \nabla u_m(s) \|_0^2 \, \| u_m(s) \|_0^{2k} \, ds \leq$$

$$\leq \frac{1}{\Theta} \frac{1}{(2+2k)\nu} \frac{(\Theta+T) C_k(\Omega,Q)}{\nu^k} \leq \frac{(1+T) C_k(\Omega,Q)}{(2+2k)\nu^{k+1}} < +\infty \ . \tag{2.8}$$

From the other hand the following estimate of the variations (1.2) of degree $q > 2$ "with respect to t" holds true if $b > \frac{n+1}{2} + 2$: for $\forall T > 0$, $m \in \mathbb{N}$, $\Theta > 1$

$$\mathscr{E}_\Theta \| u_{m,\Theta} \|_{BV_q^{-b};T} \leq C(\Omega,Q,T,\nu) < +\infty \ . \tag{2.9}$$

For the proof of (2.9) denote the first term of the right side of (1.2) by $\mathrm{Var}_{q;T}^{-b} u$. Then we get $\| u \|_{BV_q^{-b};\, T} \leq$ $\leq C(T)(\int_0^T \| u(t) \|_0^2 \, dt + 1) + 2 \, \mathrm{Var}_{q;T}^{-b} u$. Let $z_m = u_m - w_m$, then

$$\mathscr{E}_\Theta \| u_{m,\Theta} \|_{BV_q^{-b};T} \leq C(T) \frac{1}{\Theta} \int_0^\Theta \mathscr{E} [(\int_0^T \| u_m(\tau+t) \|_0^2 \, dt + 1) \ + \tag{2.10}$$

$$+ \ 2 \, \mathrm{Var}_{q;T}^{-b} (Z_m(\tau+\cdot) + w_m(\tau+\cdot))] \, d\tau \ .$$

But from (2.1) it follows

$$\| \dot{z}_m(t) \|_{-b} \le C(\Omega,\nu)(1 + \| u_m(t) \|_0^2),$$

and therefore

$$(2.11)\ \mathscr{E}\mathrm{Var}_{q;T}^{-b} z_m(\tau+\cdot) \le C(\Omega,\nu)(T+\mathscr{E}\int_\tau^{\tau+T} \| u_m(s)\|_0^2\, ds).$$

Besides for the trajectories of the Wiener process it is possible by using Levy's module to get:

$$(2.12)\ \mathscr{E}\mathrm{Var}_{q;T}^{-b} w_m(\tau+\cdot) \le C(Q,T), \forall m \in \mathbb{N}.$$

Putting (2.11) - (2.12) into (2.10) and using (2.6) one gets (2.9).

From (2.8) and (2.9) it follows for $m \in \mathbb{N}$, $\Theta > 1, \forall T > 0$: $\mathscr{E}_\Theta \| u_{m,\Theta}\|_{U;T} \le C(T) < +\infty$. Hence by Tchebychev's inequality for $\forall M(T) > 0$

$$(2.13)\ P_{m,\Theta}(\| u\|_{U;T} \ge M(T)) = \lambda x \frac{d\tau}{\Theta} (\| u_{m,\Theta}\|_{U;T} \ge M(T)) \le \frac{C(T)}{M(T)}.$$

<u>Lemma</u> 2.3. The set $\{u \in \mathscr{L}_{2;T} \equiv L_2(0,T;\mathscr{H}^0): \| u\|_{U;T} < M(T)\}$ is precompact in $\mathscr{L}_{2;T}$.

The proof of this lemma is similar to the proof of the compactness theorem in [7]. For $\forall \epsilon > 0$ we choose $M(T)$, $T \in \mathbb{N}$, such that $\sum_1^\infty \frac{C(T)}{M(T)} \le \varepsilon$. Then from (2.13) for $\mathscr{K}_\varepsilon \equiv \{u \in \mathscr{L}_2 : \|u\|_{U;T} < M(T), \forall T \in \mathbb{N}\}$ it follows that $P_{m,\Theta}(\mathscr{K}_\varepsilon) \ge 1-\varepsilon$.

But from lemma 2.3 it follows that $\mathscr{K}_\varepsilon$ has a compact closure in $\mathscr{L}_2$. Therefore by Prokhoroff's theorem we get lemma 2.2.

Proof of lemma 2.1. According to the lemma 2.2 for every $m \in \mathbb{N} \exists \{\Theta_k\} : P_{m,\Theta_k} \underset{\Theta_k\to\infty}{\to} \bar{P}_m$ where $\bar{P}_m$ is a measure on $\mathscr{L}_2$. Hence we deduce that $\bar{P}_m$ is stationary. For $\varphi \in \Pi_m \mathcal{V}^o$ from (2.1) we have (here P_m is a distribution of u_m)

$$\int \exp(i\{\mathcal{A}u,\varphi\})\bar{P}_m(du) = \lim_{\Theta_k\to\infty} \frac{1}{\Theta_k}\int_0^{\Theta_k}[\int \exp(i\{\mathcal{A}u(\tau+\cdot),$$

(2.14)

$$\varphi(\cdot)\})P_m(du)]\,d\tau = \lim_{\Theta_k\to\infty} \frac{1}{\bar{\Theta}_k}\int_0^{\Theta_k} \exp(-\int_0^{\infty} \langle Q\varphi(t-\tau),$$

$$\varphi(t-\tau)\rangle\,dt)\,d\tau = \tilde{\lambda}(-\dot{\varphi}).$$

Hence we have (2.2). The estimates (2.3) - (2.4) follows from (2.8) and (2.9) by using Fatou's lemma. We are going to deduce (2.5). For P_m restrictions $\mu_m(t,\cdot)\equiv\gamma_t^* P_m$ it follows from (2.1) the direct Kolmogoroff equation [5,6]: for $\forall v \in V_m$ $\tilde{\mu}_m(\Theta,v) - \tilde{\mu}_m(0,v) =$

$$= \int_0^{\Theta} [\int \exp(i\langle u,v\rangle)\mathcal{A}(u,v)\mu_m(\tau,du)]\,d\tau = \Theta\int \exp(i\langle u,v\rangle)\mu_{m,\Theta}(du),$$

where $\mu_{m,\Theta} = \gamma_o^* P_{m,\Theta}$. Hence, dividing by $\Theta = \Theta_{k'}$, we get (2.5) for $\mu_m = \lim_{\Theta_{k'}\to\infty} \mu_{m,\Theta_{k'}}$ instead of $\bar{\mu}_m(du)$ (the compactness of measures μ_{m,Θ_k} on $\mathcal{H}^o$ follows from (2.8)). It remains to prove, that $\mu_m = \bar{\mu}_m$ or equivalently $\tilde{\mu}_m(v)\equiv\tilde{\bar{\mu}}(v)$:

$$(2.15)\quad \lim_{\Theta_{k'}\to\infty} \int \exp(i\langle\gamma_o u,v\rangle)P_{m,\Theta_{k'}}(du) = \int \exp(i\langle\gamma_o u,v\rangle)\bar{P}_m(du), \forall\, v\in V_m .$$

We substitute here $\gamma_o u = \lim_{\varepsilon\to 0+} \gamma_o^{\varepsilon} u$ and receive from (2.1) that

$$|\int[\exp(i\langle\gamma_o^{\varepsilon}u,v\rangle) - \exp(i\langle\gamma_o u,v\rangle)]P_{m,\Theta_{k'}}(du) \le$$

$$\le \|v\|_b \int \|\gamma_o^{\varepsilon}u - \gamma_o u\|_{-b}\, P_{m,\Theta_{k'}}(du) \to 0 \text{ uniformly in } \Theta_{k'} > 1 \text{ (and}$$

$\mu \in \mathbb{N}$). Hence (2.15) follows from convergence of $P_{m,\Theta_{k'}} \to \bar{P}_m$ and continuity of $\langle\gamma_o^{\varepsilon}\cdot,v\rangle$ on $\mathcal{L}_2$.

The proof of theorem 1.1. From the estimates (2.3) -(2.4) we have the weak - compactness on $\mathcal{L}_2$ of measures $\{\bar{P}_m\}$, $m \in \mathbb{N}$, as in the proof of lemma 2.2. Hence $\exists \{m_j\}$ such that $\bar{P}_{m_j} \to \bar{P}$ as $m_j \to \infty$. Stationarity of $\bar{P}$ follows from the stationarity of $\bar{P}_{m_j}$, the identity (1.3) is the result of (2.2). The estimates (1.4) follows from (2.3). The verification of the (1.5) could be done similarly to the verification of (2.5).

3. Proof of theorem 1.2. For $v(x) = v^j e_j(x)$ we get from (1.5):

$$(3.1) \quad 0 = \int \exp(i\langle u(x), v^j e_j(x)\rangle)\mathcal{A}(u(x), v^j e_j(x)\bar{\mu}(du), v^j \in \mathbb{R} .$$

By the differentiation of (3.1) with respect to v^j, we get formally for $v^j \in \mathbb{R}$

$$(3.2) \int \exp(i\langle u, v^j e_j\rangle)[iu^j \mathcal{A}(u, v^j e_j) + \frac{\partial \mathcal{A}}{\partial v^j}(u, v^j e_j)]\bar{\mu}(du) = 0 .$$

From the estimate (1.4) for $k = 1$ it follows the uniform convergence of the integral (3.2) when $|v^j| < C$, $\forall C > 0$.

Hence (3.2) holds true. Similarly could be justified the second derivation: $\int \exp(i\langle u, v^j e_j\rangle)[(iu^j)^2 \mathcal{A} + 2iu^j \frac{\partial \mathcal{A}}{\partial v^j} + \frac{\partial^2 \mathcal{A}}{(\partial v^j)^2}]\bar{\mu}(du) = 0.$ Put here $v^j = 0$ and sum up over $j \le m$:

$$(3.3) \int [-2\langle -(u,\nabla)u + \nu\Delta u, u_m\rangle - Sp\Pi_m Q\Pi_m]\bar{\mu}(du) = 0.$$

Here $u_m \equiv \Pi_m u \underset{m \to \infty}{\to} u$ in $\overset{\circ}{\mathcal{H}}{}^1$ for $\forall u \in \overset{\circ}{\mathcal{H}}{}^1$, therefore $\langle -(u,\nabla)u + \nu\Delta u, u_m\rangle \to -\nu\langle \nabla u, \nabla u\rangle$. As $n = 2$ we deduce from the Ladyzenskaya inequality [8] the majorant for the expression in (3.3): $|\langle -(u,\nabla)u + \nu\Delta u, u_m\rangle| \le C\|u\|_1^2\|u\|_0$. But the functional $\|\cdot\|_1^2\|\cdot\|_0$ is integrable with the respect to $\bar{\mu}(du)$ in view of (1.4) for $k = 1$.

Therefore, from (3.3) as $m \to \infty$ we get $\int [2\nu\langle \nabla u, \nabla u\rangle - SpQ]\bar{\mu}(du) = 0$ by the Lebesgue theorem. Hence we have (1.6).

REFERENCES

1. Viot, M., Solution faibles d'équations aux dérivées partielles stochastiques non linéaires, Thése, Paris, 1976.
2. Višic, M.I., Komech, A. I., Infinite dimensional parabolic equations related to stochastic equations with partial derivatives. DAN, USSR, 1977, 233, No. 5, 769-772. (In Russian)
3. Višic, M.I., Komech A. I., Existence of the solutions for direct Kolmogoroff equation which correspondes to the stochastic systems Navy-Stoke type equations, paper contained in "Complex analysis and applications", Nauka, M., 1978. (In Russian)
4. Kolmogoroff, A. N., Remarks on the statistic solutions for Navy-Stoke systems, UMN, 1978, 33, No. 3, 124. (In Russian)
5. Hasminskij, R. Z., Stability of the system of differential equations with respect to regular perturbation of parameters, Nauka, M., 1969. (In Russian)
6. Mc Kean, H.P., Stochastic Integrals, Academic Press, N.Y., 1969.
7. Višic, M.I., Fursikow, A. W., Translation-invariant stochastic solutions and individual (special) solutions with infinite energy for Navy-Stoke systems, Siberian Journal of Mathematics, 1978, 19, No. 5, 1005-1031. (In Russian)
8. Ladyzenskaya, O. A., Mathematical theory of viscous incompressible flow, Nauka, M., 1970. (In Russian)

ON ABSOLUTE CONTINUITY OF PROBABILITY MEASURES FOR MARKOV-ITÔ PROCESSES

Yu. M. Kabanov, R. Sh. Liptser, A. N. Shiryayev

(Moscow)

§1. INTRODUCTION

Let $(\Omega, \mathscr{F}, F = (\mathscr{F}_t)\ t \geq 0\ \mathbb{P})$ be a filtered probability space with usual assumptions and $\xi = (\xi_t, \mathscr{F}_t, \mathbb{P})$, $\tilde{\xi} = (\tilde{\xi}_t, \tilde{\mathscr{F}}_t, \mathbb{P})$ be two Markov processes governed by the stochastic Itô equations:

$$\xi_t = \xi_o + \int_0^t a(s,\xi_{s-})dA_s + \int_0^t b(s,\xi_{s-})dm_s + \tag{1}$$

$$+ \int_0^t \int_{|u|>1} f(s,\xi_{s-},u)dp + \int_0^t \int_{|u|\leq 1} g(s,\xi_{s-},u)d(p-q),$$

$$\tilde{\xi}_t = \tilde{\xi}_o + \int_0^t \tilde{a}(s,\tilde{\xi}_{s-})dA_s + \int_0^t b(s,\tilde{\xi}_{s-})dm_s +$$

$$+ \int_0^t \int_{|u|>1} \tilde{f}(s,\tilde{\xi}_{s-},u)dp + \int_0^t \int_{|u|\leq 1} \tilde{g}(s,\tilde{\xi}_{s-},u)d(p-q). \tag{2}$$

Here $A = (A_t)t \geq 0$ is a (non-random) continuous function with the locally bounded variation $m = (m_t, \mathscr{F}_t, \mathbb{P})$ is a continuous Gaussian martingale with the characteristic $\langle m \rangle$, $dp = p\ (\omega;\ dt,\ du)$ is a well-measurable integer-valued random measure on $(\mathbb{R}_+ \times \mathbb{R}_o, \mathscr{B}_+ \otimes \mathscr{B}_o)$ with the deterministic compensator $dq = q(dt,\ du)$ $(\mathbb{R}_+ = (0,\infty), \mathbb{R}_o = \mathbb{R}\setminus\{0\}$, $\mathscr{B}_+$ and $\mathscr{B}_o$ are Borel σ-algebras on $\mathbb{R}_t$ and $\mathbb{R}_o$ respectively), [1]. We suppose that the coefficients of (1), (2) satisfy to conditions which imply the existence and uniqueness of strong solutions of (1), (2) (see [2]-[6].

Let $(\mathbf{X},\mathscr{X})$ be a measurable space of right continuous functions $X = (X_t)t \geq 0$ with left hand limits.

Denote by P and $\tilde{P}$ the distributions of the processes ξ and $\tilde{\xi}$:

$$P(\Gamma) = \mathbb{P}(\xi \in \Gamma),\ \tilde{P}(\Gamma) = \mathbb{P}(\tilde{\xi} \in \Gamma),\ \Gamma \in \mathscr{X}.$$

The purpose of the paper is to obtain sufficient conditions for absolute continuity ($\tilde{P} \ll P$) of the measure $\tilde{P}$ with respect to the measure P.

The first result in this direction has been proved by Skorokhod [2] for the case when $A_t = t$, $\langle m \rangle_t = t$, $q(dt,du) = u^{-2}dtdu$ and the coefficients are continuous. Grigelionis [7], [8] studied more general case by Girsanov's method [9] and obtained an answer which includes a condition of uniform integrability of a nonnegative martingale.

Certainly, the processes $\tilde{\xi}$ and ξ, given by (1), (2), are semimartingales and it is possible to apply the general results for semimartingales [10], [11], but arising condition of "τ_n-uniqueness" is not natural in the present setting.

Here we replace the condition of "τ -uniqueness" for the measure $\tilde{P}$ by more suitable condition which can be expressed <u>in terms of the coefficients</u> of the equation (2). Namely, we assume that for any stopping time τ the equation

$$\Theta_t = \xi_{t\wedge\tau} + \int_0^t I(\tau < s)\tilde{a}(s,\Theta_{s-})dA_s + \int_0^t I(\tau < s)b(s,\Theta_{s-})dm_s +$$

(3) $$+\int_0^t \int_{|u|>1} I(\tau < s)\tilde{f}(s,\Theta_{s-},u)p(ds,du) +$$

$$+\int_0^t \int_{|u|\le 1} I(\tau < s)\tilde{g}(s,\Theta_{s-},u)(p-q)(ds,du)$$

has the unique strong solution. (Here and further we drop variable ω).

It is worth to note that the equation (3) has this property if the functions $\tilde{a}$, b, $\tilde{g}$, satisfy the linear growth and Lipschitz conditions and $q((0,t] \times \{u: |u| \le 1\}) < \infty$, $t < \infty$.

§2. The main result.

Denote by $\tilde{P}_o$ and P_o the restrictions of $\tilde{P}$ and P on the σ-algebra $\sigma(X_o)$. Let

(4) $$q_t = q(\{t\} \times \mathbb{R}_o), \quad q_t^{(2)} = q(\{t\} \times \{u: |u| \le 1\})$$

(5) $$h(t,x,u) = I(|u| \le 1)g(t,x,u) + I(|u| > 1)f(t,x,u)$$

We use the following assumptions:

I: $\tilde{P}_0 \ll P_0$.

II: There exists the measurable function $\rho = \rho(t,x,u)$ on $\mathbb{R}_+ \times \mathbb{R} \times \mathbb{R}$ such that

a) $0 \le \rho(t,x,u) < \infty$;

b) $\hat{\rho}(t,x) = \int_{R_o} \rho(t-x,h(t,x,u))q\{t\}, du) \le 1$;

c) $q_t = 1 \Longrightarrow \hat{\rho}(t,x) = 1$;

d) for any measurable nonnegative function H(t,x,u) on $\mathbb{R}_+ \times \mathbb{R} \times \mathbb{R}$

$$\int_0^\infty \int_{|u|>1} H(t,x,\tilde{f}(t,x,u))q(dt,du) =$$

$$= \int_0^\infty \int_{|u|>1} H(t,x,f(t,x,u))\rho(t,x,f\ (t,x,u))q(dt,du);$$

e) the following functions are well-defined and finite

$$\widehat{g\rho(g)}(t,x) = \int_{|u|\le 1} g(t,x,u)\ \rho(t,x,g(t,x,u))q(\{t\},du),$$

$$\hat{\tilde{g}}(t,x) = \int_{|u|\le 1} \tilde{g}(t,x,u)q(\{t\},\ du),$$

$$\tilde{G}(t,x,u) = \begin{cases} \tilde{g}(t,x,u)-\hat{\tilde{g}}(t,x), u \ne 0, \\ -\hat{\tilde{g}}(t,x), u = 0 \end{cases}$$

and

$$G(t,x,u) = \begin{cases} g(t,x,u)-\widehat{g\rho(g)}(t,x),\ u \ne 0, \\ -\widehat{g\rho(g)}(t,x),\ u = 0; \end{cases}$$

f) for any measurable nonnegative function $H(t,x,u)$ on $\mathbb{R}_+ \times \mathbb{R} \times \mathbb{R}$

$$\int_0^\infty \int_{|u|\le 1} H(t,x,G(t,x,u))q(dt,du) + \sum_{t>0} H(t,x,G(t,x,0))(1-q_t^{(1)}) =$$

$$= \int_0^\infty \int_{|u|\le 1} H(t,x,G(t,x,u))\rho\ (t,x,g(t,x,u))q(dt,du) +$$

$$+ \sum_{t>0} H(t,x,G(t,x,0))(1-\widehat{g\rho(g)}(t,x)).$$

III: There exists the measurable function $\gamma = \gamma(t,x)$ on $\mathbb{R}_+ \times \mathbb{R}$ such that

$$\int_0^t b(s,x)\gamma(s,x)d\langle m\rangle_s = \int_0^t (\tilde{a}(s,x) - a(s,x))dA_s -$$

$$- \int_0^t \int_{|u|\le 1} g(s,x,u)(\rho(s,x,g(s,x,u)) - 1)q(ds,du).$$

Put for $X \in \mathbf{X}$

$$\mathbb{B}_t(X) = \int_0^t \gamma^2(s,X_{s-})d\,m_s +$$

$$+ \int_0^t \int_{\mathbb{R}_0} (1 - \sqrt{\rho(s,X_{s-},h(s,X_{s-},u))}\ )^2 q(ds,du) +$$

$$+ \sum_{s\le t} I(0 < q_s < 1)(1 - \sqrt{\frac{1-\hat{\rho}(s,X_{s-})}{1-q_s}})^2(1-q_s)$$

and $\mathbb{B}_\infty(X) = \lim_{t\to\infty} \mathbb{B}_t(X)$.

<u>Theorem</u>. Suppose that the equations (1) - (3) have the unique strong solutions, $|g(t,x,u)| + \tilde{g}(t,x,u)| \le K$ and conditions I - III are fulfilled.

Then

$$\mathbb{P}(\mathbb{B}_\infty(\tilde{\xi}) < \infty) = 1 \Longrightarrow \tilde{P} \ll P.$$

The proof will be given in §4. In §3 we prove an intermediate result.

§3. Auxiliary results

Let $\mathcal{P}$ be the σ-algebra of F-predictable sets on $\mathbb{R}_+ \times \Omega$ and

$\tilde{\mathscr{P}} = \mathscr{P} \otimes \mathscr{B}_o$. Assume that $\mathscr{P}$-measurable function $\beta = \beta(t,\omega)$ and $\tilde{\mathscr{P}}$-measurable function $Y = Y(t,\omega,u)$ satisfy

IV: 1) $0 \le Y < \infty$;

2) $\hat{Y}(t,\omega) = \int_{\mathbb{R}_o} Y(t,\omega,u)q(\{t\},(du) \le 1$;

3) $q_t = 1 \Longrightarrow \hat{Y}(t,\omega) = 1$;

4) for a some constant C

$$\int_0^\infty \beta^2(t,\omega)d\langle m\rangle_t + \int_0^\infty \int_{\mathbb{R}_o} (1-\sqrt{Y(t,\omega,u)})^2 q(dt,du) +$$

$$+ \sum_{t>0} I(0 < q_t < 1)\Big(1-\sqrt{\frac{1-\hat{Y}(t,\omega)}{1-q_t}}\Big)^2(1-q_t) \le C.$$

Then the following assertion is hold.

Lemma 1. ([11], Theorem 12). For any $\mathscr{F}_o$-measurable nonnegative random variable z_o with $\mathbb{E}\, z_o = 1$ the process $Z = (z_t, \mathscr{F}_t, \mathbb{P})_{t\ge 0}$ defined by

$$z_t = z_0 + \int_0^t z_{s-}\beta(s,\omega)dm_s + \tag{6}$$
$$\int_0^t \int_{\mathbb{R}_o} z_{s-}(Y(s,\omega,u)-1 + (1-q_s)^{\otimes}(\hat{Y}(s,\omega)-q_s))(p-q)(ds,du),$$

where $a^{\otimes} = a^{-1}$ if $a = 0$ and $a^{\otimes} = 0$ if $a = 0$, is a nonnegative uniformly integrable martingale with $\mathbb{E}\, Z_\infty = 1$ ($Z_\infty = \lim_{t\to\infty} Z_t$). Now we prove a weak version of our main theorem.

Lemma 2. Let the equations (1), (2) have the unique strong solutions, $|g(t,x,u)| \le K$ and conditions I-III are fulfilled.

Then

$$\mathbb{P}(\mathbb{B}_\infty(\tilde{\xi}) \le C) = 1 \Longrightarrow \tilde{P} \ll P,$$

where C is a some constant.

Proof. Put

$$z_0 = \frac{d\tilde{P}_0}{dP_0}(\xi), \quad \beta(t,\omega) = \gamma(t,\xi_{t-}), \tag{7}$$

$$Y(t,\omega,u) = \rho(t,\xi_{t-},h(t,\xi_{t-},u)).$$

Then, by virtue of Lemma 1, the process Z defined by equation (6) with z_0,β, and Y given by (7) is a uniformly integrable martingale with $Ez_\infty = 1$ ($z_\infty = \lim_{t\to\infty} z_t$).

Define on $(\Omega, \mathscr{F})$ the new probability measure $\tilde{\mathbb{P}}$ putting $d\tilde{\mathbb{P}} = z_\infty d\mathbb{P}$. According to the Theorem 10 of [11], the process $\tilde{m} = (\tilde{m}_t, \mathscr{F}_t, \tilde{\mathbb{P}})$ $t \geq 0$ with

$$\tilde{m}_t = m_t - \int_0^t \gamma(s,\xi_{s-})d\langle m\rangle_s \tag{8}$$

is a continuous local martingale with the characteristic $\langle\tilde{m}\rangle = \langle m\rangle$. Therefore, $\tilde{m}$ is a Gaussian martingale (see [12], Example 1). Further, according to the Theorem 11 of [11], and by virtue of the assumption $|g(t,x,u)| \leq K$ there exists the process $\tilde{M} = (\tilde{M}_t, \mathscr{F}_t, \tilde{\mathbb{P}})$ $t \geq 0$ with

$$\begin{aligned} \tilde{M}_t = &\int_0^t \int_{|u|\leq 1} g(s,\xi_{s-},u)(p-q)(ds,du) - \\ &- \int_0^t \int_{|u|\leq 1} g(s,\xi_{s-},u)(\rho(s,\xi_{s-})g(s,\xi_{s-},u))-1)q(ds,du). \end{aligned} \tag{9}$$

The process $\tilde{M}$ is a purely discontinuous local martingale which has the representation ($\tilde{\mathbb{P}}$-a.s.):

(10) $$\tilde{M}_t = \int_0^t \int_{|u|\le 1} g(s,\xi_{s-},u)(p-\tilde{q})(ds,du)$$

where $\tilde{q}(dt,du) = \rho(t,\xi_{t-},h(t,\xi_{t-},h(t,\xi_{t-},u))q(dt,du)$ is the compensator of the random measure $p(dt,du)$ with respect to $\tilde{\mathbb{P}}$, [6]. Note, also, that

(11) $$\begin{aligned} I(|u| \le 1)\tilde{q}(dt,du) &= I(|u| \le 1)\rho(t,\xi_{t-},g(t,\xi_{t-},u))q(dt,du) \\ I(|u| > 1)\tilde{q}(dt,du) &= I(|u| > 1)\rho(t,\xi_{t-},f(t,\xi_{t-},u))q(dt,du). \end{aligned}$$

Put

(12) $$\zeta_t = \int_0^t \int_{|u|>1} f(s,\xi_{s-},u)p(ds,du).$$

Since $\tilde{\mathbb{P}} \ll \mathbb{P}$ then the equality (1) is valid $\mathbb{P}$- and $\tilde{\mathbb{P}}$- a.s. Hence, with regard to (8), (9), (12) and (11), III we obtain $\tilde{\mathbb{P}}$-a.s.

(13) $$\xi_t = \xi_0 + \int_0^t \tilde{a}(s,\xi_{s-})dA_s + \int_0^t b(s,\xi_{s-})dm_s + \zeta_t + \tilde{M}_t.$$

Now, we shall show that there exists extended probability space

$$(\hat{\Omega},\ \hat{\mathscr{F}},\ \hat{F} = (\hat{\mathscr{F}}_t)_{t\ge 0}, \hat{\mathbb{P}}) = (\Omega \times \bar{\Omega}, \mathscr{F} \otimes \bar{\mathscr{F}},\ \hat{F}, \tilde{\mathbb{P}} \times \bar{\mathbb{P}}),\ \mathscr{F}_t \supseteq \mathscr{F}_t \otimes \{\emptyset,\Omega\},$$

with usual assumptions and a well-measurable integer-valued measure $.\hat{p}(dt,du) = \hat{p}(\omega;dt,du)$ on $(\mathbb{R}_+ \times \mathbb{R}_o, \mathscr{B}_+ \otimes \mathscr{B}_o)$ with the compensator $q(dt,du)$ such that $\hat{P}$-a.s.

(14) $$\zeta_t(\omega) = \int_0^t \int_{|u|>1} f(s,\xi_{s-}(\omega),u)\hat{p}(\hat{w};ds,du),$$

$$\tilde{M}_t(\omega) = \int_0^t \int_{|u|\le 1} g(s,\xi_{s-}(\omega),u)(\hat{p}-q)(ds,du)$$

(we denote by $\eta(\omega)$ (or η) the random variable on $(\hat{\Omega}, \hat{\mathscr{F}})$ with $\eta(\hat{\omega}) = \eta(\omega\,\bar{\omega}) \equiv \eta(\omega)$).

Let $\mu^{\zeta}(\omega;dt,dv)$ and $\mu^{\tilde{M}}(\omega;dt,dv)$ be the measures of jumps of ζ and $\tilde{M}$, i.e.

$$\mu^{\zeta}(\omega;(0,t]\times\Gamma) = \sum_{s\le t} I(\Delta\zeta_s\in\Gamma),\quad \mu^{\tilde{M}}(\omega;(0,t]\times\Gamma) =$$

$$= \sum_{s\le t} I(\Delta\tilde{M}_s\in\Gamma),\qquad \Delta\zeta_s = \zeta_s - \zeta_{s-}\,,\ \Delta\tilde{M}_s = \tilde{M}_s - \tilde{M}_{s-},\ \Gamma\in\mathscr{B}_0.$$

It is easy to see that for any nonnegative measurable function $H(t,x,v)$ on $\mathbb{R}_+ \times \mathbb{R} \times \mathbb{R}_0$ with $H(t,x,o) = 0$, $(t,x) \in \mathbb{R}_+ \times \mathbb{R}$,

$$\int_0^\infty \int_{\mathbb{R}_0} H(t,\xi_{t-},v)\mu^{\zeta}(dt,dv) = \int_0^\infty \int_{|u|>1} H(t,\xi_{t-},f(t,\xi_{t-},u))p(dt,du),$$

(15)

$$\int_0^\infty \int_{\mathbb{R}_0} H(t,\xi_{t-},v)\mu^{\tilde{M}}(dt,dv) = \int_0^\infty \int_{|u|\le 1} H(t,\xi_{t-},G(t,\xi_{t-},u))p(dt,du) +$$

$$+ \sum_{t>0} H(t,\xi_{t-},G(t,\xi_{t-},0))(1-p(\{t\}\times\{u:|u|\le 1)),$$

where the function $G(t,x,u)$ is defined in II_e).

It follows from (15) and (11) that the compensators $\nu^{\zeta}(\omega;dt,dv)$ and $\mu^{\tilde{M}}(\omega;dt,dv)$ of integer-valued random measures μ^{ζ} and $\mu^{\tilde{M}}$ with respect to $\tilde{\mathbb{P}}$ are such that

$$\int_0^\infty \int_{\mathbb{R}_0} H(t,\xi_{t-},v)\nu^{\zeta}(dt,dv) = \int_0^\infty \int_{|u|>1} H(t,\xi_{t-},f(t,\xi_{t-},u))q(dt,du)$$

(16)

$$\int_0^\infty \int_{\mathbb{R}_0} H(t,\xi_{t-},v)\nu^{\tilde{M}}(dt,dv) = \int_0^\infty \int_{|u|\le 1} H(t,\xi_{t-},G(t,\xi_{t-},u))q(dt,du) +$$

$$+ \sum_{t>0} H(t,\xi_{t-},G(t\ \xi,_{t-},0))(1-g_t^{(1)}).$$

Now, the required representation (14) follows by virtue of assumptions $II_{d)-f)}$ from the remark to Theorem 1 and Lemma 2 of [13].

Thus, (13) and (11) yield the following equality ($\hat{P}$-a.s.)

$$\xi_t(\omega) = \xi_0(\omega) + \int_0^t \tilde{a}(s,\xi_{s-}(\omega))dA_s + \int_0^t b(s',\xi_{s-}(\omega))dm_s +$$

$$+ \int_0^t \int_{|u|>1} \tilde{f}(s,\xi_{s-}(\omega),u)\hat{p}(ds,du) +$$

$$+ \int_0^t \int_{|u|\le 1} g(s,\xi_{s-}(\omega),u)(\hat{p}-q)(ds,du)$$

(cf. with (2)).

The process $\tilde{m} = (\tilde{m}, \hat{\mathscr{F}}_t, \hat{\mathbb{P}})_{t\ge 0}$ is a Gaussian martingale with $\langle\hat{m}\rangle = \langle m\rangle$. Therefore, by the uniqueness of solution of (2)

$$\hat{\mathbb{P}}\ (\xi \in A) = \mathbb{P}\,(\tilde{\xi} \in A),\ A \in \mathscr{X}\ .$$

Then for a set $A \in \mathscr{X}$

$P(A) = 0 \Longleftrightarrow \mathbb{P}\,(\xi\in A) = 0 \Longleftrightarrow \mathbb{P} \times \bar{\mathbb{P}}\ (\xi\in A) = 0 \Longrightarrow \tilde{\mathbb{P}} \times \bar{\mathbb{P}}\,(\xi\in A)=0$

$\Longleftrightarrow \hat{\mathbb{P}}\,(\xi\in A) = 0 \Longleftrightarrow \mathbb{P}\,(\tilde{\xi}\in A) = 0 \Longleftrightarrow \tilde{P}(A) = 0.$ Q.E.D.

§ 4. Proof of the Theorem

For $X \in \mathbb{X}$ define $\tau_n(X) = \inf\,(t\colon \mathbb{B}_t\,(X) > n),\ n \ge 1.$ By the definition of the function $\mathbb{B}_t(X)$ and by assumption $II_{b)}$ we have $\mathbb{B}_{\tau_{(n)}(X)}(X) \le n+2$. Since $\mathbb{P}(\mathbb{B}_\infty(\tilde{\xi}) < \infty) = 1$, it follows that

$$\lim_n \mathbb{P}(\tau_n(\tilde{\xi}) < \infty) = 0 \tag{17}$$

Let $\Theta^{(n)} = (\Theta^{(n)}, \mathscr{F}_t, \mathbb{P})_{t\ge 0}$ be defined by (3) with $\tau = \tau_n(\xi)$. Then, by definition, $\Theta^{(n)}_{t\wedge\tau_n(\xi)} = \xi_{t\wedge\tau_n(\xi)}$, $t \ge 0$. Hence,

$\tau_n(\xi) = \tau_n(\Theta^{(n)})$ and

$$\mathbb{P}(\Theta^{(n)} \in \Gamma, \tau_n(\Theta^{(n)}) = \infty) = \mathbb{P}(\xi \in \Gamma, \tau_n(\xi) = \infty) \tag{18}$$

Define the following functionals:

$$a_n(t,X) = I(\tau_n(X) \ge t)a(t,X_{t-}) + I(\tau_n(X) < t)a(t,X_{t-}) ,$$

$$f_n(t,X,u) = I(\tau_n(X) \ge t)f(t,X_{t-},u) + I(\tau_n(X) < t)f(t,X_{t-},u),$$

$$g_n(t,X,u) = I(\tau_n(X) \ge t)g(t,X_{t-},u) + I(\tau_n(X) < t)g(t,X_{t-},u),$$

$$h_n(t,X,u) = I(|u|>1)f_n(t,X,u) + I(|u| \le 1)g_n(t,X,u).$$

Then according to (1) and (3) $\Theta^{(n)}$ satisfies the equation

$$\Theta^{(n)}_t = \xi_0 + \int_0^t a_n(s,\Theta^{(n)})dA_s + \int_0^t b(s,\Theta^{(n)}_{s-})\, dm_s +$$

$$+ \int_0^t \int_{|u|>1} f_n(s,\Theta^{(n)},u)p(ds,du) +$$

$$+ \int_0^t \int_{|u|\le 1} g_n(s,\Theta^{(n)},u)(p-q)(ds,du).$$

Now put

$$\beta_n(t,\omega) + I(\tau_n(\Theta^{(n)}) \ge t)\gamma(t,\Theta^{(n)}_{t-}),$$

$$\gamma_n(t,\omega,u) = I(\tau_n(\Theta^{(n)}) \ge t)\rho(t,\Theta^{(n)}_{t-},h_n(t,\Theta^{(n)},u)) + I(\tau_n(\Theta^{(n)}) < t) ,$$

$$\hat{Y}_n(t,\omega) = \int_{\mathbb{R}_0} Y_n(t,\omega,u)q(\{t\}, du).$$

Conditions IV are evidently fulfilled and

$$\int_0^\infty \beta_n^2(t,\omega)d\langle m\rangle_t + \int_0^\infty \int_{\mathbb{R}_0} (1-\sqrt{Y_n(t,\omega,u)})^2 q(dt,du) +$$

$$+\sum_{t\geq 0} I(0 < q_t < 1)\Big(1-\sqrt{\frac{1-\hat{Y}_n(t,\omega)}{1-q_t}}\Big)^2(1-q_t) = \mathbb{B}_{\tau_n}(\Theta^{(n)}),(\Theta^{(n)}) \leq n+2.$$

Thus, by Lemma 1, the process $Z^{(n)} = (Z_t^{(n)}, \mathscr{F}_t, \mathbb{P})_{t\geq 0}$ defined by (6) with $Z_0^{(n)} = \frac{d\tilde{P}o}{dPo}(\xi)$ and $\beta = \beta_u$, $Y = Y_u$ is a uniformly integrable martingale with $EZ_\infty^{(n)} = 1$. Therefore, $\tilde{\mathbb{P}}^{(n)}$ with $d\tilde{\mathbb{P}}^{(n)} = Z_\infty^{(n)}d\mathbb{P}$ is a probability measure. As in the proof of Lemma 2, it can be shown that there exist an extended probability space $(\hat{\Omega}, \hat{\mathscr{F}}, \hat{F} = (\hat{\mathscr{F}}_t)_{t\geq 0}, \hat{\mathbb{P}}^{(n)})$, a continuous Gaussian martingale $(m_t^{(n)}(\omega), \hat{\mathscr{F}}_t, \hat{\mathbb{P}}^{(n)})_{t\geq 0}$ with the characteristic $\langle m\rangle$ and a well-measurable integer-valuable measure $\hat{p}(\hat{\omega};dt,du)$ with the compensator $q(dt,du)$ with respect to $\mathbb{P}^{(n)}$ such that $\Theta^{(n)}$ has the following representation

$$\Theta_t^{(n)}(\omega) = \xi_0(\omega) + \int_0^t \tilde{a}(s,\Theta_{s-}^{(n)}(\omega))dA_s + \int_0^t b(s,\Theta_{s-}^{(n)}(\omega))dm_s^{(n)} +$$

$$+\int_0^t \int_{|u|>1} \tilde{f}(s,\Theta_{s-}^{(n)}(\omega),u)\hat{p}(ds,du) +$$

$$+\int_0^t \int_{|u|\leq 1} \tilde{g}(s,\Theta_{s-}^{(n)}(\omega),u)(\hat{p}-q)(ds,du) \qquad (\mathbb{P}^{(n)}\text{-a.s.})$$

and

$$\hat{\mathbb{P}}^{(n)}(\Theta^{(n)} \in \Gamma) = \mathbb{P}(\xi \in \Gamma).$$

Now we can show that $\tilde{P} \ll P$. It is easy to see that

$$\mathbb{P}(\xi\in\Gamma,\ \tau_n(\xi) = \infty) = 0 \Longleftrightarrow \mathbb{P}(\Theta^{(n)}\in\Gamma,\ \tau_n(\Theta^{(n)}) = \infty) = 0 \Longrightarrow$$
$$\Longrightarrow \tilde{\mathbb{P}}^{(n)}(\Theta^{(n)} \in \Gamma,\ \tau_n(\Theta^{(n)}) = \infty) = 0 \Longleftrightarrow$$
$$\Longleftrightarrow \hat{\mathbb{P}}^{(n)}(\Theta^{(n)}\in\Gamma,\tau_n(\Theta^{(n)}) = \infty) = 0 \Longleftrightarrow \mathbb{P}(\tilde{\xi}\in\Gamma,\tau_n(\tilde{\xi}) = \infty) = 0.$$

Therefore, by virtue of (16) for the set $\Gamma \in \mathscr{X}$ with $P(\Gamma) = 0$

$$\tilde{P}(\Gamma) = \mathbb{P}(\xi\in\Gamma) = \mathbb{P}(\xi\in\Gamma,\ \tau_n(\tilde{\xi}) = \infty) + \mathbb{P}(\tilde{\xi}\in\Gamma,\ \tau_n(\tilde{\xi})<\infty) =$$
$$= \mathbb{P}(\tilde{\xi}\in\Gamma,\ \tau_n(\tilde{\xi}) < \infty) \le \mathbb{P}(\tau_n(\xi) < \infty) \to 0,\ n \to \infty. \qquad \text{Q.E.D.}$$

REFERENCES

[1] Jacod, J., Multivariante point processes: predictable projection Randon-Nikodym derivatives, representation of martingales. Z. W-theorie 31 (1975), 235-253.

[2] Skorokhod, A. V., Studies in the theory of random processes, Izdat. Kiev. Univ., Kiev, 1961; English translation, Addison-Wesley, 1965.

[3] Gikhman, I.I., Skorokhod, A.V., Stochastic differential equations, "Naukova Dumka", Kiev, 1968; English translation, Springer, 1972.

[4] Protter, Ph.D., On the existence uniqueness, convergence and explosions of solutions of stochastic integral equations, Ann. Probab., 5, 2 (1977), 243-261.

[5] Doleans-Dade, C., Existence and unicity of solutions of stochastic differential equations, Z. W-theorie 36 (1975), 93-102.

[6] Galtchouk, L. I., On existence and uniqueness of solutions of stochastic differential equations with respect to martingales and random measures. Second Vilnius Conference on Probability Theory and Mathematical Statistics, Vilnius, 1977.

[7] Grigelionis, B. I., On absolute by continuous change of measure and Markovian property of stochastic processes. Lit. math. sb. IX, 1 (1969), 57-61.

[8] Grigelionis, B. I., Random point processes and martingales. Lit. math. sb. XV 3(1975), 101-114.

[9] Girsanov, I. V., On transforming of a certain class of stochastic processes by absolutely continuous substitution of measures, Teoria Verojatn, i Primenen, V, 3 (1960), 314-330.

[10] Jacod, J., Memin, J., Caractéristiques locales et conditions de continuité absolue pour les semimartingales, Z. W-theorie 35 (1976), 1-37.

[11] Kabanov, Yu, M., Liptser, R. Š., Shiryayev, A. N., Absolute continuity and singularity of local continuous probability distributions, I, II, Math. Sbor., 107 N 3(1978), 364-415; 108 N 1 (1979), 32-61.

[12] Liptser, R. Š., On representation of local martingales, Teoria Verojatn. i Primenen. 4 (1976), 718-726.

[13] Kabanov, Yu. M., Liptser, R. Š., Shiryayev, A. N., On representation of integer-valued random measures and local martingales by random measures with deterministic compensators. To appear.

Representations of Gaussian Random Fields

C. Bromley
and
G. Kallianpur
University of Minnesota

1. Introduction. In this brief paper we present some preliminary results on the representation of multiparameter Gaussian processes (or random fields) which are equivalent to the Wiener process.

It is known that in the case of processes of a single real parameter, Gaussian processes equivalent to the Wiener process have a proper canonical representation of multiplicity one in the sense of Cramér and Hida ([2], [5]). One proof of this fact uses techniques of martingale theory and the Ito formula [6]. A second proof which exploits the Gaussian assumption and relies on a factorization theorem of Gohberg and Krein [4] has been given in [7] and [8].

The results of this paper could also be considered in the context of multiparameter martingale theory which has been extensively developed in recent years. (See, e.g. [1]). However, we do not know of results from this theory which can be readily applied to yield our results. We are thus led to extend the second method above to the case of random fields.

Another point of contact of the problems studied here is with the recent attempt by Tjøstheim to generalize the theory of canonical representations to second order random fields [9]. We show that when a non-anticipative representation in our sense exists, it is a proper canonical representation with co-ordinate multiplicity one in the sense of Tjøstheim. It is also seen from our results that random fields need not have representations of this kind. The connection between these two approaches (both of which rely crucially on the availability of commuting chains of projection operators) needs to be more fully explored. The usefulness and naturalness of the definition of non-anticipativity adopted in this paper is open to question and other formulations are possible. These and related questions will be investigated in a later paper.

For the sake of simplicity we consider processes of two parameters. The extension to a higher number of parameters is straightforward.

2. Non-anticipative representations. $W = (W_{st})$, $s,t \in D$, (D being the unit square) is a family of random variables defined on some complete probability space $(\Omega, \underline{\underline{A}}, P)$ and such that (W,P) is a standard Wiener process, i.e. , a sample continuous, Gaussian process with mean function zero and covariance

$$\Gamma_W(s,t;s',t') = \min(s,s') \min(t,t') \quad . \tag{1}$$

Let $\underline{\underline{F}}^{W,0}$ be the σ-field generated by the family (W_{st}) and suppose Q is a probability measure on $(\Omega, \underline{\underline{F}}^{W,0})$ such that (W,Q) is a Gaussian process with zero mean and a prescribed covariance function Γ_Q . We further assume that Q is absolutely continuous with respect to P relative to $\underline{\underline{F}}^{W,0}$, $(Q \ll P\, [\underline{\underline{F}}^{W,0}])$. It is well-known that, in view of the Gaussian hypothesis, $P \ll Q$, i.e. Q and P are equivalent $(Q \equiv P)$. Let $\underline{\underline{F}}^{W}_{st}$ be the σ-field generated by the family $\{W_{uv}\}$, $(0 \leq u \leq s,\ 0 \leq v \leq t)$ to which P-null sets in $\underline{\underline{F}}^{W,0}$ have been adjoined. Since $Q \equiv P$ we now assume Q to be defined on $\underline{\underline{F}}^{W}$ (where $\underline{\underline{F}}^{W} = \underline{\underline{F}}^{W}_{11}$) . Let us denote $L^2(\Omega, \underline{\underline{F}}^{W}, P)$ by L^2 . In what follows we will be primarily interested in random variables or processes as points or curves in Hilbert spaces which are subspaces of L^2 .

Let $D_{st} = [0,s] \times [0,t]$ and define $L(W;s,t)$ to be the closed linear subspace of L^2 spanned by $\{W_{uv}, (u,v) \in D_{st}\}$. Write $L(W)$ for $L(W;1,1)$. For $\Delta = (s,s'] \times (t,t']$, a rectangle in D , define $W(\Delta) = W_{s't'} - W_{s't} - W_{st'} + W_{st}$. If Δ and Δ' are two rectangles in D , (1) implies that

$$E_P[W(\Delta)W(\Delta')] = m(\Delta \cap \Delta') \tag{2}$$

where m is Lebesgue measure. Denote by P^1_s the orthogonal projection operator (orthoprojector) on $L(W)$ with range $L(W;s,1)$ and by P^2_t , the orthoprojector with range $L(W;1,t)$. Then

$$\Pi_1 = \{P^1_s\} \text{ and } \Pi_2 = \{P^2_t\} \text{ , } (0 \leq s \leq 1,\ 0 \leq t \leq 1)$$

are complete chains of orthoprojectors in the sense of [4]. The "independent increment" property (2) immediately yields the simple but important fact that Π_1 and Π_2 are <u>commuting</u> chains. Define $P^W_{st} = P^1_s P^2_t$, the orthoprojector on $L(W)$ with range $L(W;s,t)$.

A <u>representation</u> of the Gaussian process (W,Q) in terms of (W,P) is a stochastic process $Y = (Y_{st})$ defined on $(\Omega, \underline{\underline{F}}^W, P)$ and having the following properties:

(3) (Y,P) is Gaussian with mean zero and $E_P(Y_{st} Y_{s't'}) = \Gamma_Q(s,t;s',t')$.

(4) $Y_{st} \in L(W)$ for each $(s,t) \in D$.

According to this definition (which is, in fact, the same as the one given for the single parameter case and is independent of the number of parameters involved) a representation is simply a functional of a Wiener process W defined on some probability space whose distribution under P is Gaussian with mean zero and covariance Γ_Q. Condition (4) which requires the functional to be linear is natural because of the Gaussian hypothesis.

Let S be the operator on $L(W)$ occurring in Theorem A of the Appendix. It is easy to see that if F is a bounded linear operator on $L(W)$ such that

(5) $$S = F^*F ,$$

then

(6) $$Y_{st} = FW_{st}$$

defines a representation of Q . Conversely, given a representation Y of Q , there exists a unique bounded linear operator F on $L(W)$ such that (5) and (6) hold. Thus a representation corresponds to an appropriate factorization of the operator S .

In analogy with one-parameter processes we are interested in representations which are non-anticipative (NA) in the following sense:

(7) $$L(Y;s,t) \subseteq L(W;s,t) \ , \ (s,t) \in D \ .$$

(Here and elsewhere $L(Y;s,t)$ will denote the closed linear subspace of $L(W)$ spanned by $\{Y_{uv}, 0 \le u \le s \ , \ 0 \le v \le t\}$.) A necessary and sufficient condition for a NA representation to exist is that S have a factorization of the form (5) where

(8) $$FL(W;s,t) \subseteq L(W;s,t) \ , \ (s,t) \in D \ .$$

Condition (8) means that the subspaces $L(W;s,t)$ are invariant subspaces of F or, in the terminology of Gohberg and Krein, both the chains Π_1 and Π_2 are eigenchains of F .

The problem is reduced to seeking a factorization (5) in which F has the commuting chains Π_1 and Π_2 as its eigenchains. In this paper we shall confine our attention to NA representations which satisfy the additional condition

(9) $$L(Y;s,t) = L(W;s,t) \ , \ (s,t) \in D \ .$$

Such representations will be called <u>strongly non-anticipative</u>. One useful consequence of (9) is that the operator F defined in (6) by such a representation is necessarily invertible. In fact, $S^{-1}F^*F = I$, so that $F^{-1} = S^{-1}F^*$ whenever $L(Y;1,1) = L(W;1,1)$.

A <u>Gohberg-Krein</u> (G-K) representation is one in which

(10) $$Y_{st} = (I + V)W_{st}$$

where V is a Hilbert-Schmidt, Volterra operator on $L(W)$ having Π_1 and Π_2 as eigenchains. For brevity, let us denote the latter class of operators by $\mathfrak{S}_2\mathcal{V}(\Pi_1,\Pi_2)$. A G-K representation is clearly non-anticipative and in fact strongly non-anticipative. This last assertion follows since $(I+V)^{-1} = \sum_0^\infty (-1)^n V^n$ has the same invariant subspaces as $I+V$ (See [8], Lemma 2.3).

A G-K representation is equivalent to a factorization

(11) $$S = (I + V^*)(I + V)$$

where $V \in \mathcal{S}_2\mathcal{V}(\Pi_1, \Pi_2)$. The study of such special factorizations is motivated by the one-parameter theory where the analogous G-K representation always exists (See [8]).

Let us suppose now that Q admits a strongly NA representation $Y_{st} = FW_{st}$. Denote by P^y_{st} the orthoprojector on $L(W)$ with range $L(Y;s,t)$. The condition (9) means $P^y_{st} = P^W_{st}$. Let $Y'_{st} = F'W_{st}$ be a representation corresponding to an arbitrary factorization $S = (F')^*F'$. Define $J = F'F^{-1}$. This operator is an isometry of $L(W)$ into itself since

(12) $$J^*J = (F^{-1})^*(F')^*F'F^{-1} = (F^*)^{-1}SF^{-1} = I \ .$$

Notice that
$$\begin{aligned} E_P(JW_{st} \cdot JW_{s't'}) &= \langle JW_{st} , JW_{s't'}\rangle \\ &= \langle J^*JW_{st}, W_{s't'}\rangle \\ &= \Gamma_W(s,t;s',t') \ . \end{aligned}$$

Hence the process

(13) $$W'_{st} = JW_{st}$$

is again a two-parameter Wiener process on D . Also we may write

(14) $$Y'_{st} = F'W_{st} = (F'F^{-1})FW_{st} = JY_{st} \ .$$

Let $P^{W'}_{st}$ and $P^{y'}_{st}$ be the orthoprojectors onto $L(W';s,t)$ and $L(Y';s,t)$, respectively,then $P^{W'}_{st} = JP^W_{st}J^*$. Indeed, the operator on the right side is self-adjoint and indempotent. To compute its range, note that (12) implies J^* is onto and use (13) to write

$$\begin{aligned} J P^W_{st}J^*(L(W)) &= J P^W_{st}(L(W)) \\ &= J(L(W;s,t)) \\ &= L(W';s,t) \ . \end{aligned}$$

A similar argument using (14) in place of (13) shows that $P^{y'}_{st} = J\,P^{y}_{st}J^*$. Hence $P^{y'}_{st} = J\,P^{y}_{st}J^* = J\,P^{w}_{st}J^* = P^{w'}_{st}$ so that $L(Y';s,t) = L(W';s,t)$. We have thus proved the following simple result.

Theorem 1. Suppose there exists a strongly NA representation of Q in terms of (W,P) . Then every representation of Q is strongly NA in terms of some Wiener process (W',P) .

The question of when NA and, in particular, strongly NA representations exist has not yet been completely settled. For G-K representations we give an answer below.

3. Gohberg-Krein representations. We shall see that G-K representations do not always exist. This is in contrast to the one-parameter situation (See [8]). The following generalization of the factorization theorem of Gohberg and Krein (See Appendix, Theorem B) yields a class of measures Q which do not have G-K representations as defined above.

Theorem 2. (Four-fold Factorization). Let H be a separable Hilbert space and S a Hermitian, positive, invertible operator on H such that $S = I - T$ where T is Hilbert-Schmidt. Let Π_1 and Π_2 be two commuting, complete chains of orthoprojectors on H .

(a). Then there exist two operators V_1 and V_2 such that

$$S = (I+V_1^*)(I+V_2^*)(I+V_2)(I+V_1) \tag{15}$$

where V_1 and V_2 are Hilbert-Schmidt, Volterra operators on H such that V_1 has Π_1 and Π_2 as eigenchains and V_2 has $\Pi_1^\perp$ and Π_2 as eigenchains. In the notation of Section 2, $V_1 \in \mathfrak{S}_2\mathcal{V}(\Pi_1,\Pi_2)$ and $V_2 \in \mathfrak{S}_2\mathcal{V}(\Pi_1^\perp,\Pi_2)$.

(b) The factorization (15) is unique in the following sense. Suppose that

$$S = (I+X_4)(I+X_3)(I+X_2)(I+X_1)$$

where $\quad X_1 \in \mathfrak{S}_2\mathcal{V}(\Pi_1,\Pi_2)$, $X_2 \in \mathfrak{S}_2\mathcal{V}(\Pi_1^\perp,\Pi_2)$,

$$X_3 \in \mathfrak{S}_2\mathcal{V}(\Pi_1, \Pi_2^\perp) \quad \text{and} \quad X_4 \in \mathfrak{S}_2\mathcal{V}(\Pi_1^\perp, \Pi_2^\perp) \; .$$

Then

$$X_1 = V_1 \; , \; X_2 = V_2 , \; X_3 = V_2^* \quad \text{and} \quad X_4 = V_1^* \; .$$

<u>Proof</u>. Only an outline of the proof of (a) will be given here. The proof of uniqueness, which is simple, is omitted.

Since $S = I - T$ is self-adjoint, positive and invertible, Theorem B is applicable ([8], Lemma 2.1) and we have a special factorization along Π_2 :

$$(I - T)^{-1} = (I - Y_+)(I - Y_-)$$

where $Y_+ \in \mathfrak{S}_2\mathcal{V}(\Pi_2)$ and $Y_- = Y_+^* \in \mathfrak{S}_2\mathcal{V}(\Pi_2^\perp)$. Writing $Z_\pm = (I - Y_\pm)^{-1} - I$ we have

$$S = (I + Z_-)(I + Z_+)$$

where $Z_+ \in \mathfrak{S}_2\mathcal{V}(\Pi_2)$ and $Z_- \in \mathfrak{S}_2\mathcal{V}(\Pi_2^\perp)$. Now write $S_1 = (I - Y_+)^{-1}$. For each $P \in \Pi_1^\perp$ the operator PY_+P is Volterra because it is Hilbert-Schmidt and has Π_2 as an eigenchain. Hence $(I - PY_+P)^{-1}$ exists and so by Theorem B, S_1 has a special factorization along $\Pi_1^\perp$. Denote it by

$$S_1 = (I + V_2)(I + V_1)$$

where $V_1 \in \mathfrak{S}_2\mathcal{V}(\Pi_1)$ and $V_2 \in \mathfrak{S}_2\mathcal{V}(\Pi_1^\perp)$. By Theorem B, they are given by

$$V_1 = \int_{\Pi_1} dPY_+P(I - PY_+P)^{-1}$$

and

$$V_2 = \int_{\Pi_1^\perp} (I - PY_+P)^{-1}PY_+ \, dP \; .$$

The operator-valued integrals on the right converge in norm [4]. Using this integral representation and the commutativity of Π_1 and Π_2 it can be verified that Π_2 is also an eigenchain of V_1 and V_2. Thus we have $I+Z_+ = (I+V_2)(I+V_1)$ when $V_1 \in \mathcal{S}_2\mathcal{V}(\Pi_1,\Pi_2)$ and $V_2 \in \mathcal{S}_2\mathcal{V}(\Pi_1^\perp,\Pi_2)$ and the four-fold factorization (15) is obtained.

We shall refer to (15) as the $(\Pi_2,\Pi_1^\perp)$-factorization. Interchanging the roles of Π_1 and Π_2 in the proof of Theorem 2 we obtain a second factorization which we call the $(\Pi_1,\Pi_2^\perp)$-factorization

(16)
$$S = (I+W_1^*)(I+W_2^*)(I+W_2)(I+W_1)$$

where $W_1 \in \mathcal{S}_2\mathcal{V}(\Pi_1,\Pi_2)$, $W_2 \in \mathcal{S}_2\mathcal{V}(\Pi_1,\Pi_2^\perp)$, $W_2^* \in \mathcal{S}_2\mathcal{V}(\Pi_1^\perp,\Pi_2)$ and $W_1^* \in \mathcal{S}_2\mathcal{V}(\Pi_1^\perp,\Pi_2^\perp)$.

The factorization (16) is also unique, the uniqueness statement being as in (b) with the roles of Π_1 and Π_2 interchanged. A G-K representation, if one exists, could arise from either a $(\Pi_1,\Pi_2^\perp)$ or a $(\Pi_2,\Pi_1^\perp)$ factorization and it is not obvious, à-priori that the two types of factorizations yield the same G-K representation.

If a G-K representation

$$Y_{st} = (I+V)W_{st}$$

exists, then the factorization (11) can be regarded as a $(\Pi_2,\Pi_1^\perp)$-factorization and by the uniqueness of the latter implies $V_1 = V$ and $V_2 = 0$. On the other hand, (11) also coincides with the unique $(\Pi_1,\Pi_2^\perp)$ factorization, so that $W_1 = V$ and $W_2 = 0$. Thus we have the following result.

<u>Theorem 3.</u> A G-K representation exists if and only if one of the operators V_2 or W_2 vanishes. In this case both operators vanish and the two factorizations coincide.

<u>Theorem 4.</u> A G-K representation of Q is of the form

(17)
$$Y_{st} = W_{st} + \int_0^s\int_0^t \left[\int_0^u\int_0^v V(x,y;u,v)\,dW_{xy}\right] du\,dv$$

where $V(x,y;u,v)$ is the kernel of the Volterra operator V in the factorization (11). The kernel has the property

$$V(x,y;u,v) = 0 \quad \text{a.e.} \quad \text{on} \quad \{(x,y,u,v): (x,y) \notin D_{uv}\} .$$

Furthermore, the Radon-Nikodym derivative $\frac{dQ}{dP}$ relative to the σ-field $\underline{\underline{F}}^W_{st}$ is given by

$$(18) \qquad \rho_{st}(\omega) = \exp\left[- \int_0^s\int_0^t \left\{ \int_0^u\int_0^v L(x,y;u,v)\,dW_{xy}(\omega) \right\} dW_{uv}(\omega) - \frac{1}{2}\int_0^s\int_0^t \left\{ \int_0^u\int_0^v L(x,y;u,v)\,dW_{xy}(\omega) \right\}^2 du\,dv \right]$$

where $L(x,y;u,v)$ is the kernel of the Volterra operator L on $L^2(D)$ given by $L = (I+V)^{-1} - I$.

We omit the proof, the details of which can be found in [3].

<u>Remark</u>. The process (ρ_{st}) is a two-parameter exponential martingale with respect to $(\underline{\underline{F}}^W_{st}, P)$ in the sense of Cairoli and Walsh [1].

The following result yields a class of two-parameter Gaussian processes equivalent to the Wiener process for which a G-K representation cannot exist.

<u>Theorem 5</u>. Suppose that Q is a zero-mean Gaussian measure equivalent to two-parameter Wiener measure and having a covariance function whose S operator is given by

$$(19) \qquad S = (I+A^*)(I+A) , \quad A \neq 0$$

where (i) $A \in \mathcal{S}_2\mathcal{V}(\Pi_2, \Pi_1^\perp)$ or (ii) $A \in \mathcal{S}_2\mathcal{V}(\Pi_1, \Pi_2^\perp)$.
Then there exists no G-K representation of Q .

<u>Proof</u>. We consider (i), the other case being proved similarly. Equation (19) then provides a $(\Pi_2, \Pi_1^\perp)$ - factorization for S with operators $V_1 = 0$ and $V_2 = A$. If a G-K representation exists then the associated factorization (11) must coincide with (19) by Theorem 2 (b). Equating the corresponding factors would give $A = 0$.

Using the above result we have the following example.

Example. The Gaussian processes $Y = (Y_{st})$ of the types given below have no G-K representations.

(a) $$Y_{st} = W_{st} + \int_0^s\int_0^t \left[\int_0^u\int_v^1 A(x,y;u,v)dW_{xy} \right] dudv$$

(b) $$Y_{st} = W_{st} + \int_0^s\int_0^t \left[\int_u^1\int_0^v A(x,y;u,v)dW_{xy} \right] dudv .$$

$A(x,y;u,v)$ is the kernel of a Volterra operator A on $L^2(D)$ with the properties $A(x,y;u,v) = 0$ a.e. on $\{(x,y,u,v): (x,v) \notin D_{uy}\}$ in Case (a) and $A(x,y;u,v) = 0$ a.e. on $\{(x,y,u,v): (u,y) \notin D_{xv}\}$ in Case (b).

4. Canonical Representations. Recently, Tjøstheim has introduced the concept of canonical representations of second order random fields in an attempt to generalize existing theory for one-parameter processes [2]. We shall not go into details here but content ourselves with the following observation.

Theorem 6. A G-K representation of Q is a proper canonical representation in the sense of Tjøstheim, with uniform co-ordinate multiplicity one.

Proof. Suppose a G-K representation (Y_{st}) exists given by (17) of Theorem 4. From this and (9) of Sec. 2 it follows that it is a proper canonical representation in the sense of [9]. Let P^y_{st} be the orthoprojector with range $L(Y;s,t)$ and $P^y(\Delta) = P^y_{s't'} - P^y_{s't} - P^y_{st'} + P^y_{st}$ where $\Delta = (s,s'] \times (t,t']$ is any rectangle contained in D. Denote by H the self-adjoint operator determined by the spectral family $\{P^y(\Delta), \Delta \subseteq D\}$. From the proper canonical property we have $P^y(\Delta) = P^w(\Delta)$, where $P^w(\Delta) = P^w_{s't'} - P^w_{s't} - P^w_{st'} + P^w_{st}$. Hence H is a cyclic operator on $L(Y;1,1)$ and therefore, has multiplicity one. Furthermore, its spectral type which is the spectral type of a generating element W_{11} is Lebesgue. The uniform co-ordinate multiplicity as defined in [9] is then, also seen to be equal to one.

The following theorem gives a necessary and sufficient condition for a strongly NA representation to be a G-K representation.

<u>Theorem 7</u>. Let $Y = (Y_{st})$ where

$$Y_{st} = (I + L)W_{st}$$

be a strongly NA representation. Let π stand for either π_1 or π_2 and let $\mathcal{C}_\pi$ be the set of all finite partitions $\zeta = \{P_j\}_0^n$ of π. Define

$$\Phi_L(\zeta) = \sum_{j=1}^{n} \Delta P_j \, L \, \Delta P_j \tag{20}$$

where $\Delta P_j = P_j - P_{j-1}$ and $\zeta \in \mathcal{C}_\pi$. Then for Y to be a G-K representation it is necessary and sufficient that $\Phi_L(\zeta)$ converges to 0 on $\mathcal{C}_\pi$ (in the norm as defined in [4]).

<u>Proof</u>. The necessity is well-known from [4]. For the sufficiency, take $\pi = \pi_2$. From Theorem 2

$$(I + L^*)(I + L) = (I + V_1^*)(I + V_2^*)(I + V_2)(I + V_1)$$

where $V_1 \in \mathfrak{S}_2\mathcal{V}(\pi_1, \pi_2)$ and $V_2 \in \mathfrak{S}_2\mathcal{V}(\pi_1^\perp, \pi_2)$. Noting that $I + L$ is invertible we have

$$\begin{aligned}(I + L)(I + V_1)^{-1}(I + V_2)^{-1} &= (I + L^*)^{-1}(I + V_1^*)(I + V_2^*) \\ &= I + B \text{ , say .}\end{aligned} \tag{21}$$

It follows from (21) that B has both π_2 and $\pi_2^\perp$ as eigenchains and hence commutes with π_2. Also $B = L + M$ where

$$M = W_1 + W_2 + W_1W_2 + L(W_1 + W_2 + W_1W_2)$$

and $W_i = (I + V_i)^{-1} - I$ $(i = 1,2)$. Therefore M is Hilbert-Schmidt and since π_2 is a bordered chain (see [4]),

$$\Phi_M(\zeta) \to 0 \text{ in norm along } \mathcal{C}_\pi \text{ .} \tag{22}$$

Now

(23) $\Phi_B(\zeta) = \Phi_L(\zeta) + \Phi_M(\zeta) \to 0$ in norm,

using (22) and the hypothesis about $\Phi_L(\zeta)$.

But, since B commutes with Π_2 ,

$$\Phi_B(\zeta) = \sum_j \Delta P_j \, B \, \Delta P_j = \sum_j B \, \Delta P_j = B$$

for every partition ζ . Thus from (23) $B = 0$ and from (21) we get

$$(I + L) = (I + V_2)(I + V_1) \; , \quad \text{and} \quad L = V_2 + V_1 + V_1 V_2 \; .$$

The last fact shows that L is Hilbert-Schmidt and since it has Π_1 and Π_2 as eigenchains, $L \in \mathcal{S}_2\mathcal{V}(\Pi_1, \Pi_2)$. We have proved that the representation (Y_{st}) is G-K . (Incidentally, note that $V_2 = 0$ and $L = V_1$) .

Appendix

Let P and Q be the Gaussian measures introduced in Section 2. We assume the reader is familiar with the definition of the reproducing kernel Hilbert space (RKHS) of a Gaussian process. Denote by H_P the RKHS of (W,P) .

<u>Theorem A</u>. The Gaussian measure Q is equivalent to the Wiener measure P if and only if Γ_Q defines an operator S on H_P with the following properties:

(1) $\Gamma_Q(\cdot\cdot;s,t) = S\Gamma_W(\cdot\cdot;s,t)$ for each $(s,t) \in D$,

(2) S is a bounded, self-adjoint, positive, invertible operator, and

(3) $S = I - T$ where T is Hilbert-Schmidt.

There is an isometric isomorphism ψ between L(W) and H_P such that $\psi W_{st} = \Gamma_W(\cdot\cdot;s,t)$. Under the isometry ψ , S goes over into the operator $\psi^{-1}S\psi$ on L(W) . Since P is Wiener measure it is more convenient to work with the space $L^2(D)$ which is also isometrically isomorphic to H_P under the map θ which sends $f \in H_P$ into $\hat{f} \in L^2(D)$ where $f(s,t) = \int_0^s \int_0^t \hat{f}(u,v)\, du dv$. As there is no possibility of confusion we shall denote the operator $\theta S\theta^{-1}$ on $L^2(D)$ and the operator $\psi^{-1}S\psi$ on $L(W)$ by the same symbol S.

For the definitions and ideas connected with the factorization which we quote next, we refer the reader to [4].

Theorem B. Let Π be a complete chain of orthoprojectors defined on a separable Hilbert space H. Suppose that T is a Hilbert-Schmidt operator on H ($T\in \mathcal{S}_2$ for short) such that the following condition is satisfied:

(4) $\qquad I - PTP$ is invertible for each $P\in \Pi$.

Then the operator-valued integrals

$$X_+ = \int_\Pi (I - PTP)^{-1}PT\, dP\ , \quad X_- = \int_\Pi dP\, TP\, (I - PTP)^{-1}$$

converge in norm and $(I - T)^{-1}$ has a factorization (called a special factorization along Π) given by

(5) $$(I - T)^{-1} = (I + X_+)(I + X_-) \quad .$$

Furthermore, X_+ and X_- are Hilbert-Schmidt, Volterra operators having Π and $\Pi^\perp$ as eigenchains, respectively.

References

[1]. Cairoli, R. and Walsh, J.B. (1975). Stochastic integrals in the plane. Acta Math. 134, 111-183.

[2]. Cramér, H. (1961). On some classes of non-stationary stochastic processes. Proc. 4th Berkeley Symposium. Math. Stat. and Prob., 57-77.

[3]. Etemadi, N. and Kallianpur, G. (1977). Non-anticipative transformations of the two-parameter Wiener process and a Girsanov theorem. J. Mult. Analysis, 7, 28-49.

[4]. Gohberg, I.C. and Krein, M.G. (1967). Theory and Applications of Volterra Operators in Hilbert space. Vol. 24, Translations of Mathematical Monographs, American Mathematical Society. (1969).

[5]. Hida, T. (1960). Canonical representations of Gaussian processes and their applications. Mem. Coll. Sci. Univ. Kyoto Ser. A. 33, 109-155.

[6]. Hitsuda, M. (1968). Representations of Gaussian processes equivalent to Wiener process. Osaka J. Math. 5, 299-312.

[7]. Kailath, T. and Duttweiler, D. (1972). An RKHS criterion to detection and estimation problems, Part III: Generalized innovations, representations and a likelihood-ratio formula. IEEE Trans. Inform. Theor. 18, 730-745.

[8]. Kallianpur, G. and Oodaira, H. (1973). Non-anticipative representations of equivalent Gaussian processes. Annals of Probability. 1, 104-122.

[9]. Tjøstheim, D. (1977). Multiplicity theory for random fields using quantum mechanical methods, Probability Theory on Vector Spaces, Lecture Notes in Mathematics. No.656 Springer-Verlag (1978).

CONTINUOUS ADDITIVE $\mathcal{S}'$-PROCESSES

Kiyosi Itô

Research Institute for Mathematical Sciences
Kyoto University, Kyoto 606, Japan

§1. INTRODUCTION

The following fact for real-valued processes is basic and well-known [1].

Theorem 1.1. Let $\{X_t, t\in[0,\infty)\}$ be a real-valued continuous additive (= with independent increments) process. Then $\{X_t\}$ is Gaussian, $m_t=E(X_t)$ is continuous and $V_t=V(X_t)$ is continuous and increasing. Conversely for any given $\{m_t\}$ and $\{V_t\}$ satisfying the above conditions we can construct a unique (in law) continuous additive process $\{X_t\}$ such that

$$E(X_t) = m_t \quad \text{and} \quad V(X_t) = V_t.$$

The purpose of this paper is to extend this fact to $\mathcal{S}'$-valued processes, $\mathcal{S}'$ being the space of tempered distributions on $\mathbb{R}$. The same argument can be used to discuss the case of $\mathcal{S}'(\mathbb{R}^r)$ $(r=2,3,\ldots)$. In §2 we will explain some facts on the space $\mathcal{S}'$ and mention several properties of $\mathcal{S}'$-valued processes. In §3 we will discuss some properties of a Gaussian $\mathcal{S}'$-valued random variable for the later use. Our aim will be achieved in the last section.

§2. PRELIMINARIES

Throughout this paper we use the following notation.

$$H_n(x) = 2^{-n/2}(-1)^n e^{x^2} \frac{d^n(e^{-x^2})}{dx^n}, \quad n=0,1,2,\ldots$$

$$e_n(x) = \sqrt{g(x)}\, H_n\left(\frac{x}{\sqrt{2}}\right)(2^n n!\sqrt{\pi})^{-\frac{1}{2}}, \quad n=0,1,2,\ldots,$$

where

$$g(x) = \frac{1}{\sqrt{2\pi}} e^{-\frac{x^2}{2}}.$$

$H_n(x)$ and $e_n(x)$ are called respectively the Hermite polynomial of order n and the Hermite function of order n. e_n, $n=0,1,2,\ldots$, form a complete orthonormal system (abbr. CONS) in $L^2 = L^2(\mathbb{R})$. Also they are eigenfunctions of the differential operator:

$$D = \frac{d^2}{dx^2} - \frac{x^2}{4}.$$

In fact we have

$$De_n = -(n+\tfrac{1}{2})e_n, \quad n=0,1,2,\ldots.$$

Let $\mathscr{S}$ denote the space of all rapidly decreasing real functions on $\mathbb{R}$. We introduce the Hilbertian norm $\|\ \ \|_p$ on $\mathscr{S}$ by

$$\|\phi\|_p^2 = \sum_{n=0}^{\infty} (n+\tfrac{1}{2})^{2p}|\phi_n|^2, \quad \phi_n = \int_{\mathbb{R}} \phi e_n dx$$

for every $p \in \mathbb{R}$. $\mathscr{S}$ is a real pre-Hilbert space with the inner product $(\ ,\ )_p$ corresponding to $\|\ \ \|_p$. The completion of $\mathscr{S}$ with respect to $\|\ \ \|_p$ is denoted by $\mathscr{S}_p$. It is obvious that

$$L^2 \supset \mathscr{S}_p \downarrow \mathscr{S}.$$

$\mathscr{S}$ is topologized by a set of norms $\|\ \ \|_p$, $p \geqq 0$ or equivalently by $\|\ \ \|_p$, $p=0,1,2,\ldots$. Since $\|\ \ \|_p$ is dominated by $\|\ \ \|_{p+1}$ in the Hilbert-Schmidt sense, i.e.

$$\sum_n \|\phi_n\|_p^2 < \infty \quad \text{for every } \|\ \ \|_{p+1}\text{-CONS } \{\phi_n\},$$

$\mathscr{S}$ is a nuclear space with the topology introduced above.

Let $\mathscr{S}'$ denote the dual of the topological vector space $\mathscr{S}$. Define $\|\ \ \|_{-p}$ on $\mathscr{S}'$ by

$$\|x\|_{-p} = \sup_{\|\phi\|_p \leqq 1} |x(\phi)|.$$

Then the subspace of $\mathscr{S}'$ consisting of all $x \in \mathscr{S}'$ with $\|x\|_{-p} < \infty$ is denoted by $\mathscr{S}'_p$. $\mathscr{S}'_p$ is a real Hilbert space with the inner product $(\ ,\ )_{-p}$ corresponding to $\|\ \ \|_{-p}$. It is easy to see that

$$\mathscr{S}'_p \uparrow \mathscr{S}'.$$

For every $x \in \mathscr{S}'_p$ $x(\phi)(\phi \in \mathscr{S})$ can be extended onto $\mathscr{S}_p$ in the obvious way. The weak topology on $\mathcal{S}'$ and the strong topology on $\mathcal{S}'$ are defined respectively by the following sets of semi-norms:

$$\|x\|_{\phi} = |x(\phi)|, \text{ where } \phi \in \mathcal{S}$$

and

$$\|x\|_{B} = \sup_{\phi \in B} |x(\phi)|, \text{ where } B \text{ is a bounded subset of } \mathcal{S}.$$

It is obvious that the topology on $\mathcal{S}_p'$ induced from the strong topology of $\mathcal{S}'$ coincides with the $\| \quad \|_{-p}$-topology on $\mathcal{S}_p'$.

Since $\mathcal{S}'$ is a Montel space, a curve x_t, $t \in [0,\infty)$ in $\mathcal{S}'$ is strongly continuous if and only if it is weakly continuous.

The Borel structure $\mathcal{B}(\mathcal{S}')$ on $\mathcal{S}'$ is defined to be the Kolmogorov σ-algebra, i.e. the σ-algbera generated by the sets

$$\{x\colon x(\phi) < a\}, \quad \phi \in \mathcal{S}, \quad a \in \mathbb{R}.$$

It is known that the topological σ-algebra on $\mathcal{S}'$ with respect to the strong (or weak) topology coincides with the Kolmogorov σ-algebra $\mathcal{B}(\mathcal{S}')$. Measures on $\mathcal{S}'$, $\mathcal{S}'$-valued random variables (abbr. random $\mathcal{S}'$-variables), $\mathcal{S}'$-valued stochastic processes (abbr. $\mathcal{S}'$-processes) are defined with respect to the σ-algebra $\mathcal{B}(\mathcal{S}')$.

An $\mathcal{S}'$-process $\{X_t = X_{t,\omega} = X_{t,\omega}(\phi),\ t \in [0,\infty),\ \omega \in (\Omega,P)\}$ is called <u>continuous</u> if $X_{t,\omega}$ is strongly continuous in t (equivalently weakly continuous in t) for almost every ω. $\{X_t\}$ is called <u>additive</u> if it has independent increments, i.e. if $X_0(\omega) \equiv 0$ and for every $0 = t_0<t_1<\cdots<t_n$ ΔX_{t_i}, $i=1,2,\ldots,n$, are independent random $\mathcal{S}'$-variables. $\{X_t\}$ is called <u>Gaussian</u> if $^1\{X_t(\phi)\}_{t,\phi}$ is a Gaussian system. It is easy to check that $\{X_t\}$ is Gaussian if $\{X_t\}$ is additive and if for every t $\{X_t(\phi)\}_{\phi}$ is a Gaussian system.

§3. GAUSSIAN $\mathcal{S}'$-VARIABLES

An $\mathcal{S}'$-variable $X = X_\omega = X_\omega(\phi)$, $\omega \in (\Omega,P)$ is called <u>Gaussian</u> if $\{X_\omega(\phi), \phi \in \mathcal{S}\}$ is a Gaussian system. Since $X_\omega(\phi)$ is linear in ϕ, it is easy to see that X_ω is a Gaussian $\mathcal{S}'$-variable if and only if $X_\omega(\phi)$ is Gauss distributed for every $\phi \in \mathcal{S}$.

Let $X = X_\omega$ be a Gaussian $\mathcal{S}'$-variable. Then

$$m(\phi) \equiv m_X(\phi) = E(X(\phi)) \quad \text{and} \quad V(\phi) \equiv V_X(\phi) = V(X(\phi))$$

are well-defined and the characteristic functional of X takes the following form:

$$C(\phi) \equiv C_X(\phi) \equiv E(e^{iX(\phi)}) = \exp\{im(\phi)-\frac{1}{2}V(\phi)\}.$$

Observing that $C(\phi)$ is continuous in ϕ, we can check that both $m(\phi)$ and $V(\phi)$ are continuous in ϕ. Since $m(\phi)$ is clearly linear in ϕ, we obtain

$$m \in \mathcal{S}'.$$

Since $V(\phi)$ is a positive-definite quadratic functional of ϕ, the continuity of $V(\phi)$ mentioned above implies that

$$0 \leqq V(\phi) \leqq c\|\phi\|_p^2 \qquad \text{for some } p.$$

In this case $V(\phi)$ can be extended to a positive-definite quadratic functional on $\mathcal{S}_p$ in the obvious way. Since $\|\ \ \|_p$ is dominated by $\|\ \ \|_{p+1}$ in the Hilbert-Schmidt sense, the Sazonov-Minlos-Kolmogorov theorem [2] shows that

$$P\{X-m \in \mathcal{S}'_{p+1}\} = 1 ;$$

note that

$$C_{X-m}(\phi) = E(e^{i(X-m)(\phi)}) = \exp\{-\frac{1}{2}V(\phi)\}.$$

Since $m \in \mathcal{S}'$, we have $m \in \mathcal{S}'_q$ for some q. Hence

$$P\{X \in \mathcal{S}'_r\} = 1 \qquad \text{for} \qquad r = \max(p+1,q).$$

<u>Theorem 3.1</u>. If X is a Gaussian $\mathcal{S}'$-variable with $m_X=0$ such that

$$V_X(\phi) \leqq c\|\phi\|_p^2,$$

then

(i) $\quad P\{X \in \mathcal{S}'_{p+1}\} = 1,$

(ii) $\quad E(\|X\|_{-p-1}^2) = \sum_n V(\phi_n) < \infty$ for every CONS $\{\phi_n\}$ in $\mathcal{S}_{p+1}$,

(iii) $\quad E(\|X\|_{-p-1}^4) \leqq 3(E(\|X\|_{-p-1}^2))^2.$

<u>Proof</u>. The statement (i) was proved above. Since

$$V(\phi) \equiv V_X(\phi) \leqq c\|\phi\|_p^2 \leqq c\|\phi\|_{p+1}^2$$

and since $V(\phi)$ is a positive-definite quadratic functional, we have

$$V(\phi) = \|B\phi\|_{p+1}^2$$

with a bounded symmetric operator in $\mathscr{S}_{p+1}$. Let $\{\phi_n\}$ be any CONS in $\mathscr{S}_{p+1}$. Then

$$\sum_n \|B\phi_n\|_{p+1}^2 = \sum_n V(\phi_n) \leq \sum_n c\|\phi_n\|_p^2 < \infty,$$

since $\| \;\|_p$ is dominated by $\| \;\|_{p+1}$ in the Hilbert-Schmidt sense. Hence B is a Hilbert-Schmidt symmetric operator in $\mathscr{S}_{p+1}$, so it can be represented as

$$B\phi = \sum_n \sigma_n(\phi,\varepsilon_n)_{p+1}\varepsilon_n, \text{ where } \sum\sigma_n^2 < \infty$$

for a CONS $\{\varepsilon_n\}$ in $\mathscr{S}_{p+1}$. Since $X \in \mathscr{S}'_{p+1}$ a.s., $X(\varepsilon_n)$, $n=1,2,\ldots$ are well-defined and form a Gaussian system with

$$E(X(\varepsilon_n)) = 0, \quad E(X(\varepsilon_m)X(\varepsilon_n)) = (B\varepsilon_m,B\varepsilon_n)_{p+1} = \sigma_m\sigma_n\delta_{mn}.$$

Hence $X(\varepsilon_n)$, $n=1,2,\ldots$ are independent and each $X(\varepsilon_n)$ is Gauss distributed with mean 0 and variance σ_n^2. Thus we have

$$E(X(\varepsilon_n)^2) = \sigma_n^2, \qquad E(X(\varepsilon_n)^4) = 3\sigma_n^4,$$

and

$$E(X(\varepsilon_m)^2X(\varepsilon_n)^2) = E(X(\varepsilon_m)^2)E(X(\varepsilon_n)^2) = \sigma_m^2\sigma_n^2.$$

Since $X \in \mathscr{S}'_{-p-1}$ a.s. and since $\{\varepsilon_n\}$ is a CONS in $\mathscr{S}_{p+1}$, we have

$$\|X\|_{-p-1}^2 = \sum_n X(\varepsilon_n)^2 \qquad \text{a.s.},$$

so

$$\|X\|_{-p-1}^4 = \sum_n X(\varepsilon_n)^4 + 2\sum_{m<n} X(\varepsilon_m)^2X(\varepsilon_n)^2.$$

Therefore we have

$$E(\|X\|_{-p-1}^2) = \sum_n \sigma_n^2 < \infty$$

and

$$E(\|X\|_{-p-1}^4) = \sum_n 3\sigma_n^4 + 2\sum_{m<n} \sigma_m^2\sigma_n^2$$

$$\leq 3(\sum_n \sigma_n^2)^2 = 3(E(\|X\|_{-p-1}^2))^2.$$

Since B is a Hilbert-Schmidt operator in $\mathscr{S}_{p+1}$, we have

$$\sum_n V(\phi_n) = \sum_n \|B\phi_n\|^2 = \sum_n \|B\varepsilon_n\|^2 = \sum_n \sigma_n^2 = E(\|X\|_{-p-1}^2)$$

for every CONS $\{\phi_n\}$ in $\mathscr{S}'_{p+1}$. This completes the proof of our theorem.

A Gaussian $\mathscr{S}'$-variable X is called <u>standard</u> if

$$m_X = 0 \quad \text{and} \quad V_X(\phi) = \|\phi\|_0^2.$$

If Z is a standard Gaussian $\mathscr{S}'$-variable, then

$$P\{X \in \mathscr{S}_1'\} = 1.$$

Let $\phi_n = (n+\frac{1}{2})^{-1}e_n$, $n=1,2,\ldots$, where $\{e_n\}$ are the Hermite functions mentioned in §2. Then $\{\phi_n\}$ is a CONS in $\mathscr{S}_1$. Using (ii) of the theorem above, we have

$$E(\|X\|_{-1}^2) = \sum_n V(\phi_n) = \sum_n \|\phi_n\|_0^2 = \sum_n (n+\frac{1}{2})^{-2} = \frac{\pi^2}{2}.$$

§4. CONTINUOUS ADDITIVE $\mathscr{S}'$-PROCESSES

Let $\{X_t\}_t$ be a continuous additive $\mathscr{S}'$-process. Then for every $\phi \in \mathscr{S}$ the real-valued process $\{X_t(\phi)\}_t$ is a continuous additive process. Hence Theorem 1.1 claims that $\{X_t(\phi)\}_t$ is a Gaussian process satisfying

(i) $m_t(\phi) \equiv E(X_t(\phi))$ is continuous in t and $m_0(\phi) = 0$ for every $\phi \in \mathscr{S}$, and

(ii) $V_t(\phi) \equiv V(X_t(\phi))$ is continuous and increasing in t and $V_0(\phi) = 0$ for every $\phi \in \mathscr{S}$.

Since $X_t(\phi)$ is Gauss distributed and since $X_t(\phi)$ is linear in $\phi \in \mathscr{S}$, we can easily see that $\{X_t(\phi), \phi \in \mathscr{S}\}$ is a Gaussian system, so X_t is a Gaussian $\mathscr{S}'$-variable. Therefore $m_t \in \mathscr{S}'$. Since $\{m_t\}$ is a weakly continuous curve in $\mathscr{S}'$ by virtue of (i), $\{m_t\}$ is also strongly continuous. $V_t(\phi)$ is a continuous positive-definite quadratic functional of ϕ for every t. X_t-X_s is also a Gaussian $\mathscr{S}'$-variable independent of X_s and we have

$$E((X_t-X_s)(\phi)) = (m_t-m_s)(\phi)$$

and

$$V((X_t-X_s)(\phi)) = V(X_t(\phi)) - V(X_s(\phi)) = V_t(\phi) - V_s(\phi).$$

Hence $(V_t-V_s)(\phi) \equiv V_t(\phi)-V_s(\phi)$ is also a continuous positive-definite functional of $\phi \in \mathcal{S}$.

Let $m \in \mathcal{S}'$ and let $V(\phi)$ be a continuous positive-definite functional of $\phi \in \mathcal{S}$. Then there exists a unique probability measure μ on $\mathcal{S}'$ such that

$$\mathcal{F}\mu(\phi) \equiv \int_{\mathcal{S}'} e^{ix(\phi)}\mu(dx) = \exp\{im(\phi) - \frac{1}{2}V(\phi)\}.$$

This measure is called a Gaussian measure on $\mathcal{S}'$ with mean m and variance functional V and denoted by $N(m,V)$. The probability law of the Gaussian $\mathcal{S}'$-variable X_t-X_s mentioned above is $N(m_t-m_s, V_t-V_s)$.

Theorem 4.1. If $\{m_t\}$ is a continuous curve in $\mathcal{S}'$ and if $V_t(\phi)$ is a continuous positive-definite quadratic functional of ϕ for every t and is continuous in t for every ϕ, then there exists a unique (in law) continuous additive $\mathcal{S}'$-process with

$$E(X_t(\phi)) = m_t(\phi) \quad \text{and} \quad V(X_t(\phi)) = V_t(\phi).$$

Proof. Since the uniqueness is easy to see, we will prove the existence only. Without any loss of generality we can assume that $m_t = 0$. Since $\mathcal{S}'$ is a Lusin space, we can apply the Kolmogorov extension theorem to construct an additive Gaussian process $\{X_t\}$ such that $X_0 = 0$ and X_s-X_s is $N(0,V_t-V_s)$-distributed for every $s < t$. To complete the proof it is enough to check that there exists a continuous version of $\{X_t\}$ on every finite interval $0 \leq t \leq T$.

Suppose that $0 \leq t \leq T$. Then

$$E(X_t(\phi)^2) = V_t(\phi) \leq V_T(\phi) \leq c|\phi|_p^2$$

for some $c = c(T)$ and $p = p(T)$. Hence

$$P\{X_t \in \mathcal{S}'_{p+1}\} = 1$$

and

$$v_t \equiv E\{\|X_t\|_{-p-1}^2\} = \sum_n E(X_t(\phi_n)^2) = \sum_n V_t(\phi_n) < \infty$$

for any CONS $\{\phi_n\}$ in $\mathcal{S}'_{p+1}$, as we can see from (ii) of Theorem 3.1.

Since $V_t(\phi)$ is increasing in t, v_t is also increasing. We will prove that v_t is continuous in $t \in [0,T]$. Since

$$\lim_{t\to s} V_t(\phi_n) = V_s(\phi_n), \quad n=1,2,\ldots \quad (t,s \in [0,T])$$

$$0 \leqq V_t(\phi_n) \leqq V_T(\phi_n), \quad n=1,2,\ldots \quad (t \in [0,T])$$

and

$$\sum_n V_T(\phi_n) = v_T < \infty,$$

we have

$$\lim_{t\to s} v_t = \sum_n \lim_{t\to s} V_t(\phi_n) = \sum_n V_s(\phi_n) = v_s,$$

proving the continuity of v_t.

Since $\{X_t(\phi)\}$ is a Gaussian additive process with $E(X_t(\phi)) = 0$, we have

$$E[(X_t(\phi)-X_s(\phi))^2] = E[X_t(\phi)^2]-E[X_s(\phi)^2] = V_t(\phi)-V_s(\phi)$$

for $s < t \leqq T$. Hence

$$\begin{aligned} E(\|X_t-X_s\|_{-p-1}^2) &= \sum_n E[(X_t(\phi_n)-X_s(\phi_n))^2] \\ &= \sum_n (V_t(\phi_n)-V_s(\phi_n)) \\ &= v_t-v_s. \end{aligned}$$

By (iii) of Theorem 3.1 we have

$$E(\|X_t-X_s\|_{-p-1}^4) \leqq 3(E(\|X_t-X_s\|_{-p-1}^2))^2 = 3(v_t-v_s)^2.$$

Using the Kolmogorov continuous version theorem, we can easily see that $\{X_t, t \leqq T\}$ has a $\|\ \|_{-p-1}$-continuous version. Since the $\|\ \|_{-p-1}$-topology in $\mathcal{S}'_p$ coincides with the topology on $\mathcal{S}'_p$ induced from the strong topology of $\mathcal{S}'$, $\{X_t, t \in T\}$ has a continuous version in $\mathcal{S}'$.

REFERENCES

1. K. Itô, Theory of probability (in Japanese), Iwanami-Shoten, 1952.
2. I. M. Gelfand and N. Ja Vilenkin, Generalized functions IV, Academic Press, 1934.

STOCHASTIC DIFFERENTIAL EQUATION OF THE OPTIMAL NON-LINEAR FILTERING OF THE CONDITIONAL GAUSSIAN PROCESS

O.A. Glonti

I. Let $\eta = (\theta, \xi) = (\theta_t, \xi_t), 0 \le t \le T$, be a partially observable process on some probability space $(\Omega, \mathcal{F}, P)$ satisfying the following system of stochastic differential equations:

$$d\theta_t = [a_0(t,\xi) + a_1(t,\xi)\theta_t]dt + b_1(t,\xi)dw_1(t) + b_2(t,\xi)dw_2(t), \quad (1)$$

$$d\xi_t = [A_0(t,\xi) + A_1(t,\xi)\theta_t]dt + B(t,\xi)dw_2(t), \quad (2)$$

where $W_1 = (w_1(t), \mathcal{F}_t)$, $W_2 = (w_2(t), \mathcal{F}_t)$ are independent Wiener processes, $(\mathcal{F}_t), 0 \le t \le T$, is a non-decreasing family of σ-subalgebras of $\mathcal{F}$. Each of the measurable functionals $a_i(t,x), A_i(t,x), B(t,x), i=0,1, j=1,2$, is nonanticipative.

Let for each $x \in C_T$ (C_T be a space of continuous functions $x = \{x_s, s \le T\}$)

$$\int_0^T \Big(\sum_{i=0,1} \{|a_i(t,x)| + |A_i(t,x)|\} + \sum_{j=1,2} b_j^2(t,x) + B^2(t,x)\Big) dt < \infty; \quad (3)$$

$$\int_0^T [A_0^2(t,x) + A_1^2(t,x)]dt < \infty; \quad (4)$$

$$\inf_{x \in C_T} B^2(t,x) \geq C > 0, \quad 0 \leq t \leq T; \tag{5}$$

for any x and y

$$|B(t,x) - B(t,y)|^2 \leq L_1 \int_0^T |x_s - y_s|^2 dK(s) + L_2 |x_t - y_t|^2,$$

$$B^2(t,x) \leq L_1 \int_0^t (1 + x_s^2) dK(s) + L_2(1 + x_t^2), \tag{6}$$

where $K(s)$ is a non-decreasing right-continuous function, $0 \leq K(s) \leq 1$, L_1 and L_2 are some constants.

For each $x \in C_T$, $0 \leq t \leq T$,

$$|a_1(t,x)| \leq L, \quad |A_1(t,x)| \leq L; \tag{7}$$

$$\int_0^T E[a_0^4(t,\xi) + b_1^4(t,\xi) + b_2^4(t,\xi)] dt < \infty; \tag{8}$$

$$E\theta_0^4 < \infty. \tag{9}$$

The following theorem holds:

Theorem 1. Let the partially observable random process $\eta = (\theta, \xi) = (\theta_t, \xi_t), 0 \leq t \leq T$, have representations (1), (2). Then under conditions (3)-(9) and if the conditional distribution $P(\theta_0 \leq a | \xi_0)$ is Gaussian $N(m, \gamma)$, the optimal in the mean square sense filtration estimate $m_t = E(\theta_t | \mathcal{F}_t^\xi)$, $\mathcal{F}_t^\xi = \sigma\{\xi_s, s \leq t\}$ satisfies the following stochastic differential equation:

$$dm_t = [a_0(t,\xi) + a_1(t,\xi) m_t] dt +$$

$$+\{b_2(t,\xi)B(t,\xi)+A_1(t,\xi)\Big[-\langle\tilde{m}\rangle_t+e^{\int_0^t 2a_1(s,\xi)ds}\Big[\int_0^t\Big(b_1^2(s,\xi)+$$

$$+b_2^2(s,\xi)-2a_1(s,\xi)\langle\tilde{m}\rangle_s\Big)e^{-\int_0^s 2a_1(u,\xi)du}ds\Big]+\gamma\Big]B^{-2}(t,\xi)\Big[d\xi_t-$$

$$-\big(A_0(t,\xi)+A_1(t,\xi)m_t\big)dt\Big],\quad m_0=m, \tag{10}$$

where $\langle\tilde{m}\rangle_t$ is a natural increasing process from $(\tilde{m}_t^2,\mathcal{F}_t^{\xi})$-submartingal decomposition. Here

$$\tilde{m}_t=m_t-\int_0^t\big(a_0(s,\xi)+a_1(s,\xi)m_s\big)ds,$$

and

$$\langle\tilde{m}\rangle_t=\int_{-\infty}^{\infty}L(t,y)\,dy$$

whenever a local time $L(t,y)$ of the martingale $\tilde{m}_t$ exists.

Proof. For $m_t=E[\theta_t|\mathcal{F}_t^{\xi}]$ and $\delta_t=E[\theta_t^2|\mathcal{F}_t^{\xi}]$ the following representations are true (see the proof of the wellknown Lipcer-Shiryayev's theorem, Theorem 12.1 from [1]):

$$dm_t=\big[a_0(t,\xi)+a_1(t,\xi)m_t\big]dt+$$

$$+\frac{b_2(t,\xi)B(t,\xi)+\gamma_t A_1(t,\xi)}{B^2(t,\xi)}\Big[d\xi_t-\big(A_0(t,\xi)+A_1(t,\xi)m_t\big)dt\Big], \tag{11}$$

$$\delta_t=\delta_0+\int_0^t\big[2a_0(s,\xi)m_s+2a_1(s,\xi)\delta_s+b_1^2(s,\xi)+$$

$$+b_2^2(s,\xi)\big]ds+\int_0^t B^{-1}(s,\xi)\{2m_s b_2(s,\xi)B(s,\xi)+$$

$$+A_1(s,\xi)\left[E(\theta_s^2|\mathcal{F}_t^\xi)-\delta_s m_s\right]\Big\}\,d\overline{W}_s\,, \qquad (12)$$

where $\gamma_s = E[(\theta_t - m_t)^2|\mathcal{F}_t^\xi]$ and $\overline{W}_t$ is the so called innovation process.

Since $\gamma_t = \delta_t - m_t^2$, the following step in the proof is to obtain the representations for m_t^2. Here, in contrast to the proof of Theorem 12.1 from [1], we use the generalized Ito's formula from the following lemma, which is given without any proof.

__Lemma.__ (A.T. Wang, [2]). Let $(M(t), G_t, P)$ be a continuous local martingale and let f be a convex function. Then

$$f(M(t)) - f(M(0)) = \int_0^t f'(M(u))\,dM(u) + A(t), \quad \text{a.s.P} \qquad (13)$$

where $A(t)$ is a natural increasing process. Indeed,

$$A(t) = \frac{1}{2}\int_{-\infty}^{\infty} L(t,y)\,d\mu(y)$$

whenever a local time $L(t,y)$ of $M(t)$ exists, μ is a locally finite measure equal to f'' in a weak sense, namely $\int_{-\infty}^{\infty} f(x)\varphi''(x)\,dx = \int_{-\infty}^{\infty} \varphi(x)\,d\mu(x)$ for all $\varphi \in C_0^{\infty}[R]$.

Hence, using formula (13) for $\left(m_t - \int_0^t [a_0(s,\xi) + a_1(s,\xi)m_s]ds\right)^2$ after a simple transformation we have

$$m_t^2 = m_0^2 + \int_0^t 2m_s\,dm_s + \langle \tilde{m} \rangle_t \tag{14}$$

From (14) and (11) we have

$$m_t^2 = m_0^2 + \int_0^t 2m_s\,[a_0(s,\xi) + a_1(s,\xi)m_s]ds + $$

$$+ \int_0^t 2m_s \frac{b_2(s,\xi)B(s,\xi) + \gamma_s A_1(s,\xi)}{B(s,\xi)}\,d\bar{W}_s + \langle \tilde{m} \rangle_t . \tag{15}$$

Further, from (12) and (15)

$$\gamma_t = \gamma_0 + \int_0^t [2a_1(s,\xi)\gamma_s + b_1^2(s,\xi) + b_2^2(s,\xi)]ds - \langle \tilde{m} \rangle_t . \tag{16}$$

Therefore , γ_t is a solution of the following linear stochastic differential equation

$$\frac{d\gamma_t}{dt} = 2a_1(t,\xi)\gamma_t + b_1^2(t,\xi) + b_2^2(t,\xi) - \frac{d\langle \tilde{m} \rangle_t}{dt} ,$$

$$\gamma_0 = \gamma , \tag{17}$$

and it is obvious that

$$\gamma_t = e^{\int_0^t 2a_1(s,\xi)ds}\left[\int_0^t \left(b_1^2(s,\xi) + b_2^2(s,\xi) - \frac{d\langle \tilde{m} \rangle_s}{ds}\right) e^{-\int_0^s 2a_1(u,\xi)du} ds + \gamma\right].$$

After a simple transformation

$$\gamma_t = -\langle \tilde{m} \rangle_t + e^{\int_0^t 2a_1(s,\xi)ds} \left[\int_0^t \left(b_1^2(s,\xi) + b_2^2(s,\xi) - 2a_1(s,\xi)\langle \tilde{m} \rangle_s \right) e^{-\int_0^s 2a_1(u,s)du} ds + \gamma \right] \quad (18)$$

If we put (18) in (11) we obtain equation (10) and this completes the proof of the theorem.

Remark 1. In our opinion, the theorem is interesting because it suggests a single stochastic differential equation for the determination of the optimal filtration estimate instead of a traditional system of filtering equations for m_t and γ_t (see, for example, equations (12,29), (12.30) in [1]).

Remark 2. From the filtration equation (10) immediately follows the representation

$$m_t = e^{\int_0^t a_1(s,\xi)ds} \left\{ \int_0^t a_0(s,\xi)\, e^{-\int_0^s a_1(u,\xi)du} ds + \right.$$

$$\left. + \int_0^t e^{-\int_0^s a_1(u,\xi)du} d\tilde{m}_s + m \right\} \quad (19)$$

where $\tilde{m}_t$ satisfies the following stochastic equation

$$d\tilde{m}_t = \left\{ b_2(t,\xi)B(t,\xi) + A_1(t,\xi)\left[-\langle \tilde{m} \rangle_t + e^{\int_0^t 2a_1(s,\xi)ds} \left[\int_0^t \left(b_1^2(s,\xi) + \right. \right. \right. \right.$$

$$+ b_2^2(s,\xi) - 2a_1(s,\xi)\langle\tilde{m}\rangle_s\Big) e^{-\int_0^s 2a_1(u,\xi)du} ds\Big] + \gamma\Big] B^{-1}(t,\xi)\, d\bar{w}_t. \quad (20)$$

Equation (20) is naturally solved applying the method of successive approximations, taking $\tilde{m}_t^{(0)} \equiv \tilde{m}_0 = m$ as zero approximation. It is obvious that $\langle \tilde{m}^{(0)}\rangle_t \equiv 0$.

From (20) we can obtain the stochastic integro-differential equation for $\langle\tilde{m}\rangle_t$:

$$\frac{d\langle\tilde{m}\rangle_t}{dt} = \Big\{ b_2(t,\xi)B(t,\xi) + A_1(t,\xi)\Big[-\langle\tilde{m}\rangle_t + e^{\int_0^t 2a_1(s,\xi)ds}\Big[\int_0^t \Big(b_1^2(s,\xi) +$$

$$+ b_2^2(t,\xi) - 2a_1(s,\xi)\langle\tilde{m}\rangle_s\Big) e^{-\int_0^s 2a_1(u,\xi)du} ds\Big] + \gamma\Big\}^2 B^{-2}(t,\xi) .$$

Remark 3. The conditional filtration error $\gamma_t = E\left[(\theta_t - m_t)^2 | \mathcal{F}_t^\xi\right]$ can be found from representation (18), which may be effective during the investigation of control problems with squared criterion, e.g. the construction of optimal in the mean square sense schemes of transmission of Gaussian signals.

As an illustration to Theorem 1 we give the following example:

Example. Let the partially observable process $\eta = (\theta, \xi) = (\theta_t, \xi_t)$, $0 \le t \le T$, satisfy the following system of stochastic differential equations:

$$d\theta_t = b(t,\xi)\, dw_1(t) ,$$

$$d\xi_t = [A_0(t,\xi) + A_1(t,\xi)\theta_t]dt + B(t,\xi)\, dw_2(t) .$$

Then we have from (10) for the optimal filtration estimate m_t the representation

$$dm_t = \{A_1(t,\xi)[-\langle m\rangle_t + \int_0^t b^2(s,\xi)ds +$$

$$+\gamma] B^{-2}(t,\xi)[d\xi_t - (A_0(t,\xi) + A_1(t,\xi)m_t)dt], m_0 = m,$$

and for γ_t from (18) :

$$\gamma_t = -\langle m\rangle_t + \int_0^t b^2(s,\xi)ds + \gamma .$$

Here

$$\langle m\rangle_t = \int_{-\infty}^{\infty} L(t,y)\, dy$$

where $L(t,y)$ is the local time of m_t .

II. Consider the multidimensional case. Let the partially observable stochastic process $(\theta,\xi) = ((\theta_1(t),\dots,\theta_k(t)),(\xi_1(t),\dots,\xi_\ell(t))), 0\le t\le T,$ be of diffusion type, which has the representation

$$d\theta_t = [a_0(t,\xi) + a_1(t,\xi)\theta_t]dt + b_1(t,\xi)dw_1(t) + b_2(t,\xi)dw_2(t), \tag{21}$$

$$d\xi_t = [A_0(t,\xi) + A_1(t,\xi)\theta_t]dt + B_1(t,\xi)dw_1(t) + B_2(t,\xi)dw_2(t), \tag{22}$$

where $W_1 = (w_1(t),\mathcal{F}_t)$ and $W_2 = (w_2(t),\mathcal{F}_t)$ are independent of each other Wiener processes, $w_1(t) = [w_{11}(t),\dots,w_{1k}(t)]$ and $w_2(t) = [w_{21},\dots,w_{2\ell}]$; $a_0 = (a_{01},\dots,a_{0k})$,

$A_0 = (A_{01}, \ldots, A_{0\ell})$ are vector-functions , $a_1 = \|a_{ij}\|$, $A_1 = \|A_{ij}\|$, $b_1 = \|b_{ij}^{(1)}\|$, $b_2 = \|b_{ij}^{(2)}\|$, $B_1 = \|B_{ij}^{(1)}\|$, $B_2 = \|B_{ij}^{(2)}\|$ are $k\times k$, $\ell\times k$, $k\times k$, $k\times\ell$, $\ell\times k$, $\ell\times\ell$ -matrices.

Then from the proof of Theorem 1 and Theorem 12.7 [1] it is easy to see that the following theorem holds:

<u>Theorem 2.</u> Under conditions of Theorem 12.7 from [1] the optimal in the mean square sense filtration estimate $m_t = [m_1(t), \ldots, m_k(t)]$, $m_i(t) = E[\theta_i(t) | \mathcal{F}_t^{\xi}]$ of the partially observable process (θ, ξ) , satisfying (21), (22), has the representation

$$dm_t = [a_0(t,\xi) + a_1(t,\xi)m_t]dt +$$

$$+\{(b\circ B)(t,\xi) + \int_0^t \varphi(t,s;\xi)\left[(b\circ b)(s,\xi) - \frac{d\langle\tilde m\rangle_s}{ds}\right]\varphi^*(t,s;\xi)ds +$$

$$+\varphi(t,0;\xi)\gamma\varphi^*(t,0;\xi)\}(B\circ B)^{-1}(t,\xi)[d\xi_t - (A_0(t,\xi) + A_1(t,\xi)m_t)dt],$$

where $(b\circ b) = b_1 b_1^* + b_2 b_2^*$, $(b\circ B) = b_1 B_1^* + b_2 B_2^*$, $(B\circ B) = B_1 B_1^* + B_2 B_2^*$, C^* is the matrix conjugate of C and $\langle \tilde m\rangle_t = \langle \tilde m, \tilde m\rangle_t$ is a $k\times k$ matrix with components $\langle \tilde m_{i_1}, \tilde m_{i_2}\rangle$, which is a process from the decomposition of the product $\tilde m_{i_1}\tilde m_{i_2}$ of square integrable martingales $\tilde m_i(t) = m_i(t) - \int_0^t [a_{0i}(s,\xi) + \sum_j a_{ij}(s,\xi)m_j]ds$, $i=\overline{1,k}$; $\varphi(t,s;\xi)$ is the solution of the matrix equation

$$\frac{d\varphi(t,s;\xi)}{dt} = a_1(t,\xi)\varphi(t,s;\xi), \quad \varphi(s,s;\xi) = E_{(k\times k)}.$$

R e f e r e n c e s

1. Lipcer R. S.,Shiryayev A.N.Statistics of Random Processes, (Rassian),Moscow,1974.

2. Wang A.T.Generalized Ito's Formula and Additive Functionals of Brownian Motions,Z.Warhscheinlichkeitstheorie Verw. Gebiete,1977,41,153-159.

THE MAXIMUM RATE OF CONVERGENCE OF DISCRETE APPROXIMATIONS FOR STOCHASTIC DIFFERENTIAL EQUATIONS

J.M.C. Clark and R.J. Cameron
Department of Computing and Control
Imperial College of Science and Technology
London SW7 2BZ, UK

INTRODUCTION AND SUMMARY

This paper presents some results on the best rate of convergence that can be attained by methods of approximating the solution of stochastic differential equations that depend on "discrete" evaluations. Suppose $(x_t)_{t\geq 0}$ is a vector process solving the Ito equation

(1) $$dx_t = f(x_t)dt + g(x_t)dw_t \qquad x_0 = \alpha, \text{ a constant}$$

where (w_t) is a vector Brownian motion and the components of f and g are uniformly Lipschitz. Let $T > 0$ be fixed. The type of method that we are concerned with produces for each regular partition $(0,T/n,2T/n,\ldots)$ of the positive real line an approximation to x_T that depends on the values of the forcing Brownian motion only at the dividing points of the partition; in other words the method produces a P_n-measurable approximation, where P_n denotes the "partition" σ-field generated by $(w_{iT/n}: i=0,1,2,\ldots)$. The natural stochastic generalisations of the traditional difference methods of numerical analysis are of this type.

The error of approximation will be measured by the L_2 norm $E[(\cdot)^2]^{\frac{1}{2}}$. Since the conditional mean $E[x_T|P_n]$ is the best approximation of x_T with respect to this norm among all P_n-measurable approximations, the sequence of conditional means has the best rate of convergence of all (P_n)-adapted sequences. Our results are therefor couched in terms of conditional-mean sequences.

Lower bounds on the maximum rate of convergence have been known for some time. For example the stochastic version of the Cauchy-Euler method is known ([1] p. 240 Lemma 2) to produce approximations with errors (at most) of the order $(\frac{1}{n})^{\frac{1}{2}}$. Another method due to McShane ([2] p. 205) has errors of the order of $\frac{1}{n}$ for an interesting class of equations that satisfy the following commutativity condition: the column vectors g_i of the matrix coefficient g commute in the sense that

$$(2) \qquad \sum_k g_i^k \frac{\partial g_j^m}{\partial x^k} - \sum_k g_j^k \frac{\partial g_i^m}{\partial x^k} = 0 \text{ for all } i, j \text{ and } m \text{ in their ranges.}$$

This condition is satisfied trivially if g has constant components or is scalar, but it is also satisfied by equations with non-constant vector g_i that arise in the theory of nonlinear filtering.

The results of this paper show that these bounds are sharp in the sense that for at least one equation the "semilinear" Cauchy-Euler rate is the best rate that can be achieved and that for others the "linear" rate of McShane's method is the best. Roughly speaking, these two rates are "generically" the best rates, the semilinear rate in the class of all equations and the linear rate among equations satisfying the commutativity condition. In this short paper this question is considered only for an illustrative class of scalar equations satisfying (2). It turns out that if the class is endowed with a natural "uniform C^3" topology the linear rate is indeed generically the best.

It should be emphasised that these results apply only to discrete methods generating P_n-measurable approximations. Wagner and Platen [3]* have recently developed a hierarchy of approximation methods, that can legitimately be called "discrete", with arbitrarily fast rates of convergence. The distinction is that the approximations given by these methods depend in a more complicated fashion on the forcing Brownian motion and are not P_n-measurable. The methods of Wagner and Platen are particularly important in their application to the simulation of diffusion processes; the requirement of "P_n-measurability" considered here is relevant mainly to problems of approximation in nonlinear filtering.

AN EXAMPLE OF A SLOW BEST RATE

Suppose (v_t), (w_t), $t \geq 0$ are independent continuous Brownian motions, zero at $t = 0$, defined on a probability space (Ω,F,P). Suppose (x_t,y_t) solves

$$(3) \qquad dx_t = y_t dv_t \qquad x_0 = y_0 = 0$$

$$dy_t = dw_t$$

We note that the Brownian motion coefficients $\binom{y}{0}$, $\binom{0}{1}$ do not commute in the sense of the introduction. Fix $T > 0$ and for integral n let P_n denote the partition

---

* Some of these results are described in the paper of Platen in this volume.

σ-field generated by $(v_{iT/n}, w_{iT/N})_{i\in N}$. We assert that the minimum error norm

(4) $$E[(x_T - E[x_T|P_n])^2]^{1/2} = \frac{T}{2}(\frac{1}{n})^{1/2}$$

To prove this assertion it is convenient to represent the Brownian motions as sums of piecewise-linear P_n-measurable processes and Brownian-bridge processes: for brevity, let h denote $\frac{T}{n}$; let

$$\hat{v}_t = (i+1-\frac{t}{h})v_{ih} + (\frac{t}{n}-i)v_{(i+1)h} \quad \text{for } ih \le t \le (i+1)h,\ i=0,1,2,\ldots$$

and let $\tilde{v}_t$ denote $v_t - \hat{v}_t$. Define $\hat{w}_t$ and $\tilde{w}_t$ similarly. Then it can be easily verified that the four processes $\hat{v}_t$, $\tilde{v}_t$, $\hat{w}_t$ and $\tilde{w}_t$ are independent continuous Gaussian processes and that $\tilde{v}_t$ and $\tilde{w}_t$ are Brownian-bridge processes pinned to zero at t = 0, h, 2h, ..., with zero means and covariances

(5) $$E[\tilde{v}_s\tilde{v}_t] = E[\tilde{w}_s\tilde{w}_t] = \frac{1}{h}(s-ih)((i+1)h-t) \quad \text{for } ih \le s \le t \le (i+1)h,\ i=0,1,\ldots$$
$$= 0 \qquad \text{for } s < t \text{ otherwise}$$

Furthermore, since the process defined by

$$M_t \equiv M_{ih} + \tilde{v}_t + \int_{ih}^{t} \frac{1}{ih+h-s}\tilde{v}_s ds, \quad ih < t \le (i+1)h$$
$$= 0,\ t = 0$$

is a Brownian motion, $\tilde{v}_t$ is a semimartingale on its canonical σ-fields and therefore a semimartingale on the larger family (G_t), where G_t is generated both by $(\tilde{v}_s)_{0\le s\le t}$ and $(\hat{v}_s, w_s)_{s\ge 0}$. It follows that $\tilde{v}_t$, $\hat{v}_t$ and therefore v_t are all (G_t)-semimartingales and that the Ito integral $\int_0^T w_t dv_t$ can be identified with a (G_t)-semimartingale integral. Hence

$$x_T = \int_0^T w_t dv_t = \int_0^T (\hat{w}d\hat{v}_t + \hat{w}_t d\tilde{v}_t + \tilde{w}_t d\hat{v}_t + \tilde{w}_t d\tilde{v}_t)$$

and from mutual independence

$$\hat{x}_T \equiv E[x_T|P_n] = \int_0^T \hat{w}_t d\hat{v}_t, \quad \tilde{x}_T \equiv x_T - \hat{x}_t = \int_0^T \hat{w}_t d\tilde{v}_t + \int_0^T \tilde{w}_t dv_t$$

Also $\int_0^T \hat{w}d\tilde{v} = -\int_0^T \tilde{v}d\hat{w}$ since $\tilde{v}_T = 0$, and the product of the two integrals in the equation for $\tilde{x}_T$ turns out to have zero expectation. Consequently

$$E[\tilde{x}_T^2] = E[(-\int_0^T \tilde{v}_t d\hat{w}_t)^2] + \int_0^T E\tilde{w}_t^2 dt$$

The second integral reduces to $n\int_0^h \frac{s}{h}(h-s)ds = \frac{T^2}{6n}$; the first becomes

$$\sum_{i=0}^{n-1} \int_{ih}^{(i+1)h} \int_{ih}^{(i+1)h} E[\tilde{v}_s\tilde{v}_t]\frac{1}{h^2} E\Delta_i w^2 dsdt,$$ where $\Delta_i w = w_{(i+1)h} - w_{ih}$, and this reduces to $\frac{T^2}{12n}$. The assertion then follows from this equation.

The Cauchy-Euler method applied to (2) gives the following approximation for x_T: $\bar{x}_{T,n} = \sum_{i=0}^{n-1} w_{iT/n}(v_{(i+1)T/n} - v_{iT/n})$. Direct computation shows that

$$E[(x_T - \bar{x}_{T,n})^2]^{\frac{1}{2}} = T(\frac{1}{2n})^{\frac{1}{2}}$$

which is of the same order as the minimum error (4). We note, however, that for this example the Cauchy-Euler method is "inefficient" in that this error is $2^{\frac{1}{2}}$ times the minimum error.

SYSTEMS WITH ADDITIVE DISTURBANCES

As before, let $(w_t)_{t\geq 0}$, $w_0 = 0$, be a continuous Brownian motion on (Ω,F,P), but now let P_n be the σ-field generated by $(w_{iT/n})_{i\in N}$ for some fixed $T > 0$.

Theorem 1

If (x_t) is the continuous solution of

(6) $$dx_t = a(x_t)dt + dw_t, \quad x_0 = \beta, \text{ a constant}$$

where a is of class C^3 with bounded first, second and third order derivatives then

(7) $$E[(x_T - E[x_T|P_n])^2]^{\frac{1}{2}} = (\frac{T^3}{12}\int_0^T E[\exp(2\int_t^T a'(x_s)ds)a'(x_t)^2]dt)^{\frac{1}{2}}\frac{1}{n} + R_n$$

where $nR_n \to 0$ as $n \to \infty$.

In particular it follows from this that for the special case where a is of the form $k_1 + k_2x$ with $k_2 \neq 0$ the best rate is linear.

Proof

Arguments based on Gronwall's inequality, similar to, but simpler than, those that follow, establish that all moments of x_t, and therefore of $a(x_t)$, exist. Let $\hat{w}_t$ and $\tilde{w}_t$ be defined as in the previous section and set $h = \frac{T}{n}$. It follows from (6) that the process $\hat{x}_t = \beta + \int_0^t E[a(x_s)|P_n]ds + \hat{w}_t$ is a modification of $E[x_t|P_n]$ with continuous sample paths. We shall denote $x_t - \hat{x}_t$ by $\tilde{x}_t$, but we shall use NX to denote the projection $X - E[X|P_n]$ for other variables X. Expanding a in (6) we have

$$x_t = \beta + \int_0^t (a(\hat{x}_s) + a'(\hat{x}_s)\tilde{x}_s + \tfrac{1}{2}a_s^{**}\tilde{x}_s^2)ds + w_t$$

where a_s^{**} is the coefficient $\int_0^1 a''(\hat{x}_s + \Theta\tilde{x}_s)d\Theta$; it is clear from the form of a_s^{**} that it is jointly measurable in (s,ω).

Let $A_s = \tfrac{1}{2}N[a_s^{**}\tilde{x}_s^2]$. If the operator N is applied to both sides of the previous equation it follows that

$$\tilde{x}_t = \int_0^t (a'(\hat{x}_s)\tilde{x}_s + A_s)ds + \tilde{w}_t$$

and therefore that the process $\tilde{x}_t - \tilde{w}_t$ solves the equation

$$\dot{\xi} = a'(\hat{x}_t)\xi + a'(\hat{x}_t)\tilde{w}_t + A_t, \quad \xi_0 = 0$$

Hence, if $F_{s,t}(\hat{x})$ denotes the fundamental solution $\exp \int_0^t a'(\hat{x}_s)ds$ of $\xi = a'(\hat{x}_t)\xi$,

$$\tilde{x}_t = \tilde{w}_t + \int_0^t F_{s,t}(\hat{x})a'(\hat{x}_s)\tilde{w}_s ds + \int_0^t F_{s,t}(\hat{x})A_s ds \tag{8}$$

The asymptotic form of $E\tilde{x}_T^2$ derived from this equation will give the statement of the theorem. Most of the remaining argument is to show that the second integral in (8) is of negligible order. Suppose that $|a'|$, $|a''|$ and $|a^{(3)}|$ are less than L. Since $x_t = \beta + \int_0^t (a(\hat{x}_s) + a_s^* \tilde{x}_s)ds + w_t$ where $a_s^* = \int_0^1 a'(\hat{x}_s + \theta\tilde{x}_s)d\theta$ we have

$$\tilde{x}_t = \int_0^t N[a_s^* \tilde{x}_s]ds + \tilde{w}_t$$

The following inequality, which will be used repeatedly,

$$E[(NX)^{2p}] \leq E[(2X)^{2p}] \text{ for positive integral } p \tag{9}$$

and Jensen's inequality then yield

$$E\tilde{x}_t^8 \leq (2t)^7 \int_0^t (2L)^8 E\tilde{x}_s^8 ds + 2^7 E\tilde{w}_t^8$$

Since $\tilde{w}_t$ is normal and $E\tilde{w}_t^2 \leq \frac{h}{4}$ (see equation (5)) it follows that $E\tilde{w}_t^8 = 15(E\tilde{w}_t^2)^4 \leq \frac{15}{64}h^4$. Consequently by Gronwall's lemma there is a positive constant K_1 such that for $t : 0 \leq t \leq T$,

$$E\tilde{x}_t^8 \leq K_1 h^4 \tag{10}$$

We now show that Ex_{ih}^4 for integral i is also of order h^4. For any independent random variables $X_0, X_1, \ldots X_{i-1}$ with zero means and finite fourth moments $E[(\sum_j X_j)^4] = \sum_j EX_j^4 + 6\sum_{jk} EX_j^2 EX_k^2$. An application of this formula to the P_n-conditionally independent variables $X_j = \int_{jh}^{(s+1)h} F_{s,ih}(\hat{x})\tilde{w}_s ds$, together with Jensen's inequality and the inequalities $|F_{s,t}(\hat{x})| < e^{LT}$ for $s, t \leq T$, $E\tilde{w}_s^2 \leq \frac{h}{4}$ and $E\tilde{w}_s^4 \leq \frac{3}{16}h^2$ and the fact that $\tilde{w}_r$ and $\tilde{w}_{jh+r}$ are identically distributed shows that the first integral in (8) satisfies

$$E[(\int_0^{ih} F_{s,ih}(\hat{x})\tilde{w}_s ds)^4 | P_n] \leq e^{4LT}(ih^3\int_0^h E\tilde{w}_s^4 ds + 6i^2(h\int_0^h E\tilde{w}_s^2 ds)^2)$$

$$\leq K_2 h^4$$

for $K_2 = (9T^2/16)e^{4LT}$. Since, by the inequality (9), $EA_s^4 \le E[a_s^{**4}\tilde{x}_s^8] \le L^4K_1h^4$, the second integral in (8) satisfies

$$E[(\int_0^t F_{s,t}(\hat{x})A_s ds)^4] \le e^{4LT}t^3\int_0^t EA_s^4 ds \le K_3h^4$$

For suitable K_3. As $\tilde{w}_{ih} = 0$, it then follows from (8) and these two inequalities that for $i \le n$

(11) $$E\tilde{x}_{ih}^4 \le 8(K_2+K_3)h^4 \equiv K_4h^4$$

We now obtain a sharper estimate of the second integral in (8) and show it is of negligible order. First, note that, since $a \in C^3$, the remainder term

$$\tfrac{1}{2}a^{**}\tilde{x}_t^2 = \tfrac{1}{2}a''(\hat{x}_t)\tilde{x}_t^2 + \frac{1}{6}a_t^{3*}\tilde{x}_t^3$$

where a_t^{3*} denotes $\int_0^1 a^{(3)}(\hat{x}_t+\Theta\tilde{x}_t)d\Theta$. Second, for $ih \le t < (i+1)h$ it follows from (6) that $\tilde{x}_t$ can be expressed as

$$\tilde{x}_t = \tilde{x}_{ih} + \int_{ih}^t Na(x_s)ds + \tilde{w}_t \equiv R_t + \tilde{w}_t$$

But $E(\int_{ih}^t Nads)^4 \le (2h)^4 \max_{0\le t\le T}E[a^4]$; this together with (11) implies that $ER_t^4 \le K_5h^4$ for some positive K_5. Furthermore $E\tilde{w}_t^4 \le \frac{3h^2}{16}$. Hence if B_t denotes $N[R_t^2+2R_t\tilde{w}_t]$ Schwarz's inequality implies that $EB_t^2 \le K_6h^3$ for some K_6. Let $C_t = \frac{1}{6}N[a^{3*}x_t^3]$. Then the inequalities (9) and (11) show that $EC_t^2 \le K_7h^3$ for $K_7 = \frac{L}{6}K_1^{\frac{3}{4}}$. Since $N\tilde{x}_t^2 = B_t + N\tilde{w}_t^2$ we have the expression

$$A_t = \tfrac{1}{2}a''(\hat{x}_t)(N\tilde{w}_t^2+B_t) + C_t$$

Therefore, since $|F_{s,T}| \le e^{LT}$,

$$E[(\int_0^T F_{s,T}(\hat{x})A_s ds)^2] = E[E[(\dots)^2|P_n]]$$
$$\le 3e^{2LT}(\frac{L^2}{4}\int_0^T\int_0^T E[N\tilde{w}_s^2N\tilde{w}_t^2]dsdt+K_8h^3)$$

where $K_8 = T(\tfrac{1}{4}L^2K_6+K_7)$. But $N\tilde{w}_s^2$ and $N\tilde{w}_t^2$ are orthogonal if $s \le ih \le t$ for some i; hence

$$\int_0^T\int_0^T |E[N\tilde{w}_s^2N\tilde{w}_t^2]|dsdt = \sum_{i=0}^{n-1}\int_{ih}^{(i+1)h}\int_{ih}^{(i+1)h} |E[N\tilde{w}_s^2N\tilde{w}_t^2]|dsdt$$

Schwarz's inequality, (9) and (6) show that $E[N\tilde{w}_s^2,N\tilde{w}_t^2] \le \frac{3h^2}{4}$ and therefore that the right-hand sum is bounded by $\tfrac{3}{4}Th^3$. Combining these inequalities we find that for suitable K_9

$$E[(\int_0^T F_{s,T}(\hat{x})A_s ds)^2] \le K_9h^3 = K_9(\frac{T}{n})^3$$

Consequently the second integral in (8) is of order $(\frac{1}{n})^{3/2}$ and the theorem is proved if we show that

(12) $$\min_{n\to\infty} n^2 E[(\int_0^T F_{s,T}(\hat{x})a'(\hat{x}_s)\tilde{w}_s ds)^2] = \frac{T^3}{12}\int_0^T E[F_{s,T}^2(x)a'(x_s)^2]ds$$

Now, on taking the conditional expectation with respect to P_n before the total expectation we find that

$$E[(\int_0^T F_{s,T}(\hat{x})a'(\hat{x}_s)\tilde{w}_s ds)^2] = \int_0^T\int_0^T E[G_{s,t}(\hat{x})]E[\tilde{w}_s\tilde{w}_t]dsdt$$

where $G_{s,t}(\hat{x})$ is short for $F_{s,T}(\hat{x})F_{t,T}(\hat{x})a'(\hat{x}_s)a'(\hat{x}_t)$. But given the boundedness and continuity of a' and a" and the convergence to zero of $\sup_{0\le t\le T}E[(x_t-\hat{x}_t)^2]$, which follows from (10), it is straightforward to show that $E[G_{s,t}(\hat{x})]$ is continuous on $[0,T]^2$ and converges uniformly to the continuous function $E[G_{s,t}(x)]$. Furthermore the covariance measure $n^2E[\tilde{w}_s\tilde{w}_t]dsdt$ has as its support the string of squares $[ih,(i+1)h]^2$ lying on the diagonal of $[0,T]^2$; it is easy to verify from formula (6) that its mass is $\frac{T^3}{12}$ and that it converges weakly to a measure uniformly distributed on the diagonal of $[0,T]^2$. Equation (12) then follows and the proof of the theorem is complete.

A GENERIC BEST RATE

The following theorem concerns a class of scalar stochastic equations endowed with a topology. The definition of this topology is more simply expressed in terms of the coefficients of the Stratonovich form of equation rather than in terms of those of the corresponding Ito equation. It is unnecessary to repeat here Stratonovich's original definition of his integral [4]; it is sufficient simply to point out that the correspondence between an Ito integral and a Stratonovich integral is such that a solution (x_t) of the scalar Ito equation

(13_I) $$x_t = \alpha + \int_0^t f(x_s)ds + \int_0^t g_1(x_s)dw_s$$

where (w_t) is a scalar Brownian motion and g_1 is of class C^1, is also a solution of the equation containing a Stratonovich integral

(13_S) $$x_t = \alpha + \int_0^t g_0(x_s)ds + \oint_0^t g_1(x_s)dw_s$$

where $f = g_0 + \frac{1}{2}g_1'g_1$. For our purposes the advantage of the Stratonovich form is that under a change of coordinates the coefficients transform according to the usual rules of calculus. We shall now characterise the equations (13) by the triple (α,g_0,g_1).

Let $C_b^3(R)$ be the class of C^3-functions on R for which the function and its first three derivatives are bounded in moduli. The space $E = R \times C_b^3(R) \times C_b^3(R)$ can be identified with the class of equations (α, g_0, g_1) for which the coefficients $g_0, g_1 \in C_b^3(R)$. Endowed with the norm

$$||(\alpha, g_0, g_1)||_E = \max \{|\alpha|, \sup_x |g_i^{(j)}(x)| : i=0,1; j=0,1,2,3\}$$

E is a Banach space and the non-empty open sets of the norm topology are of the second category of Baire. We note that the norm topology is sufficiently strong for the map $E \to L_2(\Omega, F, P)$ of an equation into its solution x_T at a fixed time T to be continuous. Let E_+ be the subset of equations (α, g_0, g_1) in E for which $\inf_x |g_1(x)| > 0$. It is clear that E_+ is open (and non-empty) in E; therefore open subsets in the relative norm topology on E_+ are also open in the Banach space E and hence of the second category both in the norm topology on E and in the relative norm topology on E_+.

Theorem 2

Suppose (x_t) is the solution of equations (13) for which $(\alpha, g_0, g_1) \in E_+$. Let P_n be the partition σ-field of the previous section. Then the rate of convergence of $E[x_T|P_n]$ to x_T is linear in the sense that

$$\lim_{n\to\infty} |\tfrac{1}{2} \log E[(x_T - E[x_T|P_n])^2]/\log n| = 1$$

unless $g_0 = Kg_1$ for some constant K. This rate is generic in E_+ in that the set of equations $A = \{(\alpha, g_0, g_1) : g_0/g_1$ is not constant$\}$ on which it occurs is open, of the second category, and dense in E_+ in the relative norm topology on E_+. On the complementary set of equations in E_+, x_T is P_n-measurable for all n.

The following is McShane's approximation procedure in a form appropriate for the solution of equations (13). Let g_{ij} denote $g_i' g_j$, i, j = 0, 1. For $0 \le i \le n-1$ define $\bar{x}_{n,i}$ by: $\bar{x}_{n,0} = \alpha$,

$$\bar{x}_{n,i+1} = \bar{x}_{n,i} + g_0 \frac{T}{n} + g_1 \Delta_i w + \tfrac{1}{2}[g_{00}(\tfrac{T}{n})^2 + (g_{01}+g_{10})\tfrac{T}{n} \Delta_i w + g_{11}(\Delta_i w)^2] + \frac{1}{6} \frac{dg_{11}}{dx} (\Delta_i w)^3$$

where all the functions are evaluated at $\bar{x}_{n,i}$ and $\Delta_i w = w_{(i+1)T/n} - w_{iT/n}$. Then it follows from ([2], p. 205 Theorem 4.8) that for equations in the class E

$$E[(\bar{x}_{n,n} - x_T)^2]^{\frac{1}{2}} \le \frac{K}{n}$$

for some constant K. Consequently Theorem 2 demonstrates that McShane's approximation sequence $(\bar{x}_{n,n})$ converges generically at the best rate possible at least for the subclass of equations E_+.

<u>Proof</u>

For (α,g_0,g_1) in E_+ let $L = ||(\alpha,g_0,g_1)||$ and $M = \sup_x |1/g_1(x)|$. Let $\phi(x)$ denote $\int_0^x \frac{1}{g_1(z)}dz$. Then $\frac{1}{L} \leq |\phi'| \leq M$ and ϕ is strictly monotonic. If $\hat{\cdot}$ denotes $E[\cdot|P_n]$ it follows from the minimizing property of conditional expectations that

$$E(y_T-\hat{y}_T)^2 \leq E(\phi(\hat{x}_T)-\phi(\hat{x}_T))^2 \leq M^2E(x_T-\hat{x}_T)^2$$

and similarly that $E(x_T-\hat{x}_T)^2 \leq L^2E(y_T-\hat{y}_T)^2$. Hence $y_T - \hat{y}_T$ and $x_T - \hat{x}_T$ both converge to zero at the same rate. By the usual formula for transformation of coordinates applied to (13_S)

$$y_t = \beta + \int_0^t a(y_s)ds + w_t$$

where $\beta = \phi(\alpha)$ and $a = g_0 \circ \phi^{-1}/g_1 \circ \phi^{-1}$. It is easy to verify that the first three derivatives of a are bounded in moduli and that Theorem 1 applies to this equation. Hence $E[y_T|P_n]$ converges linearly to y_T if the coefficient $\int_0^T E[\exp(2\int_t^T a'(y_s)ds)a'(y_t)^2]dt$ is positive. But this coefficient is zero if and only if $a'(y_t) = 0$ a.e. $dt \times dP$ on $[0,T]\times R$. However by, for instance, Girsanov's theorem the support of the distribution of y_t for fixed t is the same as that for w_t, namely R. The continuity of a' then implies that the condition $a'(y_t) = 0$ a.e. is equivalent to a' being identically zero. If this is so $a \equiv K$ for some constant K, $y_T = \beta + KT + w_T$ and $x_T = \phi^{-1}(\beta+KT+w_T)$, which is P_n-measurable.

It remains to verify that the linear rate is generic. If $e = (\alpha,g_0,g_1) \in E_+$ and (e_n) is a sequence in A^c such that $||e_n-e||_E \to 0$ then as $|g_1| > 0$ $g_0/g_1 = \lim_n g_{0,n}/g_{1,n}$ which is constant. So A^c is closed relative to E_+ and A is open. Finally for $\delta > 0$ sufficiently small the δ-neighbourhood $\{e : ||e-\bar{e}||_E < \delta\}$ of a point $\bar{e} = (\alpha,K\bar{g}_1,\bar{g}_1)$ in A^c lies in E_+ and contains the point $(\alpha,K\bar{g}_1 + \frac{\delta}{2}f,\bar{g}_1)$ for any f with the moduli of f and its first three derivatives less than one. If f/g_1 is not constant this also lies in A, and so A is dense in E_+. This completes the proof of the theorem.

REFERENCES

[1] I.I. Gihman, A.V. Skorohod, Stochastic Differential Equations, Springer-Verlag, Berlin, 1972.

[2] E.J. McShane, Stochastic Calculus and Stochastic Models, Academic Press, New York, 1974.

[3] W. Wagner, E. Platen, Approximation of Ito integral equations, internal publication of Zentralinstitut für Mathematik und Mechanik der Akademie der Wissenschaffen der DDR.

[4] R.L. Stratonovich, A new representation for stochastic integrals and equations, J. SIAM Control, 4, pp. 362-371 (1966). Originally published in Russian in Vestnik Moskov. Univ. Ser. I. Mat. Meh., 1, pp. 3-12 (1964).

APPROXIMATION OF ITÔ INTEGRAL EQUATIONS

E. Platen (Berlin, GDR)

We consider Itô integral equations of the form

$$x_t = x_{t_0} + \int_{t_0}^{t} a(u,x_u)\,du + \int_{t_0}^{t} b(u,x_u)\,dw_u \ , \qquad (1)$$

$t \in [t_0, T]$, where $x_t = \{x_t^{(i)}\}_{i=1}^{m}$ and $w_t = \{w_t^{(j)}\}_{j=1}^{n}$ is a n-dimensional standard Wiener process with independent components. Drift and diffusion are continuous functions which fulfil a Lip schitz condition. Further we assume $E\|x_{t_0}\|^2 < \infty$, such that existence and uniqueness of the solution of (1) is given.

Itô integral equations are of increasing importance in many fields of application. But only a few papers exist dealing with methods for numerical solution of problems connected with Itô integral equations. Effective and universal applicable methods seems to be simulation algorithms which generate step by step approximate realizations on a digital computer.

In papers of Mil'štein [3] and Rao, Borwankar, Ramakrishna [5] such algorithms were considered. Here the mean square difference between approximation and original process or a quite similar functional converges to zero, if the maximum step size tends to zero. Further Gichman, Skorochod [1] and Kushner [2] studied weak convergence of such step by step approximations.

In the following we show mean square convergence of a given order and weak convergence for a special class of step by step approximations.

Let

$M := \{ (j_1,\dots,j_k) : k \in \{1,2,\dots\},\ j_i \in \{0,1,\dots,n\} \text{ for } i \in \{1,\dots,k\}\} \cup \{v\}$

denote the set of row vectors $\alpha = (j_1,\dots,j_k)$ with finite length $l(\alpha) := k$ where $l(v) = 0$. We write $-\alpha$ respectively $\alpha-$, if we delete the first respectively last component of α.

Further we define stochastic integrals

$$J_\alpha(s,t) := \begin{cases} 1 & \text{for } \alpha = v \\ \int_s^t J_{\alpha-}(s,u)\, dw_u^{(j_k)} & \text{for } \alpha = (j_1,\dots,j_k) \end{cases}$$

for all $\alpha \in M$, $s,t \in [t_0,T]$, $s \le t$, where $dw_u^{(0)} = du$.

Let D denote the set of all continuous functions $f \mid [t_0,T] \times R^m \to R^m$ with continuous partial derivatives of first order $\frac{\partial}{\partial t} f$, $\frac{\partial}{\partial x^{(i)}} f$ and second order $\frac{\partial^2}{\partial x^{(i)} \partial x^{(r)}} f$, $i,r \in \{1,2,\dots,m\}$. We introduce the following differential operators defined for $f \in D$, $k \in \{0,1,\dots,n\}$

$$L^{(k)} f := \begin{cases} \frac{\partial}{\partial t} f + \sum_{i=1}^m \left(\frac{\partial}{\partial x^{(i)}} f\right) a^{(i)} + \frac{1}{2} \sum_{i,r=1}^m \left(\frac{\partial^2}{\partial x^{(i)} \partial x^{(r)}} f\right) \sum_{j=1}^n b^{(i,j)} b^{(r,j)} & \text{for } k = 0 \\ \sum_{i=1}^m \left(\frac{\partial}{\partial x^{(i)}} f\right) b^{(i,k)} & \text{for } k \in \{1,\dots,n\}. \end{cases}$$

Further we use functions $f_\alpha \mid [t_0,T] \times R^m \to R^m$, $\alpha \in M$, $f_\alpha = \{ f_\alpha^{(i)} \}_{i=1}^m$ where

$$f_\alpha^{(i)} := \begin{cases} 0 & \text{for } \alpha = v \\ a^{(i)} & \text{for } \alpha = (0) \\ b^{(i,j)} & \text{for } \alpha = (j),\ j \in \{1,\dots,n\} \\ L^{(j_1)} f_{-\alpha}^{(i)} & \text{for } \alpha = (j_1,\dots,j_k),\ k \ge 2. \end{cases}$$

Let $n(\alpha)$ denote the number of components of $\alpha \in M$ which are zero.

For a given natural number $\varepsilon \in \{1,2,\dots\}$ we set

$A_\varepsilon := \{ \alpha \in M : l(\alpha) + n(\alpha) \leq \varepsilon \quad \text{or} \quad l(\alpha) = n(\alpha) = (\varepsilon+1)/2 \}$.

In the following we assume for all $\alpha \in A_\varepsilon$, that $f_\alpha \in D$ and a Lipschitz condition holds for f_α. Further we suppose for all $s \in [t_0, T]$ and $\beta \in \{\alpha \in M \setminus A_\varepsilon : -\alpha \in A_\varepsilon\}$

$$E \| f_\beta (s, X_s) \|^2 \leq K_1 .$$

Finally Γ denotes the set of all finite discretizations $\tau = \{t_i\}_{i=0}^{N}$, $t_0 < t_1 < \dots < t_N = T$, of the intervall $[t_0, T]$ and

$$h(\tau) := \max_{i \in \{0,1,\dots,N-1\}} (t_{i+1} - t_i)$$

Then the following assertion is proved in Wagner, Platen [6] .

<u>Theorem 1:</u>

Let for a given $\varepsilon \in \{1,2,\dots\}$ and $\tau \in \Gamma$ the approximate process $\{X_\tau^\varepsilon(t)\}_{t \in [t_0,T]}$ defined by

$$X_\tau^\varepsilon(t_0) := X_{t_0}$$

and

$$X_\tau^\varepsilon(t) := X_\tau^\varepsilon(t_i) + \sum_{\alpha \in A_\varepsilon} f_\alpha(t_i, X_\tau^\varepsilon(t_i))\, J_\alpha(t_i, t) \qquad (2)$$

for all $i \in \{0,1,\dots,N-1\}$ and $t \in [t_i, t_{i+1}]$, then

$$\sup_{t \in [t_0,T]} E \| X_t - X_\tau^\varepsilon(t) \|^2 \leq K_2\, h(\tau)^\varepsilon ,$$

where K_2 is a constant not depending on τ .

The proof is basing on a representation of the increment of a diffusion process which can be obtained by iterated application of the Itô formula.

For practial purposes it is interesting to have weak

convergence of such or more simpler approximations. Therefore we consider now continuous interpolated versions of the approximation above.

Let for a given $\varepsilon \in \{1, 2, \ldots\}$ and $\tau \in \Gamma$ the approximate process $\{\bar{X}^{\varepsilon}_{\tau}(t)\}_{t \in [t_0, T]}$ defined in the discretization points by

$$\bar{X}^{\varepsilon}_{\tau}(t_0) := X_{t_0}$$

and

$$\bar{X}^{\varepsilon}_{\tau}(t_{i+1}) := \bar{X}^{\varepsilon}_{\tau}(t_i) + \sum_{\alpha \in A_{\varepsilon}} f_{\alpha}(t_i, \bar{X}^{\varepsilon}_{\tau}(t_i)) \, J_{\alpha}(t_i, t_{i+1}) \tag{3}$$

for all $i \in \{0, 1, \ldots, N-1\}$. In the remaining time $\{\bar{X}^{\varepsilon}_{\tau}(t)\}_{t \in [t_0, T]}$ shall be continuous interpolated, such that for all $i \in \{0, 1, \ldots N-1\}$, $t, s \in [t_i, t_{i+1}]$, $t \geq s$,

$$E \| \bar{X}^{\varepsilon}_{\tau}(t) - \bar{X}^{\varepsilon}_{\tau}(s) \|^4 \leq K_3 (1 + E \| \bar{X}^{\varepsilon}_{\tau}(t_i) \|^4)(t-s)^2 .$$

For instance, this interpolation condition is fulfilled for $\{X^{\varepsilon}_{\tau}(t)\}_{t \in [t_0, T]}$ and in the case of usual linear interpolation. Then the following theorem is proved in Platen [4] .

Theorem 2:

If $E \| X_{t_0} \|^4 < \infty$ and $h(\tau)$ tends to zero, then for all $\varepsilon \in \{1, 2, \ldots\}$ the family of probability measures adapted to $\{\bar{X}^{\varepsilon}_{\tau}(t)\}_{t \in [t_0, T]}$, $\tau \in \Gamma$, converges weakly to that probability measure adapted to $\{X_t\}_{t \in [t_0, T]}$.

In the proof it is shown at first, that the finite dimensional distributions converge. This follows by the mean square convergence proved in Theorem 1. Then the kriterion

$$\sup_{\tau \in \Gamma} E \| \bar{X}^{\varepsilon}_{\tau}(t+\Delta) - \bar{X}^{\varepsilon}_{\tau}(t) \|^4 \leq K_4 \Delta^2$$

is proved for all $t, t+\Delta \in [t_0, T]$ and weak convergence follows by well known theorems.

References

1. Gichman, I.I.; Skorochod, A.W.: Introduction to the theory of stochastic processes, Moscow (1977) (in Russian)
2. Kushner, H.J.: On the weak convergence of interpolated Markov chains to a diffusion. Ann. Probability 2, (1974), 40-50
3. Mil'štein, G.N.: Approximate integration of stochastic differential equations, Torija verjatn. primen. XIX, (1974), 3, 583-588
4. Platen, E.: Weak convergence of approximations of Itô integral equations. Preprint ZIMM, AdW der DDR, Berlin (1978)
5. Rao, N.J.; Borwankar, J.D.; Ramakrishna, D.: Numerical solution of Itô integral equations. SIAM J. Control, 12, 1, (1974), 124-139.
6. Wagner, W.; Platen, E.: Approximation of Itô integral equations. Preprint ZIMM, AdW der DDR, Berlin (1978).

A PROBABILISTIC APPROACH TO THE REPRESENTATION PROBLEM OF MARTINGALES AS STOCHASTIC INTEGRAL

by

H.J. ENGELBERT and JULIANE HEß

UNIVERSITY OF JENA

GDR

1. INTRODUCTION

It is well known that every martingale of a Wiener process can be represented as a stochastic integral of the Wiener process. We consider the case when this Wiener process is replaced by a local continuous martingale. To describe the problem exactly we use the following notations.

Let $(\Omega, \underline{\underline{F}}, P)$ be a complete probability space and $(\underline{\underline{F}}_t)_{t\geq 0}$ an increasing family of sub-σ-algebras of $\underline{\underline{F}}$ such that $\underline{\underline{F}}_0$ is containing all events from $\underline{\underline{F}}$ having probability zero. By $(\underline{\underline{G}}_t)_{t\geq 0}$ we denote the smallest right continuous family of σ-algebras containing $(\underline{\underline{F}}_t)_{t\geq 0}$, that means $\underline{\underline{G}}_t = \bigcap_{\varepsilon>0} \underline{\underline{F}}_{t+\varepsilon}$ for all $t\geq 0$. Let $X = (X_t, \underline{\underline{F}}_t, t\geq 0)$ be a local continuous martingale on $(\Omega, \underline{\underline{F}}, P)$ where X_t is $\underline{\underline{F}}_t$-measurable for all $t\geq 0$. Denote by $\underline{\underline{F}}_t^X$ the σ-algebra generated by $\{X_s, s\leq t\}$ and completed with respect to P in the usual manner. Introduse the family $(\underline{\underline{G}}_t^X)_{t\geq 0}$ of σ-algebras by $\underline{\underline{G}}_t^X = \bigcap_{\varepsilon>0} \underline{\underline{F}}_{t+\varepsilon}^X$ for all $t\geq 0$.

We say that the representation property for X holds if every martingale $Y = (Y_t, \underline{\underline{F}}_t^X, t\geq 0)$ can be represented as a stochastic integral of X,

$$Y_t = Y_0 + \int_0^t f(s,\omega)\, dX_s,$$

where $f(s,\omega)$ is a previsible process with respect to $(F_t^X)_{t\geq 0}$.

Let $\underline{\underline{M}}^c_{loc}(X)$ be the set of all probability measures Q on $\underline{\underline{F}}^X_\infty$ such that X is a local continuous martingale. Denote by $\langle X\rangle$ the unique increasing previsible process such that $X^2-\langle X\rangle$ is a local martingale. For two local martingales Y and Z we set $\langle Y,Z\rangle = \frac{1}{4}(\langle Y+Z\rangle - \langle Y-Z\rangle)$. This is the uniquely determined process of bounded variation such that $YZ - \langle Y,Z\rangle$ is a local martingale.

Then in papers of LIPCER /8/ and of JACOD and YOR /4/ the following equivalent properties can be found:

i) The representation property holds for X.

ii) P is an extremal point of the convex set $\underline{\underline{M}}^c_{loc}(X)$.

iii) For all $Q \in \underline{\underline{M}}^c_{loc}(X)$ the condition $Q \ll P$ (i.e. Q is absolutely continuous with respect to P) implies that $Q = P$.

iv) For every uniformly bounded martingale $Z = (Z_t,\underline{\underline{F}}^X_t,t\geq 0)$ satisfying the properties $Z_0 = 0$ and $\langle Z,X\rangle = 0$ it follows $Z_t \equiv 0$ for all $t\geq 0$.

The equivalence of these properties shows us that the representation property does not hold for all continuous martingales. An example for this fact we can find in the following way:

Let P_1 and P_2 be two different extremal points of $\underline{\underline{M}}^c_{loc}(X)$. (Their existence is guaranteed by the examples in section 3. Let, for instance, C_{R_+} be the space of continuous functions x defined on $R_+ = [0,+\infty)$, supplied with the σ-algebra $\underline{\underline{B}}$ generated by the cylinder sets, and let X_t be the coordinate mapping $X_t(x) = x(t)$ for all $x \in C_{R_+}$ and $t \in R_+$. Consider the Wiener measure P_1 on $(C_{R_+},\underline{\underline{B}})$ and the image law P_2 of P_1 by the mapping H with $(Hx)(t) = x(2t)$ for all $x \in C_{R_+}$ and $t \in R_+$.) Set $P = qP_1 + (1-q)P_2$ where $0 < q < 1$. Then we have $P \in \underline{\underline{M}}^c_{loc}(X)$, but P does not possess the representation property.

Further examples are given by MAISONNEUVE /5/ and in section 3 of this paper.

In this connection another problem arises, namely, to describe the set of all martingales Y which can be represented as a stochastic integral of a fixed local continuous martingale X. This problem will be treated in the following section.

Furthermore we get sufficient conditions on X which guarantee that X has the representation property.

In section 3 we consider examples of martingales possessing this property. A special example the investigation of which provides a condition for the uniqueness of the solution of stochastic differential equations is treated in this section, too.

In the last section we give an application to the n-dimensional time change for reducing a certain class of n-dimensional martingales to the Wiener process.

2. SUFFICIENT CONDITIONS FOR THE REPRESENTABILITY

Since the representation property holds for the Wiener process we use that any local continuous martingale can be reduced to the Wiener process in order to treat the general case. This result is due to KUNITA and WATANABE /7/ and is based on the method of random time change.

Denote by $(A_t)_{t\geq 0}$ the increasing previsible process of X, i.e. $A_t = \langle X\rangle_t$ for all $t\geq 0$. We define a family of stopping times $(T_t)_{t\geq 0}$ with respect to $(\underline{G}^X_t)_{t\geq 0}$ by

$$T_t = \inf \{ u \geq 0 : A_u > t \}$$

where we set $T_t = +\infty$ if the set $\{u\geq 0 : A_u > t\}$ is empty. Then the process $W = (W_t, \underline{G}_{T_t}, t\geq 0)$ with +) $W_t = X_{T_t}$ is a Wiener process if

---

+) Here the convention $X_\infty = \limsup_{s\to\infty} X_s$ is used. Note that $\lim_{s\to\infty} X_s$ exists on the set $\{A_\infty < +\infty\}$ and, in particular, on $\{T_t = +\infty\}$.

$A_\infty = \lim_{t\to\infty} A_t = \infty$ holds. Because, in general, W is only a Wiener process stopped in the randomized stopping time A_∞ we treat the general case in the following way to get a Wiener process on $[0,\infty)$: Let $W' = (W'_t, \underline{\underline{F}}_t^{W'}, t\geq 0)$ be a Wiener process on a second probability space $(\Omega', \underline{\underline{F}}', P')$. Then we set

$$\tilde{W}_t(\omega,\omega') = W_t(\omega) + W'_t(\omega') - W'_{t\wedge A_\infty(\omega)}(\omega').$$

This is a martingale on $(\Omega\times\Omega', \underline{\underline{F}} \otimes \underline{\underline{F}}', P\times P')$ with respect to $(\underline{\underline{G}}_{T_t} \otimes \underline{\underline{F}}_t^{W'})_{t\geq 0}$ because every term in the sum is a martingale with respect to this family of σ-algebras. Since $\langle\tilde{W}\rangle_t = t$ we get that $\tilde{W}$ is a Wiener process. In the sequel we consider $\tilde{W} = (\tilde{W}_t, \underline{\underline{F}}_t^{\tilde{W}}, t\geq 0)$.

If A_∞ is a stopping time with respect to $(\underline{\underline{F}}_t^{\tilde{W}})_{t\geq 0}$ we get W from $\tilde{W}$ by stopping in A_∞. The condition is valid if A_∞ is a previsible stopping time with respect to $(\underline{\underline{F}}_t^{W})_{t\geq 0}$. Under this assumption we can verify the inclusion $\underline{\underline{F}}_\infty^{W} \subseteq \underline{\underline{F}}_\infty^{\tilde{W}}$ which is important for the proof of the following theorem.

THEOREM 1. <u>Let</u> y <u>be a</u> F_∞^W <u>-measurable integrable random variable and suppose that</u> A_∞ <u>is a previsible stopping time with respect to</u> $(F_t^W)_{t\geq 0}$. Then the martingale

$$Y_t = E(y \,/\, \underline{\underline{G}}_t)$$

<u>can be represented as a stochastic integral of</u> X

$$Y_t = Y_0 + \int_0^t f(s,\omega)\, dX_s.$$

We give a sketch of the proof of this theorem in the case $A_\infty = \infty$. The general case can be treated quite similar on $(\Omega\times\Omega', \underline{\underline{F}} \otimes \underline{\underline{F}}', P\times P')$ replacing W by $\tilde{W}$ because from the condition that A_∞ is a previsible stopping time the $\underline{\underline{F}}_\infty^{W}$-measurability of y implies the $\underline{\underline{F}}_\infty^{\tilde{W}}$-measurability of y. Since Y is in fact independent of ω' we get the representation on $(\Omega, \underline{\underline{F}}, P)$. Suppose now that $A_\infty = \infty$. From the assumptions of the theorem follows that Y_{T_t} is $\underline{\underline{F}}_t^W$-measurable. Using the representation theorem for the Wiener process we can therefore derive

$$Y_{T_t} = \int_0^t g(s,\omega)\, dW_s.$$

To make the time change in the opposite direction is easy if A_t has no intervals of constancy. In the general case we can show that these intervals are stochastic intervals of the form $[S_n, T_n]$ for n=1,2,... where S_n and T_n are stopping times with respect to $(\underline{\underline{G}}_t)_{t\geq 0}$. Using the martingale property and the fact that X_t and A_t have the same intervals of constancy we get the statement of the theorem.

REMARKS. i) The assumption that A_∞ is previsible with respect to $(\underline{\underline{F}}^W_t)_{t\geq 0}$ holds, for example, if $A_\infty = \infty$ or if A_∞ is equal to a constante $c < \infty$.

ii) In the case where the A_t are stopping times with respect to $(\underline{\underline{F}}^W_s)_{s\geq 0}$ it is necessary and sufficient for A_∞ being previsible that

$$S_\infty = \inf \{ t \geq 0 : A_t = A_{t+\varepsilon} \text{ for all } \varepsilon > 0 \}$$

is previsible with respect to $(\underline{\underline{G}}^X_t)_{t\geq 0}$. The random time S_∞ describes the first moment after which the process X is constant. Especially we obtain from $S_\infty = \infty$ that A_∞ is previsible. The condition $S_\infty = \infty$ includes the case that A_t is strictly increasing.

As a corollary from theorem 1 we get the following proposition.

PROPOSITION 1. Let $Y = (Y_t, \underline{\underline{G}}_t, t\geq 0)$ be a martingale such that Y_{T_t} is F^W_t-measurable for all $t \geq 0$ and suppose again that A_∞ is previsible with respect to $(\underline{\underline{F}}^W_t)_{t\geq 0}$. Then Y can be represented as a stochastic integral of X.

A simular result as in proposition 1 is given in a paper of AL-HUSSAINI /3/.[+)] He consideres the representation of martingales Y

+) We wish to thank R.S. Liptzer for sending us this paper after the days of the conference, until thistime it escaped our notice.

on the interval (R_0, S_∞) where

$$R_0 = \inf \{t \geq 0 : X_t \neq 0\}$$

and S_∞ is defined as above, under the assumption that Y_{T_t} is $\underline{\underline{F}}_t^W$-measurable for all $t \geq 0$ but without the restriction that A_∞ is previsible and without any restriction to S_∞. We can give an example which shows that the assertion of proposition 1 does not hold without assuming the previsibility of A_∞.

In connection with proposition 1 we get another condition which guarantees that the representation property for X holds.

PROPOSITION 2. Suppose that A_∞ is previsible. Then the condition $\underline{\underline{F}}_\infty^X = \underline{\underline{F}}_\infty^W$ implies that X possesses the representation property.

As a corollary from proposition 2 we can give some other sufficient conditions for the representation property if we assume again that A_∞ is previsible.

LEMMA 1. The following conditions are equivalent:

i) $\underline{\underline{F}}_\infty^W = \underline{\underline{F}}_\infty^X$.

ii) $\underline{\underline{F}}_t^W = \underline{\underline{G}}_{T_t}^X$ for all $t \geq 0$.

iii) T_t is $\underline{\underline{F}}_t^W$-measurable for all $t \geq 0$.

iv) T_t is $\underline{\underline{F}}_\infty^W$-measurable for all $t \geq 0$.

v) A_t is a stopping time with respect to $(\underline{\underline{F}}_s^W)_{s \geq 0}$ for all $t \geq 0$.

3. EXAMPLES FOR MARTINGALES POSSESSING THE REPRESENTATION PROPERTY

Using the results from the last section we consider now some classes of martingales having the representation property.

GAUSSIAN MARTINGALES

Suppose that $\langle X \rangle_t = m(t)$ is an increasing continuous function on $[0, \infty)$ which does not depend on $\omega \in \Omega$. Then the representation property for X holds by point v) of lemma 1.

Martingales of this kind are continuous processes with independent increments, i.e. continuous Gaussian martingales.

STOPPED MARTINGALES

We consider a continuous martingale $(X_t, \underline{\underline{G}}_t, t \geq 0)$, and suppose that $(X_t, \underline{\underline{G}}_t^X, t \geq 0)$ has the representation property. Let T be a stopping time of $(\underline{\underline{G}}_t)_{t\geq 0}$. Then we define Y by $Y_t = X_{t \wedge T}$ for all $t \geq 0$. Thus Y is a martingale with respect to $(\underline{\underline{G}}_t)_{t\geq 0}$.

PROPOSITION 3. <u>The following conditions are equivalent</u>:

i) $(Y_t, \underline{\underline{G}}_t^Y, t\geq 0)$ <u>possesses the representation property</u>.

ii) T <u>is a previsible stopping time of</u> $(\underline{\underline{G}}_t^Y)_{t\geq 0}$.

iii) T <u>is a previsible stopping time of</u> $(\underline{\underline{G}}_t^X)_{t\geq 0}$.

iv) T <u>is a stopping time of</u> $(\underline{\underline{G}}_t^X)_{t\geq 0}$.

Using this result, we can construct examples for martingales which do not have the representation property. For simplicity we use the Wiener process but it is possible to use any other martingale satisfying the representation property.

Let (W^1, W^2) be a 2-dimensional Wiener process and $\underline{\underline{F}}_t^{(W^1,W^2)} = \sigma(W_s^1, W_s^2, s \leq t)$. Suppose that $s < t$ and $A \in \underline{\underline{F}}_s^{W^2}$ with the property $0 < P(A) < 1$ are fixed and set

$$T = \begin{cases} t & \omega \in A \\ s & \omega \in A^c \end{cases} .$$

Then T is a stopping time of $(\underline{\underline{F}}_t^{W^2})_{t\geq 0}$ and consequently of $(\underline{\underline{F}}_t^{(W^1,W^2)})_{t\geq 0}$, too. Therefore $(W^1_{t\wedge T}, \underline{\underline{F}}_t^{(W^1,W^2)}, t\geq 0)$ is a martingale and also $W^T = (W^1_{t\wedge T}, \underline{\underline{G}}_t^T, t\geq 0)$ where $\underline{\underline{G}}_t^T$ denotes the σ-algebra generated by $\{W^1_{s\wedge T}, s\leq t\}$.

If we assume that every martingale $Y = (Y_t, \underline{\underline{G}}_t^T, t\geq 0)$ can be represented as a stochastic integral of W^T, we can show that any stopping time of $(\underline{\underline{G}}_t^T)_{t\geq 0}$ is previsible. Consequently T has to be previsible, too, and there exists a sequence of stopping times

$(T_n)_{n \in N}$ of $(\underline{\underline{G}}_t^T)_{t \geq 0}$ with $T_n < T$ and $T_n \uparrow T$. Since $T_n < T$, T_n are also stopping times of $(\underline{\underline{F}}_t^{W^1})_{t \geq 0}$. This means that T is a stopping time of $(\underline{\underline{F}}_t^{W^1})_{t \geq 0}$ and, in particular, that T is $\underline{\underline{F}}_\infty^{W^1}$-measurable. But T is independent of $\underline{\underline{F}}_\infty^{W^1}$ by its construction and hence the assumption must be false, i.e. W^T does not have the representation property.

MARTINGALES CONSTRUCTED FROM A WIENER PROCESS

Another method to get martingales possessing the representation property is given in the next proposition.

PROPOSITION 4. _Let_ $W = (W_t, \underline{\underline{F}}_t^W, t \geq 0)$ _be a Wiener process and_ $(A_t)_{t \geq 0}$ _be an increasing family of stopping times of_ $(\underline{\underline{F}}_t^W)_{t \geq 0}$, _where_ $A_t(\omega)$ _is continuous a.s. as a function of_ t. _Define_ $X_t = W_{A_t}$ _for all_ $t \geq 0$. _Then_ $X = (X_t, \underline{\underline{F}}_t^X, t \geq 0)$ _is a local continuous martingale satisfying the representation property_.

The statement of this proposition is true because A is the increasing process of X and A_t is a previsible stopping time of $(\underline{\underline{F}}_s^W)_{s \geq 0}$ for all $t \geq 0$. Thus A_∞ is previsible, too, and the assertion follows from point v) of lemma 1.

WEAK SOLUTIONS OF STOCHASTIC DIFFERENTIAL EQUATIONS

The following example is of importance for itself, too, because we get from it an answer to the question for the uniqueness of weak solutions of stochastic differential equations.

THEOREM 2. _Let_ $(\Omega, F, P, (\underline{\underline{F}}_t)_{t \geq 0}, X, W^*)$ _be a weak solution of the equation_

$$dX_t = \sigma(X_t)\, dW_t^* , \quad X_0 = 0$$

where W^* _is a Wiener process and_ σ _is any measurable function on_ R_1 _such that_

$$\sigma^2(X_t(\omega)) > 0 \qquad P \times l - \text{a.s.}$$

where l _denotes the Lebesgue measure on_ R_1. _Then the local continuous martingale_ X _has the representation property_.

The main idea of the proof is to show that T_t is $\underline{\underline{F}}_t^W$-measurable for all t. From $dX_t = \sigma(X_t)\, dW_t^*$ follows that

$$dA_t = \sigma^2(X_t)\, dt.$$

Therefore we get

$$t = \int_0^t \sigma^{-2}(X_s)\, dA_s$$

and hence

$$T_t = \int_0^{T_t} \sigma^{-2}(X_s)\, dA_s.$$

In order to show the $\underline{\underline{F}}_t^W$-measurability of T_t we have to transform this integral such that it only depends on functions of W_s, $s \leq t$. This is possible using the method of time change, for instance, in the set-up given by DELLACHERIE in /1/. From this we get

$$T_t = \int_0^{T_{A_t}} \sigma^{-2}(X_{T_s})\, ds.$$

But in view of the condition $\sigma^2(X_t) > 0$ we know that A_t strictly increases and therefore T_t is continuous in t for $t < A_\infty$. Furthermore we get

$$T_{A_{\overline{\infty}}} = T_{A_\infty} = \infty \ .$$

By the continuity of T_t we also obtain that

$$T_{A_t} = t$$

and, consequently,

$$T_t = \int_0^t \sigma^{-2}(X_{T_s})\, ds = \int_0^t \sigma^{-2}(W_s)\, ds$$

for $t < A_\infty$, i.e. $T_t < \infty$.

For $t \uparrow A_\infty$ we can now derive

$$+\infty = T_{A_{\overline{\infty}}} = T_{A_\infty} = \int_0^{A_\infty} \sigma^{-2}\,(W_s)\, ds$$

from which follows that

$$T_t = \int_0^t \sigma^{-2}\,(W_s)\, ds$$

for $t \geq A_\infty$ and, therefore, for all $t \geq 0$.

Hence T_t is $\underline{\underline{F}}_t^W$-measurable for all $t \geq 0$.

The condition that A_∞ is previsible is valid because A_t is strictly increasing.

COROLLARY. <u>The weak solution of</u>

$$dX_t = \sigma(X_t)\, dW_t^*,$$

<u>if it exists, is unique under the condition of theorem 2, i.e. for two weak solutions</u>

$$(\Omega^i, \underline{\underline{F}}^i, P^i, (\underline{\underline{F}}_t^i)_{t\geq 0}, Z^i, W^i)$$

<u>with</u>

$$\sigma^2(Z_t^i) > 0, \quad P^i \times 1 - \text{a.s.} \quad \text{for } i=1,2$$

<u>we have</u>

$$P_{Z^1}(A) = P_{Z^2}(A)$$

<u>for all</u> $A \in \underline{\underline{\mathcal{B}}}$ <u>where</u> $\underline{\underline{\mathcal{B}}}$ <u>denotes the</u> σ<u>-algebra on the space</u> C_{R_+} <u>of continuous functions</u> x <u>on</u> R_+ <u>generated by the coordinate mappings</u> X_t <u>defined by</u> $X_t(x) = x(t)$ <u>for</u> $x \in C_{R_+}$ <u>and all</u> $t \geq 0$, <u>and</u>

$$P_{Z^i}(A) = P^i(\omega^i : Z^i(\omega^i) \in A)$$

<u>for</u> $i=1,2$,.

The corollary follows from the fact that

$$P = qP_{Z^1} + (1-q)P_{Z^2}$$

with $0 < q < 1$ represents a weak solution. To verify this we have to show that there exists a Wiener process $(B_t)_{t\geq 0}$ on $(C_{R_+}, \underline{\underline{\mathcal{B}}}, P)$ such that

$$dX_t = \sigma(X_t)\, dB_t$$

where again $X_t(x) = x(t)$. The existence of a Wiener process $(B_t)_{t\geq 0}$ satisfying this condition is provided by a theorem due to DOOB /9/: Because X is a martingale with respect to P such that $\langle X\rangle_t = \int_0^t \sigma^2(X_s)\, ds$ with $\sigma^2(X_t) > 0$ $P \times 1$ - a.s. there exists a Wiener process $(B_t)_{t\geq 0}$ satisfying $X_t = \int_0^t \sigma(X_s)\, dB_s$. Then according to theorem 2 P has the representation property and, consequently, P is an extremal point of $\underline{\underline{M}}^c_{loc}$ (X). But this contradicts the representation of P as a mixture of extremal points

$$P = qP_{Z^1} + (1-q)\, P_{Z^2}$$

with $P_{Z^1} \neq P_{Z^2}$, i.e. we get $P_{Z^1} = P_{Z^2}$.

REMARK. Suppose that $\sigma^2(x) > 0$ for all $x \in R_1$. Then we obtain the uniqueness of a weak solution of

$$dX_t = \sigma(X_t)\, dW^*_t \quad .$$

A similar result can also be found in GIHMAN and SKOROHOD /2/.

4. AN APPLICATION TO THE n-DIMENSIONAL TIME CHANGE

It is known that any n-dimensional martingale $X = (X^1_t,\dots,X^n_t)_{t\geq 0}$ with continuous orthogonal components can be reduced to a n-dimensional Wiener process by random time change. This fact was proved by KUNITA and WATANABE /7/ in the case where the increasing processes $(A^k_t)_{t\geq 0}$ are independent of k. Later F.B. KNIGHT /6/ has shown the result without this restriction on the increasing processes. However, KNIGHT's proof is very complicated and, in fact, he proved a somewhat weaker theorem, namely, that there exists a larger probability space on which the result holds.

Using the statement of theorem 1 we can give a simple proof for this fact. For simplicity we suppose $A^k_\infty = \infty$ for all $k \leq n$.

THEOREM 3. <u>Let</u> $X = (X^1_t,\dots,X^n_t)_{t\geq 0}$ <u>be a martingale</u> +) <u>with continuous components such that</u>

$$\langle X^k, X^l \rangle = 0$$

+) With respect to an increasing family $(\underline{\underline{G}}_t)_{t\geq 0}$ of σ-algebras.

for $k \neq 1$. Set

$$T_t^k = \inf \{ s \geq 0 : A_s^k > t \}$$

and

$$W_t^k = X^k_{T_t^k} \quad .$$

Then $W = (W_t^1, \ldots, W_t^n)$ is a Wiener process.

For the proof we use the fact that any n-dimensional martingale with continuous components $(X_t^k)_{t \geq 0}$ such that

$$\langle X^k, X^l \rangle_t = \delta_{ij} \cdot t$$

holds for all $t \geq 0$ and $i,k \leq n$ is a Wiener process. Suppose, for example, that n=2. The crucial step is then to show that +) $(X_t^1, \underline{\underline{G}}_t \vee \underline{\underline{F}}_\infty^{W^2}, t \geq 0)$ is a martingale, because from this easily follows that (W^1, W^2) and $W^1 . W^2$ are martingales with respect to $(\underline{\underline{F}}_t^{(W^1, W^2)})_{t \geq 0}$. But for showing that $(X_t^1)_{t \geq 0}$ is a martingale with respect to $(\underline{\underline{G}}_t \vee \underline{\underline{F}}_\infty^{W^2})_{t \geq 0}$ it is sufficient to show that X^1 is orthogonal to any martingale Y defined by $Y_t = E(y/\underline{\underline{G}}_t)$ with y $\underline{\underline{F}}_\infty^{W^2}$-measurable and bounded. In view of theorem 1 we get for such martingales

$$Y_t = Y_o + \int_0^t f(s, \omega) \, dX_s^1$$

and consequently

$$\langle Y, X^1 \rangle_t = \int_0^t f(s, \omega) \, d \langle X^1, X^2 \rangle_s = 0.$$

+) For two σ-algebras $\underline{\underline{G}}_1$ and $\underline{\underline{G}}_2$, by $\underline{\underline{G}}_1 \vee \underline{\underline{G}}_2$ we denote the smallest σ-algebra containing $\underline{\underline{G}}_1$ and $\underline{\underline{G}}_2$.

REFERENCES

/1/ C. DELLACHERIE, Capacités et processus stochastiques, Springer Verlag Berlin, Heidelberg, New York (1972)

/2/ I.I. GIHMAN, A.N. SKOROHOD, The theory of stochastic processes, vol. III, Moscow, Nauka (1975), (in Russian)

/3/ A.N. AL-HUSSAINI, Stochastic integral representations of some martingales, J. Math. Anal. Appl. 58, 637 - 646 (1977)

/4/ J. JACOD, M. YOR, Étude des solutions extrémales et représentation integrale des solutions pour certain problèmes de martingales, Z. Wahrscheinlichkeitstheorie und Verw. Gebiete 38, 83 - 125 (1977)

/5/ L.J. GALČUK, Representation of some martingales, Teor. Verojatnost. i Primenen. XXI 3, 613 - 620 (1976), (in Russian)

/6/ F.B. KNIGHT, A reduction of continuous square-integrable martingales to Brownian motion, Lecture Notes in Mathematics 190, 19 - 31 (1970)

/7/ H. KUNITA, S. WATANABE, On square-integrable martingales, Nagoya Math. J. 30, 209 - 245 (1967)

/8/ R.S. LIPCER, On the representation of local martingales, Teor. Verojatnost. i Primenen. XXI 4, 718 - 726 (1976) (in Russian)

/9/ R.S. LIPCER, A.N. SHIRYAYEV, Statistics of stochastic processes, Moscow, Nauka (1974) (in Russian)

DIFFUSION IN REGIONS WITH MANY SMALL HOLES

G. C. Papanicolaou* and S. R. S. Varadhan**

Courant Institute of Mathematical Sciences, New York University

1. FORMULATION AND STATEMENT OF MAIN THEOREM

Let D be a bounded open set containing the origin, having C^2 boundary and with diameter less than or equal to one. For each N = 1,2,..., let $y_1^{(N)}, y_2^{(N)} \dots y_N^{(N)}$ be points in R^3 and define sets $D_i^{(N)}$ by

(1.1) $$D_i^{(N)} = \{x \in R^3 \mid N(x - y_i^{(N)}) \in D\}, \quad i = 1,2,\dots,N.$$

We shall call the set $D_i^{(N)}$ the hole centered at $y_i^{(N)}$ with diameter less than or equal to N^{-1}. Let $G^{(N)}$ denote the region

(1.2) $$G^{(N)} = R^3 - \bigcup_{i=1}^{N} D_i^{(N)}$$

which is R^3 with holes of diameter $\leq N^{-1}$ centered at $y_1^{(N)},\dots,y_N^{(N)}$.

We shall analyze the asymptotic behavior of $u^{(N)}(x,t)$ as $N \to \infty$ which is the solution of

(1.3)
$$\frac{\partial u^{(N)}(x,t)}{\partial t} = \frac{1}{2}\Delta u^{(N)}(x,t), \qquad t > 0,\ x \in G^{(N)}$$
$$u^{(N)}(x,t) = 0, \qquad t > 0,\ x \in \partial G^{(N)} = \bigcup_{i=1}^{N} \partial D_i^{(N)},$$
$$u^{(N)}(x,0) = f(x), \qquad x \in G^{(N)},$$

with f(x) a given bounded continuous function with compact support in R^3. The precise meaning in which problem (1.3) is taken is as follows.

Let P_x be the Brownian motion measure on continuous trajectories on R^3 starting from $x \in R^3$; its transition probability density p(t,x,y) is given by

(1.4) $$p(t,x,y) = (2\pi t)^{-3/2} \exp\left\{-\frac{|x-y|^2}{2t}\right\}.$$

Let $E_x\{\cdot\}$ denote expectation with respect to P_x. Then we set by definition

(1.5) $$u^{(N)}(x,t) = E_x\left\{f(x(t))\ \chi_{\{\tau^{(N)} > t\}}\right\}.$$

---

* Research supported under Grant AFOSR-78-3668 with the U. S. Air Force.
**Research supported under Grant No. NSF-MCS-77-02687 with the National Science Fndn.

Here χ_A is the indicator function of a set A of trajectories and equals one if the trajectory is in A and zero if not. The stopping time $\tau^{(N)}$ is defined by

$$\tau^{(N)} = \min\left\{\tau_1^{(N)}, \tau_2^{(N)}, \ldots, \tau_N^{(N)}\right\} \tag{1.6}$$

where $\tau_j^{(N)}$ is the first time the trajectory reaches $\partial D_j^{(N)}$.

For the theorem that follows we shall need the following two hypotheses regarding the location of the hole centers $y_1^{(N)}, \ldots, y_N^{(N)}$.

First, there exists a bounded continuous function of compact support $V(x) \geq 0$ such that for each $\phi \in C_0(R^3)$ (continuous and with compact support)

$$\lim_{N\uparrow\infty} \frac{1}{N} \sum_{i=1}^{N} \phi(y_i^{(N)}) = \int_{R^3} \phi(x)\, V(x)\, dx . \tag{1.7}$$

Second, we assume that

$$\lim_{N\uparrow\infty} \frac{1}{N^2} \sum_{\substack{i,j=1\\ i\neq j}}^{N} \frac{1}{|y_i^{(N)} - y_j^{(N)}|} = \int_{R^3}\int_{R^3} \frac{V(x)\, V(y)}{|x-y|}\, dx\, dy \tag{1.8}$$

Let α denote the capacitance of the set D. Since ∂D is C^2, α is given by

$$\alpha = \int_{\partial D} \frac{\partial w(x)}{\partial \hat{n}(x)}\, dS(x) \tag{1.9}$$

where $\hat{n}(x)$ is the unit inner normal at $x \in \partial D$ and dS denotes element of surface area. The capacitory potential $w(x)$ is the solution of

$$\begin{aligned} \Delta w(x) &= 0 , & x &\in R^3 - D, \\ w(x) &= 1 , & x &\in \partial D , \\ w(x) &\to 0 \quad \text{as} & |x| &\to \infty . \end{aligned} \tag{1.10}$$

THEOREM 1. <u>Under hypotheses</u> (1.7) <u>and</u> (1.8), <u>given</u> $\varepsilon > 0$ <u>and</u> $T < \infty$ <u>there exists an integer</u> N_0 <u>such that for all</u> $N \geq N_0$ <u>there is a set</u> $G_\varepsilon^{(N)} \subset G^{(N)}$ <u>such that</u>

$$\sup_{0\leq t\leq T}\ \sup_{x\in G_\varepsilon^{(N)}} |u^{(N)}(x,t) - u(x,t)| < \varepsilon \tag{1.11}$$

<u>and</u>

$$\mathrm{vol}(R^3 - G_\varepsilon^{(N)}) < \varepsilon . \tag{1.12}$$

Here $u(x,t)$ is given by

$$u(x,t) = E_x\left\{f\big(x(t)\big)\ \exp\left[-\alpha\int_0^t V\big(x(s)\big)\,ds\right]\right\}. \tag{1.13}$$

Remarks.

1. The limit function $u(x,t)$ is the solution of

$$\begin{aligned} u_t &= \frac{1}{2}\Delta u - \alpha V u\,, && t > 0 \\ u(0,x) &= f(x)\,, && x \in R^3, \end{aligned} \tag{1.14}$$

by the Feynman-Kac formula.

2. The above theorem implies that

$$\sup_{0\le t\le T} \int_{G^{(N)}} |u^{(N)}(x,t) - u(x,t)|^2\,dx \to 0 \tag{1.15}$$

as $N \to \infty$ for each $T < \infty$.

3. Theorems valid in bounded regions with various boundary conditions can also be obtained but are not discussed further here.

If we think of $u^{(N)}(x,t)$ as the temperature at x at time t of a solid occupying $G^{(N)}$ of conductivity $1/2$ then, the effect of keeping the holes $D_i^{(N)}$ in $G^{(N)}$ at temperature zero appears in the limit $N \to \infty$ as a continuous heat removal rate $\alpha V(x)$ which is the limiting capacitance per unit volume. The scaling is such that this quantity has a finite limit. It is clear that condition (1.7) is a natural one; it gives the limiting density of hole centers $V(x)$. The meaning of condition (1.8) is less immediate although it is clear that it puts a restriction on how close hole centers can be from each other.

We shall consider next two particular cases in which (1.8) follows from simpler conditions.

Suppose first that the holes $D_i^{(N)}$ are spheres with centers at $y_1^{(N)},\dots,y_N^{(N)}$ and diameters equal to N^{-1}. Suppose that there is a constant $\gamma > 0$ such that for all $N = 1,2,\dots$

$$|y_i^{(N)} - y_j^{(N)}| \ge \gamma\, N^{-1/3}\,, \qquad i,j = 1,2,\dots,N. \tag{1.16}$$

We assume that((1.7) holds and note that now the capacitance $\alpha = 2\pi$ (the capacitance of a sphere of unit diameter). <u>We shall show that</u> (1.16) <u>implies</u> (1.8).

Let $B_j^{(N)}$ denote the sphere of radius $\gamma N^{-1/3}/2$ with center at $y_i^{(N)}$. For a fixed $y_i^{(N)}$ satisfying (1.16) the function $|y_i^{(N)} - y|^{-1}$ is harmonic for $y \neq y_i^{(N)}$. Thus by the mean value theorem

$$(1.17)\qquad |y_i^{(N)} - y_j^{(N)}|^{-1} = \frac{1}{b^{(N)}} \int_{B_j^{(N)}} \frac{dy}{|y_i^{(N)} - y|}\ , \qquad j \neq i$$

where $b^{(N)} = \pi\gamma^3/6N$ is the volume of $B_j^{(N)}$. With y fixed in $B_j^{(N)}$ we may use (1.17) again for $|y_i^{(N)} - y|^{-1}$. This yields

$$|y_i^{(N)} - y_j^{(N)}|^{-1} = \frac{1}{(b^{(N)})^2} \int_{B_i^{(N)}} \int_{B_j^{(N)}} \frac{dx\ dy}{|x - y|}$$

and hence

$$(1.18)\qquad \frac{1}{N^2} \sum_{\substack{i,j=1\\ i\neq j}}^{N} \frac{1}{|y_i^{(N)} - y_j^{(N)}|} = \sum_{\substack{i,j=1\\ i\neq j}}^{N} \frac{1}{(Nb^{(N)})^2} \int_{B_i^{(N)}} \int_{B_j^{(N)}} \frac{dx\ dy}{|x - y|}$$

To prove that (1.7) and (1.16) imply (1.8) it follows from (1.18) that we must show that

$$(1.19)\qquad \lim_{N\uparrow\infty} \frac{1}{N^2} \sum_{i,j=1}^{N} \frac{1}{(Nb^{(N)})^2} \int_{B_i^{(N)}} \int_{B_j^{(N)}} \frac{dx\ dy}{|x - y|} = \int\int \frac{V(x)\ V(y)}{|x - y|}\, dx\ dy$$

The addition of the diagonal terms on the right side of (1.19) makes a negligible contribution for N large. To show (1.19) define

$$(1.20)\qquad \psi^{(N)}(x) = \frac{1}{N} \sum_{i=1}^{N} \frac{1}{b^{(N)}} \chi_{B_i^{(N)}}(x)$$

and note that $\psi^{(N)}(x)$ is uniformly bounded by a constant independent of N and x. In terms of $\psi^{(N)}$ we rewrite (1.19) as

$$(1.21)\qquad \lim_{N\uparrow\infty} \int\int \frac{\psi^{(N)}(x)\ \psi^{(N)}(y)}{|x - y|}\, dx\ dy = \int\int \frac{V(x)\ V(y)}{|x - y|}\, dx\ dy$$

which we prove as follows.

Let $\pi^{(N)}$ be a measure defined on the Borel sets of R^3 by

$$(1.22)\qquad \pi^{(N)}(A) = \frac{1}{N}\ [\#\ \text{of}\ i = 1,2,\ldots,N\ \text{for which}\ y_i^{(N)} \in A]\ .$$

Hypothesis (1.7) says that $\pi^{(N)}$ converges weakly as $N \to \infty$ to the measure π whose density with respect to Lebesgue measure is $V(x)$. Hence the product measure $\pi^{(N)} \times \pi^{(N)}$ on the Borel sets of R^6 converges weakly to $\pi \times \pi$. Let $\sigma^{(N)}$ be the uniform measure on the ball in R^3 centered at the origin with radius $\gamma N^{-1/3}/2$. Then, if $*$ denotes convolution and $\pi \cdot \phi$ stands for the integral of ϕ with respect to π, we have

$$(1.23) \quad \left(\pi^{(N)} * \sigma^{(N)}\right) \times \left(\pi^{(N)} * \sigma^{(N)}\right) \cdot \phi = \int\int \psi^{(N)}(x) \; \psi^{(N)}(y) \; \phi(x,y) \; dx \; dy$$

for any $\phi \in C_0(R^6)$. It is easily verified that if $\pi^{(N)} \to \pi$ weakly then $\pi^{(N)} * \sigma^{(N)} \to \pi$ weakly also. Hence by (1.23) we have

$$(1.24) \quad \lim_{N\to\infty} \iint \psi^{(N)}(x) \; \psi^{(N)}(y) \; \phi(x,y) \; dx \; dy = \iint V(x) \; V(y) \; \phi(x,y) \; dx \; dy$$

for each $\phi \in C_0(R^6)$. Now we approximate the function $|x - y|^{-1}$ by $(|x-y|+\delta)^{-1}$ with $\delta > 0$. In order that (1.21) be valid (hence (1.19) and hence (1.8)) it suffices to note that the $\psi^{(N)}$ are bounded uniformly in N and x and for any compact set $K \subset R^3$

$$\int_K \int_K \left(\frac{1}{|x - y|} - \frac{1}{|x - y| + \delta} \right) dx \; dy \to 0 \quad \text{as} \quad \delta \downarrow 0 .$$

We have thus shown that (1.7) and (1.16) imply (1.8).

The second case we consider in which (1.8) can be verified from other conditions is when <u>the hole centers are distributed randomly</u> as follows.

For each $N = 1,2,\ldots$, let $y_1^{(N)}, y_2^{(N)}, \ldots, y_N^{(N)}$ be independent identically distributed random variables in R^3 with density $V(x)$ that is,

$$(1.25) \quad \text{Prob}\left\{ y_i^{(N)} \in A \right\} = \int_A V(x) \; dx$$

for any Borel subset A in R^3. We assume that $V(x)$ is bounded, continuous, has compact support and with no loss in generality

$$\int_{R^3} V(x) \; dx = 1$$

We denote expectation with respect to (1.25) by $< \cdot >$ to distinguish it from expectation relative to P_x. Thus, for any $\phi \in C_0(R^3)$

$$<\phi(y_i^{(N)})> = \int \phi(y)\ V(y)\ dy\ , \qquad i = 1,2,\ldots,N. \tag{1.26}$$

and by the strong law of large numbers

$$\lim_{N\to\infty} \frac{1}{N} \sum_{i=1}^{N} \phi(y_i^{(N)}) = \int V(y)\ \phi(y)\ dy \tag{1.27}$$

with probability one. Therefore (1.7) holds with probability one.

Condition (1.8) holds now in probability. To see this we note that for any $1 \leq i < j \leq N$

$$<|y_i^{(N)} - y_j^{(N)}|^{-1}> = \int\int \frac{V(x)\ V(y)}{|x - y|}\ dx\ dy \tag{1.28}$$

and

$$<|y_i^{(N)} - y_j^{(N)}|^{-2}> = \iint \frac{V(x)\ V(y)}{|x - y|^2}\ dx\ dy < \infty \tag{1.29}$$

Let

$$S_N = \frac{2}{N(N-1)} \sum_{1\leq i<j\leq N} \left\{ |y_i^{(N)} - y_j^{(N)}|^{-1} - \int\int \frac{V(x)\ V(y)}{|x - y|}\ dx\ dy \right\} .$$

Using (1.28) and (1.29) we verify that $<S_N^2> \to 0$ as $N \to \infty$. This implies that for each $\varepsilon > 0$ and $\delta > 0$ there is an N_0 such that for all $N \geq N_0$

$$\text{Prob}\left\{ |S_N| > \varepsilon \right\} < \delta\ , \tag{1.30}$$

that is, (1.8) holds in probability.

Let $(\Omega^{(N)}, F^{(N)}, P^{(N)})$ be a probability space on which $y_1^{(N)},\ldots,y_N^{(N)}$ are defined. If $\omega \in \Omega^{(N)}$ then $u^{(N)}(x,t) = u^{(N)}(x,t,\omega)$ in this case. Theorem 1 takes the the following form.

THEOREM 2. <u>When the hole centers are independent identically distributed random variables with density</u> $V(x)$ <u>then given</u> $\varepsilon > 0$, $\delta > 0$ <u>and</u> $T < \infty$ <u>there is an</u> N_0 <u>such that for all</u> $N \geq N_0$ <u>there is a set</u> $\Omega_{\varepsilon,\delta}^{(N)} \subset \Omega^{(N)}$ <u>and a set</u> $G_\varepsilon^{(N)}(\omega) \subset G^{(N)}(\omega)$, $\omega \in \Omega_{\varepsilon,\delta}^{(N)}$, <u>such that</u>

$$\sup_{0\leq t\leq T}\ \sup_{x\in G_\varepsilon^{(N)}(\omega)} |u^{(N)}(x,t,\omega) - u(x,t)| < \varepsilon\ , \qquad \omega \in \Omega_{\varepsilon,\delta}^{(N)}\ , \tag{1.31}$$

and

$$\text{(1.32)} \qquad \mathrm{vol}\left[R^3 - G_\varepsilon^{(N)}(\omega)\right] < \varepsilon , \qquad \omega \in \Omega_{\varepsilon,\delta}^{(N)} ,$$

while

$$\text{(1.33)} \qquad \mathrm{Prob}\left\{\Omega^{(N)} - \Omega_{\varepsilon,\delta}^{(N)}\right\} < \delta .$$

Here $u(x,t)$ is given by (1.13) as in Theorem 1.

Remarks.

1. This theorem implies that (1.15) holds in probability, that is, for each $\delta > 0$ and $T < \infty$,

$$\text{(1.34)} \qquad \mathrm{Prob}\left\{\sup_{0\le t\le T} \int_{G^{(N)}} |u^{(N)}(x,t) - u(x,t)|^2 \, dx > \delta\right\} \to 0 \quad \text{as} \quad N \to \infty,$$

where $u^{(N)}(x,t) = u^{(N)}(x,t,\omega)$, $\omega \in \Omega^{(N)}$.

2. The proof of Theorem 2 is immediate from Theorem 1 and the validity of (1.8) in probability. The set $\Omega_{\varepsilon,\delta}^{(N)}$ is selected so that $|S_N|$ (cf. (1.30)) is suitably small and thereafter randomness plays no role whatsoever. The estimates are obtained for each $\omega \in \Omega_{\varepsilon,\delta}^{(N)}$ as in the deterministic case.

The content of Theorem 2, in the form (1.34), was obtained by Kac [1] using properties of the Wiener sausage. It was also obtained by Rauch and Taylor in [2].

A comprehensive analysis of problems of the form considered here is given in [3] using analytical methods.

2. PROOF OF THEOREM 1

We shall show the following fact. Given $\varepsilon > 0$ and $t_0 < \infty$ there exists an integer N_0 such that for all $N \ge N_0$ there is an open set $G_\varepsilon^{(N)} \subset G^{(N)}$ such that

$$\text{(2.1)} \qquad \left|P_x\left\{\tau^{(N)} \le t\right\} - \alpha \int_0^t \int V(y)\, p(s,x,y)\, dy\, ds\right| < \varepsilon + g(t)$$

for all $x \in G_\varepsilon^{(N)}$ and $0 \le t \le t_0$ and moreover

$$\text{(2.2)} \qquad \mathrm{vol}\left(R^3 - G_\varepsilon^{(N)}\right) < \varepsilon ,$$

and

$$\overline{\lim_{t\downarrow 0}}\, t^{-1} g(t) = 0 . \tag{2.3}$$

Before proving this we shall use it to complete the proof of Theorem 1.

We define on $\hat{C}(R^3)$, the bounded continuous functions on R^3 that vanish at infinity, three semigroups as follows.

$$(T_t f)(x) = E_x\left\{f(x(t))\right\} \tag{2.4}$$

$$(T_t^{(N)} f)(x) = E_x\left\{f\big(x(t)\big)\ \chi_{\{\tau_t^{(N)} > t\}}\right\} \tag{2.5}$$

$$(T_t^V f)(x) = E_x\left\{f\big(x(t)\big)\ \exp\left(-\alpha \int_0^t V\big(x(s)\big)\, ds\right)\right\} . \tag{2.6}$$

Let $\Delta > 0$ be fixed; it will be chosen appropriately later. We rewrite the left side of (1.11).

$$\begin{aligned} I_\varepsilon^{(N)} &\equiv \sup_{0\le t\le t_0}\ \sup_{x\in G_\varepsilon^{(N)}} \left|T_t^{(N)} f(x) - T_t^V f(x)\right| \\ &= \sup_{0\le t\le t_0}\ \sup_{x\in G_\varepsilon^{(N)}} \Bigg\{\Bigg|\sum_{k=1}^{[t/\Delta]} T_{(k-1)\Delta}^{(N)} [T_\Delta^{(N)} - T_\Delta^V] T^V_{([t/\Delta]-k)\Delta} f(x) \\ &\quad + \left(T_t^{(N)} - T^{(N)}_{\Delta[t/\Delta]}\right) f(x) + \left(T_t^V - T^V_{\Delta[t/\Delta]}\right) f(x)\Bigg|\Bigg\} \end{aligned} \tag{2.7}$$

Here $[t/\Delta]$ denotes the integer part of t/Δ. Since both $T_t^{(N)}$ and T_t^V are contractions, (2.7) can be estimated as follows.

$$\begin{aligned} I_\varepsilon^{(N)} &\le 2\|f\| \sum_{k=2}^{[t_0/\Delta]} \sup_{x\in G_\varepsilon^{(N)}} T_{(k-1)\Delta}\, \chi_{[R^3 - G_\varepsilon^{(N)}]}(x) \\ &\quad + \frac{t_0}{\Delta} \sup_{0\le t\le t_0}\ \sup_{x\in G_\varepsilon^{(N)}} \left|\left(T^{(N)} - T^V\right) T_t^V f(x)\right| + 2 \sup_{0\le t\le\Delta}\ \sup_{x\in G_\varepsilon^{(N)}} \left|T_t^{(N)} f(x) - f(x)\right| \\ &\quad + 2\|f\| \sup_{\Delta\le t\le t_0}\ \sup_{x\in G_\varepsilon^{(N)}} T_{\Delta[t/\Delta]}\, \chi_{[R^3 - G_\varepsilon^{(N)}]}(x) + \sup_{0\le t\le\Delta}\ \sup_{x\in G} \left|T_t^V f(x) - f(x)\right| . \end{aligned} \tag{2.8}$$

Here $\|f\| = \sup_x |f(x)|$. Using the fact that $T_t^{(N)} g \le T_t g$ for $g \ge 0$ we may estimate (2.8) further by

$$(2.9)\quad I_\varepsilon^{(N)} \le 2\|f\|\left(\frac{t_0}{\Delta}+1\right)\sup_{\Delta\le t\le t_0}\sup_{x\in G_\varepsilon^{(N)}} T_t\,\chi_{[R^3-G_\varepsilon^{(N)}]}(x)$$

$$+\frac{t_0}{\Delta}\sup_{0\le t\le t_0}\sup_{x\in G_\varepsilon^{(N)}}\left|\left(T_\Delta^{(N)}-T_\Delta^V\right)T_t^V f(x)\right| + 2\sup_{0\le t\le\Delta}\sup_{x\in G_\varepsilon^{(N)}}\left|T_t^{(N)}f(x)-f(x)\right|$$

$$+\sup_{0\le t\le\Delta}\sup_{x\in G}\left|T_t^V f(x)-f(x)\right| .$$

We estimate next each term on the right side of (2.9) starting with the first term. We have

$$(2.10)\quad \sup_{\Delta\le t\le t_0}\sup_{x\in G_\varepsilon^{(N)}} T_t\,\chi_{[R^3-G_\varepsilon^{(N)}]}(x) \le \theta(\Delta)\ \mathrm{vol}\,(R^3-G_\varepsilon^{(N)})$$

where $\theta(\Delta)\to\infty$ as $\Delta\to\infty$ but is finite for $\Delta>0$ and fixed. Thus the first term on the right of (2.9) can be controlled for $\Delta>0$ fixed by letting N be large and using (2.2).

Let us consider the second term on the right side of (2.9). Let $V(x)\le C$. Then

$$(2.11)\quad \sup_{x\in G_\varepsilon^{(N)}}\left|T_\Delta^{(N)}f(x)-T_\Delta^V f(x)\right| =$$

$$=\sup_{x\in G_\varepsilon^{(N)}}\left|E_x\{f(x(\Delta))\,\chi_{\{\tau^{(N)}\le\Delta\}}\} - E_x\left\{\left(1-\exp\left(-\alpha\int_0^\Delta V(x(s))\,ds\right)\right)f(x(\Delta))\right\}\right|$$

$$\le \sup_{x\in G_\varepsilon^{(N)}}\left|E_x\left\{\left(1-\exp\left(-\alpha\int_0^\Delta V(x(s))\,ds\right)-\alpha\int_0^\Delta V(x(s))\,ds\right)f(x(\Delta))\right\}\right|$$

$$+\sup_{x\in G_\varepsilon^{(N)}}\left|E_x\left\{f(x(\Delta))\,\chi_{\{\tau^{(N)}\le\Delta\}}\right\} - E_x\left\{\alpha\int_0^\Delta V(x(s))\,ds\, f(x(\Delta))\right\}\right|$$

$$\le \alpha C\Delta^2\|f\| + \|f\|\sup_{x\in G_\varepsilon^{(N)}}\left|P_x\left\{\tau^{(N)}\le\Delta\right\} - E_x\left\{\int_0^\Delta \alpha V(x(s))\,ds\right\}\right|$$

$$+\sup_{x\in G_\varepsilon^{(N)}}\left|E_x\left\{\left(f(x(\Delta))-f(x)\right)\chi_{\{\tau^{(N)}\le\Delta\}}\right\}\right| + \alpha C\Delta\sup_x E_x\left\{\left|f(x(\Delta))-f(x)\right|\right\}.$$

Let $\omega(f,\delta)$ be the modulus of continuity of f,

$$(2.12)\quad \omega(f,\delta) = \sup_{|x-y|\le\delta}|f(x)-f(y)|$$

and note that $\omega(f,\delta)\to 0$ as $\delta\to 0$ since $f\in\hat{C}(R^3)$. For each $\delta>0$ we have

$$(2.13)\quad \sup_{x\in G_\varepsilon^{(N)}} \left| E_x\left\{\left(f\left(x(\Delta)\right) - f(x)\right) \chi_{\{\tau^{(N)} \leq \Delta\}}\right\}\right|$$

$$\leq \omega(f,\delta) \sup_{x\in G_\varepsilon^{(N)}} \left| P_x\left\{\tau^{(N)} \leq \Delta\right\} - E_x\left\{ \int_0^\Delta \alpha V(x(s))\, ds\right\}\right|$$

$$+ \alpha C\Delta\ \omega(f,\delta) + 2\|f\|\ P_x\left\{\left|x(\Delta) - x\right| > \delta\right\}$$

and similarly

$$(2.14)\qquad E_x\left\{\left|f\left(x(\Delta)\right) - f(x)\right|\right\} \leq \omega(f,\delta) + 2\|f\|\ P_x\left\{|x(\Delta) - x| > \delta\right\} .$$

Using (2.13) and (2.14) in (2.11) we see that

$$(2.15)\quad \sup_{x\in G_\varepsilon^{(N)}} \left|T_\Delta^{(N)} f(x) - T_\Delta^V f(x)\right|$$

$$\leq \alpha C\Delta^2 \|f\| + \left(\|f\| + \omega(\delta,f)\right) \sup_{x\in G_\varepsilon^{(N)}} \left|P_x\left\{\tau^{(N)} \leq \Delta\right\} - E_x\left\{\alpha \int_0^\Delta V(x(s))\, ds \right\}\right|$$

$$+ 2\alpha C\Delta\omega(\delta,f) + 2\left(1 + \alpha C\Delta\right)\|f\|\ P_x\left\{|x(\Delta) - x| > \delta\right\}.$$

Now we note that the third term on the right side of (2.9) is precisely the one estimated by (2.13). We also note that since $f \in \hat{C}(R^3)$ and V is bounded continuous and of compact support in R^3.

$$(2.16)\qquad \sup_{0\leq t\leq t_0} \omega(T_t^V f,\delta) \to 0 \quad \text{as} \quad \delta \to 0$$

and

$$(2.17)\qquad \sup_{0\leq t\leq \Delta}\ \sup_x \left|T_t^V f(x) - f(x)\right| \to 0 \quad \text{as} \quad \Delta \to 0 .$$

In addition, since P_x is the Brownian motion measure

$$(2.18)\qquad \sup_x \Delta^{-1} P_x\left\{|x(\Delta) - x| > \delta\right\} \to 0 \quad \text{as} \quad \Delta \downarrow 0 \quad \text{for each} \quad \delta > 0.$$

If we replace f by $T_t^V f$ in (2.15) use (2.13), (2.16), (2.17), (2.18) and in particular (2.1) and (2.3) we see that all terms on the right side of (2.9) can be made small by first choosing δ small and fixing it, next choosing Δ small enough and fixing it and finally choosing N large enough.

This completes the proof of Theorem 1 given the validity of (2.1), (2.2), and (2.3).

We turn now to the proof of (2.1), (2.2) and (2.3) which is given in a series of lemmas. The first lemma is a "one hole" estimate, the second gives the upper half of (2.1) and the third lemma gives the lower half of (2.1).

LEMMA 1. Let D be a bounded open set in R^3 containing the origin, having C^2 boundary and capacitance α. Let

$$D^{(N)} = \{x \in R^3 \mid N(x-y) \in D\}\ , \qquad y \in R^3$$

and let $\sigma^{(N)}$ be the first time the Brownian motion reaches $\partial D^{(N)}$ starting from $x \in R^3 - D^{(N)}$. Then

$$N\, P_x\left\{\sigma^{(N)} \le t\right\} \to \alpha \int_0^t p(s,x,y)\, ds \quad \text{as} \quad N \to \infty \tag{2.19}$$

uniformly on compact sets of x and y with $x \neq y$ where $p(s,x,y)$ is given by (1.4).

Proof: We may take the hole center y to be the origin. Let

$$v^{(N)}(x,t) = NP_x\left\{\sigma^{(N)} \le t\right\} \tag{2.20}$$

and

$$\hat{v}^{(N)}(x,\lambda) = \int_0^\infty e^{-\lambda t}\, v^{(N)}(x,t)\, dt\ , \qquad \lambda > 0\ . \tag{2.21}$$

The result (2.19) is equivalent to showing that

$$\hat{v}^{(N)}(x,\lambda) \to \frac{\alpha}{\lambda\, 4\pi|x|} \exp\left(-\sqrt{2\lambda}\ |x|\right) \tag{2.22}$$

for each $\lambda > 0$, uniformly on compact sets of x excluding $x = 0$. By change of variables we find that

$$\hat{v}^{(N)}(x,\lambda) = \frac{N}{\lambda}\, w\left(Nx, \frac{\lambda}{N^2}\right) \tag{2.23}$$

where $w(x,\mu)$ satisfies

$$\begin{aligned} 2\mu w - \Delta w &= 0 \quad \text{in} \quad R^3 - D \\ w(x,\mu) &= 1 \quad \text{for} \quad x \in \partial D\ . \end{aligned} \tag{2.24}$$

If $\hat{n}$ denotes the unit inner normal to D at the boundary, Green's identity gives

$$\frac{N}{\lambda}\, w\left(Nx, \frac{\lambda}{N^2}\right) = \frac{1}{4\pi\lambda} \int_{\partial D} \frac{\exp(-\sqrt{2\lambda}|x-N^{-1}y|)}{|x-N^{-1}y|} \frac{\partial w(y,N^{-2}\lambda)}{\partial\hat{n}}\, dS \tag{2.25}$$

$$\to \frac{1}{4\pi\lambda} \frac{\exp\{-\sqrt{2\lambda}|x|\}}{|x|} \int_{\partial D} \frac{\partial w(y,0)}{\partial \hat{n}}\, dS = \frac{\alpha}{4\pi\lambda} \frac{\exp(-\sqrt{2\lambda}\ |x|)}{|x|}$$

uniformly on compact x sets excluding $x = 0$. Note that we have used (1.9). The proof of Lemma 1 is complete.

LEMMA 2. <u>Given</u> $\varepsilon > 0$ <u>and</u> $t_0 < \infty$ <u>there exists an</u> N_0 <u>such that for all</u> $N \geq N_0$ <u>there is a set</u> $G_\varepsilon^{(N)} \subset G^{(N)}$ <u>such that</u>

$$P_x\{\tau^{(N)} \leq t\} \leq \int_0^t \int \alpha V(y)\, p(s,x,y)\, dy\, ds + \varepsilon \tag{2.26}$$

<u>for all</u> $x \in G_\varepsilon^{(N)}$ <u>and</u> $0 \leq t \leq t_0$ <u>and</u>

$$\mathrm{vol}\,(R^3 - G_\varepsilon^{(N)}) < \varepsilon. \tag{2.27}$$

<u>Proof:</u> Let $\varepsilon > 0$ be given and fix $\delta > 0$ to be chosen later. We write

$$\begin{aligned} P_x\left\{\tau^{(N)} \leq t\right\} &\leq \sum_{i=1}^{N} P_x\left\{\tau_i^{(N)} \leq t\right\} \\ &= \sum_{\substack{i=1 \\ |x-y_i^{(N)}|>\delta}}^{N} P_x\left\{\tau_i^{(N)} \leq t\right\} + \sum_{\substack{i=1 \\ |x-y_i^{(N)}|\leq\delta}}^{N} P_x\left\{\tau_i^{(N)} \leq t\right\}. \end{aligned} \tag{2.28}$$

We estimate first the second term on the right of (2.28).

From Lemma 1 and the estimate

$$\int_0^t p(s,x,y)\, ds \leq c_2\, |x - y|^{-1} \exp\left(- |x - y|^2/4t\right)$$

we conclude that there is a constant c_1 independent of x, $y_i^{(N)}$ and N such that

$$P_x\left\{\tau_i^{(N)} \leq t\right\} \leq c_1\, N^{-1} |x - y_i^{(N)}|^{-1} \exp\left(- |x - y_i^{(N)}|^2/4t\right) \tag{2.29}$$

for all $t < \infty$. Using (2.29) we obtain for all t

$$\sum_{\substack{i=1 \\ |x-y_i^{(N)}|\leq\delta}}^{N} P_x\left\{\tau^{(N)} \leq t\right\} \leq \frac{c_1}{N} \sum_{\substack{i=1 \\ |x-y_i^{(N)}|\leq\delta}}^{N} |x - y_i^{(N)}|^{-1} \tag{2.30}$$

Now we choose $\delta = \delta(\varepsilon)$ so that

$$\int_{|z|\leq\delta} |z|^{-1}\, dz \leq \frac{\varepsilon^2}{2c_1} \tag{2.31}$$

Then

$$(2.32)\quad \int dx\, \frac{c_1}{N} \sum_{\substack{i=1 \\ |x-y_i^{(N)}|\le\delta}}^{N} |x-y_i^{(N)}|^{-1} = \frac{c_1}{N} \sum_{i=1}^{N} \int dx\, |x-y_i^{(N)}|^{-1} \chi_{\{|x-y_i^{(N)}|\le\delta\}} \le \frac{\varepsilon^2}{2}$$

We define the set $G_\varepsilon^{(N)} \subset G^{(N)}$ by

$$(2.33)\quad G_\varepsilon^{(N)} = \left\{ x \in R^3 \;\middle|\; \frac{c_1}{N} \sum_{\substack{i=1 \\ |x-y_i^{(N)}|\le\delta(\varepsilon)}}^{N} |x - y_i^{(N)}|^{-1} \le \frac{\varepsilon}{2} \right\}$$

By Tchebychev's inequality

$$(2.34)\quad \text{vol}\,(R^3 - G_\varepsilon^{(N)}) \le \varepsilon$$

and by definition of $G_\varepsilon^{(N)}$ and (2.28)

$$(2.35)\quad P_x\left\{\tau^{(N)} \le t\right\} \le \sum_{\substack{i=1 \\ |x-y_i^{(N)}|>\delta(\varepsilon)}}^{N} P_x\left\{\tau_i^{(N)} \le t\right\} + \frac{\varepsilon}{2}$$

for all $x \in G^{(N)}$ and all $t \ge 0$.

We apply Lemma 1 to the first term on the right of (2.35) to obtain

$$(2.36)\quad \sum_{\substack{i=1 \\ |x-y_i^{(N)}|>\delta(\varepsilon)}}^{N} P_x\left\{\tau_i^{(N)} \le t\right\} = \frac{1}{N} \sum_{i=1}^{N} \alpha \int_0^t p(s,x,y_i^{(N)})\, ds\, \chi_{\{|x-y_i^{(N)}|>\delta\}} + h_N$$

where $h_N \to 0$ as $N \to \infty$ uniformly in $x \in G_\varepsilon^{(N)}$ and $t \in [0,t_0]$. Replacing $\chi_{\{|z|>\delta\}}$ by a smooth bounded function $\phi_\delta(z)$ that vanishes for $|z| \le \delta/2$ and using hypothesis (1.7) we obtain (2.26). The uniformity in t and $x \in G_\varepsilon^{(N)}$ is obtained by approximating the function

$$\alpha \int_0^t p(s,x,y)\, ds\, \phi_\delta(x - y) ,$$

which is uniformly continuous in $[0,t_0] \times R^3 \times R^3$, by finite sums of products of functions of one variable t, x and y at a time and then using (1.7). After this is done we let $\delta \downarrow 0$ and this gives (2.26). The proof is complete.

LEMMA 3. <u>Given</u> $\varepsilon > 0$ <u>and</u> $t_0 < \infty$ <u>there exists an</u> N_0 <u>such that for all</u> $N \ge N_0$

$$(2.37)\quad \int_0^t \alpha\, V(y)\, p(s,x,y)\, dy\, ds - \varepsilon - g(t) \le P_x\left\{\tau^{(N)} \le t\right\}$$

<u>for all</u> $x \in R^3$ <u>and</u> $0 \le t \le t_0$ <u>where</u> $g(t) \ge 0$ <u>and</u>

$$(2.38)\quad \overline{\lim_{t\downarrow 0}}\, \frac{1}{t}\, g(t) = 0 .$$

<u>Proof</u>: Let $\hat{\tau}^{(N)}$ be the minimum of $\tau_i^{(N)}$ (cf. (1.6)) over those $i = 1,2,\ldots,N$ for which $|y_i^{(N)} - y_j^{(N)}| \geq 3/N$ for all $j \neq i$ and $|x - y_i^{(N)}| \geq \delta$ with $\delta > 0$ to be chosen later. We have

$$(2.39)\quad P_x\left\{\tau^{(N)} \leq t\right\} \geq P_x\left\{\hat{\tau}^{(N)} \leq t\right\}$$

$$\geq \sum_{\substack{i=1 \\ |x - y_i^{(N)}| \geq \delta \\ |y_i^{(N)} - y_j^{(N)}| \geq 3/N}}^{N} P_x\left\{\tau_i^{(N)} \leq t\right\} - \frac{1}{2} \sum_{\substack{i \neq j \\ |x - y_i^{(N)}| \geq \delta \\ |y_i^{(N)} - y_j^{(N)}| \geq 3/N}} P_x\left\{\tau_i^{(N)} \leq t,\ \tau_j^{(N)} \leq t\right\}$$

where we have used the principle of inclusion and exclusion for estimating the probability of the union of sets. We look first at the single sum on the right of (2.39). We have the estimate

$$(2.40)\quad \sum_{\substack{i=1 \\ |x - y_i^{(N)}| \geq \delta \\ |y_i^{(N)} - y_j^{(N)}| \geq 3/N}}^{N} P_x\left\{\tau_i^{(N)} \leq t\right\} = \sum_{\substack{i=1 \\ |x - y_i^{(N)}| \geq \delta}}^{N} P_x\left\{\tau_i^{(N)} \leq t\right\} - \sum_{\substack{i=1 \\ |y_i^{(N)} - y_j^{(N)}| \leq 3/N \\ |x - y_i^{(N)}| \geq \delta}}^{N} P_x\left\{\tau_i^{(N)} \leq t\right\}$$

where the second sum on the right is over all $i = 1,2,\ldots,N$ for which $|y_i^{(N)} - y_j^{(N)}| \leq \frac{3}{N}$ for some $j \neq i$. The first sum on the right side of (2.40) is estimated exactly as was done in Lemma 2.

We must show that the second sums on the right of (2.39) and of (2.40) are small. We use (2.29) with $t = \infty$ to obtain

$$(2.41)\quad \sum_{\substack{i=1 \\ |y_i^{(N)} - y_j^{(N)}| \leq 3/N \\ |x - y_i^{(N)}| \geq \delta}}^{N} P_x\left\{\tau_i^{(N)} \leq t\right\} \leq \frac{c_1}{N} \sum_{\substack{i=1 \\ |y_i^{(N)} - y_j^{(N)}| \leq 3/N}}^{N} |x - y_i^{(N)}|^{-1} \chi_{\{|x - y_i^{(N)}| \geq \delta\}}$$

We then note that we also have the estimate

$$(2.42)\quad \frac{c_1}{N} \sum_{\substack{i=1 \\ |y_i^{(N)} - y_j^{(N)}| \leq \frac{3}{N}}}^{N} |x - y_i^{(N)}|^{-1} \chi_{\{|x - y_i^{(N)}| \geq \delta\}} \leq \frac{3c_1}{N^2} \sum_{i \neq j} \frac{\phi_\delta(|x - y_i^{(N)}|)}{|x - y_k^{(N)}|} \frac{\psi_\gamma(|y_i^{(N)} - y_j^{(N)}|)}{|y_i^{(N)} - y_j^{(N)}|}$$

for all sufficiently large N. The function $\psi_\gamma(s)$, $s \geq 0$, is defined for each $\gamma > 0$ so that it is smooth and it equals one for $s \leq \gamma$ and zero for $s \geq 2\gamma$. The function $\phi_\delta(s) \geq \chi_{\{s \geq \delta\}}$ is a smooth function that equals one when $s \geq \delta$ and zero when $s \leq \delta/2$.

LEMMA 4. *Let* μ_n *be a sequence of measures on a separable metric space* X *and suppose that the* μ_n *converge to a measure* μ *weakly as* $n \to \infty$. *Let* $f \geq 0$ *be a function on* X *such that*

$$\int_X f(x)\ \mu_n(dx) \to \int_X f(x)\ \mu(dx) \quad \text{as} \quad n \to \infty \tag{2.43}$$

and assume there exists a sequence of bounded continuous functions $f_k(x) \geq 0$ *on* X, $k = 1,2,\ldots$ *such that*

$$f_k(x) \uparrow f(x) \quad \text{as} \quad k \to \infty \tag{2.44}$$

for each $x \in X$. *Then the measures* $f\mu_n$ *converge weakly to* $f\mu$ *as* $n \to \infty$, *i.e., for any bounded continuous function* ϕ *on* X

$$\int_X \phi(x)\ f(x)\ \mu_n(dx) \to \int_X \phi(x)\ f(x)\ \mu(dx)\ . \tag{2.45}$$

We may apply this lemma to the right side of (2.42) by using hypothesis (1.8). We conclude that the right side of (2.42) tends as $N \to \infty$, uniformly in x, to

$$3c_1 \int_{R^3} \int_{R^3} \frac{\phi_\delta(|x-y|)\ \psi_\gamma(|y-z|)}{|x-y|\ |y-z|} V(y)\ V(z)\ dy\ dz\ .$$

Since $\gamma > 0$ can be chosen as small as desired we conclude that the second term on the right of (2.40) can be made arbitrarily small for all t and x by choosing N large. Moreover this estimate is independent of $\delta > 0$.

We look next at the double sum on the right of (2.39). We have that since $|y_i^{(N)} - y_j^{(N)}| \geq \frac{3}{N}$,

$$\begin{aligned} P_x\left\{\tau_i^{(N)} \leq t,\ \tau_j^{(N)} \leq t\right\} &= \int_0^t \int_{\partial D_j^{(N)}} P_x\left\{\tau_i^{(N)} \in ds,\ x(\tau_i^{(N)}) \in dz\right\} P_z\left\{\tau_j^{(N)} \leq t-s\right\} \\ &\leq P_x\left\{\tau_i^{(N)} \leq t\right\} \sup_{z \in D_i^{(N)}} P_z\left\{\tau_j^{(N)} \leq t\right\} \\ &\leq \frac{c_1 \hat{c}_1}{N^2} \frac{\exp\{-|x-y_i^{(N)}|^2/4t\}}{|x-y_i^{(N)}|} \frac{\exp\{-|y_i^{(N)}-y_j^{(N)}|^2/4t\}}{|y_i^{(N)}-y_j^{(N)}|} \end{aligned}$$

In the last inequality we make use of the fact that $|x - y_i^{(N)}| \geq \delta > 0$ and (2.29) with $\hat{c}_1$ a slightly bigger constant to take care of the sup over $z \in D_i^{(N)}$. Thus

$$\sum_{\substack{i \neq j \\ |x-y_i^{(N)}| \geq \delta \\ |y_i^{(N)}-y_j^{(N)}| \geq \frac{3}{N}}} P_x\left\{\tau_i^{(N)} \leq t,\ \tau_j^{(N)} \leq t\right\} \leq \frac{c_1\hat{c}_1}{N^2} \sum_{i \neq j} \frac{\exp(-|x-y_i^{(N)}|^2/4t)}{|x - y_i^{(N)}|} \cdot \frac{\exp(-|y_i^{(N)}-y_j^{(N)}|/4t)}{|y_i^{(N)} - y_j^{(N)}|} \phi_\delta(|x - y_i^{(N)}|) .$$

Applying Lemma 4 we see that the double sum is bounded independently of N by a constant times

$$c_1\hat{c}_1 \iint \frac{\exp(-|x-y|^2/4t)}{|x-y|} \frac{\exp(-|y-z|^2/4t)}{|y-z|} V(y)\ V(z)\ dy\ dz$$

and this is in turn less than a constant times t^2 uniformly in x.

Returning to the statement of our Lemma 3 we see that we may take for g(t) a constant times t^2 and hence (2.38) holds. This proves Lemma 3.

The proof of Theorem 1 is complete, since Lemmas 2 and 3 give (2.1), (2.2), and (2.3), except for the real variables Lemma 4. We prove this next.

Proof of Lemma 4: Let ϕ be continuous and assume with no loss in generality that $0 \leq \phi \leq 1$.

We have

$$\int \phi f \mu_n(dx) \geq \int \phi f_k \mu_n(dx) \quad \text{for } k = 1,2,\ldots$$

and hence by the weak convergence of μ_n

$$\underline{\lim}_{n\uparrow\infty} \int \phi f \mu_n(dx) \geq \int \phi f_k \mu(dx) .$$

Letting $k \to \infty$ we have

$$\underline{\lim}_{n\uparrow\infty} \int \phi f \mu_n(dx) \geq \int \phi f \mu(dx) .$$

Next we repeat the above with ϕ replaced by $1-\phi$. We have

$$\int f\mu_n(dx) - \int \phi f \mu_n(dx) \geq \int f_k \mu_n(dx) - \int \phi f_k \mu_n(dx)$$

and hence for each $k = 1,2,\ldots$, by the weak convergence of μ_n and (2.43),

$$\int f\mu(dx) - \overline{\lim_{n\uparrow\infty}} \int \phi f\mu_n(dx) \geq \int f_k\mu(dx) - \int \phi f_n\mu(dx)$$

By the monotone convergence theorem as $k \to \infty$ this gives

$$\overline{\lim_{n\uparrow\infty}} \int \phi f\mu_n(dx) \leq \int \phi f\mu(dx)$$

and the proof is complete.

REFERENCES

[1] M. Kac, Probabilistic methods in some problems of scattering theory, Rocky Mountain J. Math. 4 (1974) pp. 511-538.

[2] J. Rauch and M. Taylor, Potential and scattering theory on wildly perturbed domains, J. Funct. Anal. 18 (1975) pp. 27-59.

[3] E. I. Khruslov and V. A. Marchenko, Boundary value problems in regions with fine-grained boundaries, Naukova Dumka, Kiev, 1974.

EXTERIOR DIRICHLET PROBLEMS AND THE ASYMPTOTIC BEHAVIOR OF DIFFUSIONS

by

Michael Cranston, Steven Orey, Uwe Rösler*
University of Minnesota

Let L be the differential operator

$$Lu(x) = \frac{1}{2} \sum_{i,j=1}^{d} a_{ij}(x)u(x) + \sum_{i=1}^{d} b_i(x) \frac{\partial u}{\partial x_i}(x) .$$

Assume that L is uniformly elliptic and that the coefficients satisfy suitable regularity and growth conditions. If $D' \subseteq R^d$ is the complement of a compact set, with smooth boundary ∂D, one may consider the exterior Dirichlet problem

$$(0.1) \qquad Lu = 0 \text{ in } D' , \quad u = \varphi \text{ on } \partial D ,$$

where φ is a continuous function. We will be interested in bounded solutions of (0.1) .

We study this problem and related ones by means of the diffusion process X with differential generator L . Our approach is somewhat related to that of Freidlin [4]. In contrast to [4], however, our emphasis is on obtaining <u>all</u> bounded solutions, in certain special cases.

In Section 1 we establish some general propositions. Among these is a reduction of our problem to the case $D' = R^d$. So the problem becomes that of representing all bounded functions u satisfying $Lu = 0$ throughout R^d ; such functions we call bounded <u>harmonic</u>. We also discuss the bounded solutions $h(x,t)$ of

$$\frac{\partial h}{\partial t}(x,t) + Lh(x,t) = 0 , \quad x \in R^d , -\infty < t < \infty .$$

These are the bounded <u>parabolic functions</u>.

---

* This work was partially supported by the National Science Foundation, and the Deutsche Forschangsgemeinschaft.

In Section 2 we study the special case where $d = 2$, (a_{ij}) is the identity and $b_i(x) = \Sigma_{k=1}^{2} B_{ik}x_k$. For this very special class we obtain complete solutions. This problem already reveals many interesting features. The asymptotic behavior of X is closely related to that of the dynamical system

$$\dot{x} = Bx$$

where x is a function of t , $\dot{x}$ is the derivative with respect to t , and $B = (B_{ij})$ is the 2×2 matrix of coefficients.

Our procedure will be to give a complete description of the tail σ-field of X , and of the invariant σ-field. The discussion is by cases, depending on the nature of the eigen-values of B . In all cases this leads to very concrete "ideal boundaries" on which the bounded parabolic or harmonic functions can be represented by means of a suitable "Poisson kernel". Such representations imply convergence theorems of the Fatou type; we give only one very simple one that follows immediately from our results.

Let us anticipate two natural questions. The work of Section 2 can be extended to treat operators of the same form in R^d . However it does not completely solve the problems in R^d . In two dimensions we use the fact that if both eigen-values of B are pure imaginary X is recurrent. There is no analogous fact in higher dimensions. Our methods solve the problem in R^d provided none of the eigen-values of B have zero real part.

The second question concerns the relation between our ideal boundary and the Martin boundary. In particular, is our boundary actually the Martin boundary or is it only part of the Martin boundary? If it does agree with the Martin boundary we would have at hand a representation for all positive harmonic functions. We hope to address this problem in another paper.

1. Generalities. Let X be a diffusion process in R^d with differential generator

$$L = \frac{1}{2} \sum_{i,j=1}^{d} a_{ij}(x) \frac{\partial^2}{\partial x_i \partial x_j} + \sum_{i=1}^{d} b_i(x) \frac{\partial}{\partial x_i}$$

where it is assumed that $(a_{ij}(x))$ is uniformly positive definite, all coefficients are uniformly Hölder continuous, and $|a_{ij}(x)| \leq k_0(|x|+1)$, $|b_i(x)| \leq k_0|x|+1$, for $i,j = 1,2 \ldots d$, $x \in R^d$ and k_0 is a positive number.

Let D be a bounded open set in R^d, with smooth boundary ∂D; the case $D = \emptyset$ is <u>not</u> excluded. If φ is a continuous function on ∂D, $D' = R^d \setminus (D \cup \partial D)$, the Dirichlet problem

$$(1.1) \qquad Lu = 0 \ \text{ on } \ D', \quad u = \varphi \ \text{ on } \ \partial D$$

has a solution

$$(1.2) \qquad u(x) = E^x \varphi(X_T), \quad T = \inf\{t: t > 0, X_t \in \partial D\}$$

where as usual $\varphi(X_T) = 0$ on $T = \infty$. If X is recurrent (1.2) is the unique bounded solution. Let $\mathcal{H}_b(D)$ denote the class of all bounded harmonic functions in D', i.e. functions which are defined and twice continuously differentiable in D' and satisfy $Lu = 0$ there. Let $\mathcal{H}_b^0(D)$ be the set of $h \in \mathcal{H}_b(D)$ satisfying $h(x) \to 0$ as $x \to \partial D$. Evidently any bounded solution of (1.1) is obtained from (1.2) by adding an element of $\mathcal{H}_b^0(D)$. When $D = \emptyset$, $\mathcal{H}_b^0(\emptyset)$ and $H_b(\emptyset)$ coincide and we write simply $\mathcal{H}_b$.

Our purpose now is to show that if X is transient there exists a one-one correspondence

$$(1.3) \qquad \mathcal{H}_b^0(D) \longleftrightarrow \mathcal{H}_b$$

which (i) is linear, (ii) preserves positivity, (iii) preserves bounded pointwise convergence. Once this is shown it follows that finding the general solution of (1.1) is equivalent to specifying $\mathcal{H}_b$.

The notation $\mathcal{F}(U_t, t \in J)$ will be used to denote the smallest σ-field with respect to which the random variables U_t, $t \in J$, are measurable.

Let $X(D)$ be "the part of X on D'" in the terminology of Dynkin [1], that is the process obtained from the original X by killing at the time T defined in (1.2) above. This process lives on D'; there it behaves locally like X , so that it still has differential generator L . Let θ_t , $0 \leq t < \infty$ be the shift operators associated with the Markov process X . An X(D) measurable random variable ξ is called <u>invariant</u> if $\xi \circ \theta_t = \xi$ on $[0 \leq t < T]$. If ξ' is another X(D) measurable random variable it is <u>equivalent</u> to ξ if $P^x[\xi = \xi'] = 1$ for all $x \in D'$. What is of interest is not the invariant random variables ξ but their equivalence classes $[\xi]$. We exploit the basic correspondence, observed originally by Blackwell [1] and independently by Feller [3] between bounded invariant random variables and bounded harmonic functions. For the details relevant here see Dynkin [2], Chapter XII. It is shown that if h is bounded and harmonic on D' , then

(1.4) $$\lim_{t \uparrow T} h(X_t(D)) = H$$

exists P^x - a.s., for $x \in D'$, and H is invariant. Conversely if H is bounded and invariant

(1.5) $$h(x) = E^x H , \quad x \in D'$$

is bounded and harmonic on D' . From (1.4) and (1.5) one obtains a one-one correspondence

(1.6) $$\mathcal{H}_b(D) \longleftrightarrow \mathcal{I}_b(D)$$

where

$$\mathcal{I}_b(D) = \{[H]: H \text{ is bounded and invariant for } X(D)\} .$$

If H is invariant and $H = I_\Lambda$ is the indicator of an event, Λ is called an <u>invariant event</u>, and

$$\mathcal{I}(D) = \{[\Lambda]: \Lambda \text{ is invariant}\}$$

is known as the <u>invariant σ-field</u>. Note that H is invariant if and only if for every λ the event $[H < \lambda]$ is invariant. So describing $\mathcal{I}_b(D)$ or $\mathcal{I}(D)$ is really the same problem.

Let $\mathcal{H}_b^0(D)$ consist of those $h \in \mathcal{H}_b(D)$ satisfying $h(x) \to 0$ as $x \to \partial D$, and let $\mathcal{J}_b^0(D)$ consist of those $H \in \mathcal{J}_b(D)$ satisfying $H = 0$ on $[T < \infty]$, P^x - a.s. for all $x \in D'$. By restricting (1.6) one obtains the correspondence

$$\mathcal{H}_b^0(D) <\!\!-\!\!> \mathcal{J}_b^0(D) \quad .$$

From this we obtain (1.3) by making a suitable correspondence

$$\mathcal{J}_b^0(D) <\!\!-\!\!> \mathcal{J}_b \ , \quad \mathcal{J}_b = \{[H]: \text{H is bounded and invariant for X}\} \ ,$$

as follows: given $H \in \mathcal{J}_b$ define

$$(1.7) \qquad H^0 = \begin{cases} H \ , & T = \infty \\ 0 \ , & \text{otherwise} \end{cases}$$

and note $H^0 \in \mathcal{J}_b^0(D)$. Conversely starting with $H^0 \in \mathcal{J}_b^0(D)$, let

$$H(\omega) = H^0(\theta_t \omega)$$

where t is chosen so large that $T(\theta_t\omega) = \infty$; on the exceptional set where no such t exists let $H(\omega) = 0$. Because of the transience of X the exceptional set is indeed a P^x-null set for all x. This establishes (1.3).

Associated with X is the space-time process $\tilde{X}$, where $\tilde{X}_t = (X_t, \xi_t)$, $\xi_t = \xi_0 + t$, $0 \le t < \infty$, $\xi_0 \in R^1$. This process has differential generator $(\partial/\partial t + L)$. Thus knowledge of the invariant σ-field of $\tilde{X}$ amounts to knowing all bounded parabolic functions, i.e. functions satisfying $\partial u/\partial t + Lu = 0$, $x \in R^d$, $-\infty < t < \infty$. The invariant random variables for $\tilde{X}$ are exactly the <u>tail random variables</u> of the X process, that is those measurable with respect to $\mathcal{F}^\infty$, where

$$\mathcal{F}^\infty = \bigcap_t \mathcal{F}^t \ , \quad \mathcal{F}^t = \mathcal{F}(X_s, \ s \ge t) \quad .$$

When more than one process is being discussed the notations will specify which one is meant: thus $\mathcal{J}(X)$ is the invariant σ-field of X, $\mathcal{J}(\tilde{X})$ the invariant σ-field of $\tilde{X}$. Since

$$\mathcal{J}(\tilde{X}) = \{[\Lambda]: \Lambda \in \mathfrak{F}^{\infty}(X)\}$$

$\mathcal{J}(\tilde{X})$ is also called the <u>tail</u> σ-<u>field</u> of X. Evidently

$$\mathcal{J}(\tilde{X}) \supseteq \mathcal{J}(X)$$

though the inclusion need not be proper. Indeed equality is the more normal state of affairs. However in the examples we investigate in the next section there will be proper inclusion. It will turn out that the easiest way to find $\mathcal{J}(X)$ is to first find $\mathcal{J}(\tilde{X})$.

It is important to note that the diffusions introduced at the beginning of this section have transition probability densities $p(x,s;y,t)$ such that

(1.8) $$p(x,s;u,t) > 0 \quad \text{for} \quad t > s .$$

This implies that if Λ is a tail event, and the natural versions of the conditional probabilities are chosen

(1.9) $$P^{\mu}(\Lambda)=1 \text{ if and only if } P[\Lambda|X_s = x] = 1 \text{ for all } x \text{, and all } s \geq 0 .$$

<u>Proposition 1</u>. Let $U = (U_t)$ be a Markov process, not necessarily with stationary transition probabilities, with state space a metric space. Assume there exists a random variable U_{∞} such that as $t \to \infty$, for each initial distribution μ

(1.10) $$P^{\mu}[U_t \to U_{\infty}] = 1$$

and also

(1.11) $$P^{\mu}[U_t \in \cdot\,] \to P^{\mu}[U_{\infty} \in \cdot\,] \text{ in total variation norm .}$$

(1.12) $$\mathfrak{F}^t(U) = \mathfrak{F}(U_s,\ s \geq t) , \quad \mathfrak{F}^{\infty}(U) = \bigcap_{t>0} \mathfrak{F}^t(U) .$$

Then for each μ and each $\Lambda \in \mathfrak{F}^{\infty}(U)$ there exists $\Lambda' \in \mathfrak{F}(U_{\infty})$ such that

(1.13) $$P^{\mu}[\Lambda = \Lambda'] = 1 .$$

If (1.9) holds Λ' will not depend on μ , and the conclusion becomes: U_∞ generates the tail σ-field.

<u>Proof</u>. Let

$$(1.14) \qquad \mathfrak{F}_t(U) = \mathfrak{F}(U_s,\ s \leq t);\ \mathfrak{F}_\infty(U) = \text{least } \sigma\text{-field including } \mathfrak{F}_t,\ 0 \leq t < \infty .$$

For $s < \infty$, $\Lambda \in \mathfrak{F}_s(U)$, the Markov property and martingale convergence theorem give, for $t \to \infty$,

$$E^\mu[\Lambda | \mathfrak{F}^t(U)] = E^\mu[\Lambda | U_t] \to E^\mu[\Lambda | \mathfrak{F}^\infty(U)]$$

in $L_1(P^\mu)$. We will show that the last member equals $E^\mu[\Lambda | U_\infty]$ P^μ- a.s. It follows then that for every $\Lambda \in \mathfrak{F}_\infty(U)$

$$E^\mu[\Lambda | \mathfrak{F}^\infty(U)] = E^\mu[\Lambda | U_\infty] \quad P^\mu\text{- a.s.}$$

Observe that the proposition follows from this at once.

So let $s < \infty$, $\Lambda \in \mathfrak{F}_s$, and let B be a measurable set in the state space of the U process. Let g be a measurable function such that

$$E^\mu[\Lambda | U_\infty] = g(U_\infty) , \quad 0 \leq g(z) \leq 1 \quad \text{for all } z .$$

Choose $t > s$. By conditioning on $\mathfrak{F}_s(U)$ and using the assumption that $P^\nu[U_t \in \cdot]$ converges in variation to $P^\nu[U_\infty \in \cdot]$ for every ν one obtains that, as $t \to \infty$,

$$\int E^\mu[\Lambda | U_t] I_B(U_t) dP^\mu = P^\mu[\Lambda, U_t \in B] \to P^\mu[\Lambda, U_\infty \in B]$$

$$= \int g(U_\infty) I_B(U_\infty) dP^\mu$$

the convergence being uniform in B . Also

$$\int g(U_t) I_B(U_t) dP^\mu \to \int g(U_\infty) I_B(U_\infty) dP^\mu$$

uniformly in B . The last two relations imply

$$\int |E^{\mu}[\wedge|U_t] - g(U_t)|dP^{\mu} \to 0 \quad .$$

So writing

$$\int |E^{\mu}[\wedge|U_t] - g(U_{\infty})|dP^{\mu} \leq \int |E^{\mu}[\wedge|U_t] - g(U_t)|dP^{\mu} + \int |g(U_t) - g(U_{\infty})|dP^{\mu}$$

the first term on the right goes to zero. To handle the last term use (1.11). By Lusin's theorem there exists for $\epsilon > 0$ a set B such that $P^{\mu}[U_{\infty} \in B] > 1 - \epsilon$ and g is continuous when restricted to B. In the last integral decompose the domain of integration into three disjoint sets $\Gamma_1(t)$, $\Gamma_2(t)$, $\Gamma_3(t)$ defined by

$$\Gamma_1(t) = [U_{\infty} \in B, U_t \in B], \; \Gamma_2(t) = [U_{\infty} \notin B], \; \Gamma_3(t) = [U_{\infty} \in B, \; U_t \notin B] \; .$$

As $t \to \infty$, $\int_{\Gamma_1(t)} \to 0$ by Lebesgue's bounded convergence theorem. Actually $\Gamma_2(t)$ does not depend on t, and $\int_{\Gamma_2(t)} \leq 2\, P^{\mu}[U_{\infty} \notin B] \leq 2\epsilon$. Finally $\int_{\Gamma_3(t)} \leq 2P[U_t \notin B] \leq 3\epsilon$ for t sufficiently big, because $P^{\mu}[U_t \in B] \to P^{\mu}[U_{\infty} \in B]$. Since $\epsilon > 0$ is arbitrary the proposition is proved.

2. <u>The Bounded Solutions</u>. We return to the case

$$Lu(x) = \frac{1}{2}\,\Delta u(x) + Bx \cdot \nabla u(x) \;, \quad x \in R^2 \;.$$

Here B is a constant 2×2 matrix.

Let $X = (X^1, X^2)$ be the diffusion with differential generator L. Then

$$W_t = (X_t - X_0) - \int_0^t BX_s \, ds \;, \quad 0 \leq t < \infty$$

defines a Wiener process $W = (W^1, W^2)$. One has then

(2.1) $$dX_t = BX_t dt + dW_t$$

and this can be solved to obtain

(2.2) $$X_t = e^{Bt}[X_0 + \int_0^t e^{-Bs}dW_s] \quad .$$

Not surprisingly the asymptotic behavior of X is closely related to that of the dynamical system

(2.3) $$\dot{x} = Bx$$

where x is a function of t and $\dot{x}$ is the derivative with respect to t. Often it will be convenient to work with $Z_t = e^{-Bt}X_t$, given by

(2.4) $$Z_t = X_0 + \int_0^t e^{-Bs}dW_s \quad .$$

We proceed to solve the problems mentioned in the introduction. The discussion will be by cases, depending on the eigenvalues λ_1 and λ_2 of B.

THE TAIL σ-FIELD:

Case i): Both eigenvalues have non-positive real part. In this case, and in this case only X is recurrent. If $h \in \mathcal{H}_b$, $h(X_t)$ is a bounded martingale and must converge, so $h \equiv$ constant easily follows. Standard arguments extend this to show $\tilde{h} \in \mathcal{H}_b(\tilde{X})$, that is $\tilde{h}$ bounded parabolic, implies $\tilde{h} \equiv$ constant. So the tail σ-field is trivial.

Case ii): Both eigen-values of B have positive real parts. Then as $t \to \infty$

(2.5) $$Z_t \to Z_\infty , \quad P^\mu \text{ - a.s. for each } \mu \quad .$$

From (2.4) we see that Z_t and Z_∞ are Gaussian, and the density of Z_t converges to that of Z_∞. So Proposition 1 applies to give that Z_∞ generates the tail σ-field of X. In terms of the space-time process $\tilde{X}_t = (X_t, \xi_t)$

$$e^{-B\xi_t} X_t \to e^{-B\xi_0} Z_\infty$$

generates $\mathcal{J}(\tilde{X})$. Note that, given $\tilde{X}_0 = (x,s)$, $e^{-B\xi_0} Z_\infty$ is Gaussian with mean $e^{-Bs} x$ and covariance depending on s but not on x. So the probability density of the limiting random variable is given by a Gaussian kernel $k(s,x-z)$.

Case iii): $\lambda_2 \le 0 < \lambda_1$. By an orthogonal change of coordinates $Y_t = CX_t$, with C an orthogonal matrix, one obtains a process Y satisfying an equation like (2.1), but with B replaced by a matrix in triangular form. The fact that an orthogonal transformation preserves the Wiener process is used. In other words, in the present case we may as well assume that B has the form

$$B = \begin{pmatrix} \lambda_1 & 0 \\ \alpha & \lambda_2 \end{pmatrix} .$$

Now the first coordinate X^1 is a one-dimensional diffusion and proceeding as in case ii) one sees that as $t \to \infty$

$$(2.6) \qquad e^{-\lambda_1 t} X_t^1 \to \hat{Z}_\infty$$

and $\hat{Z}_\infty$ generates the tail σ-field of X^1 . Again $e^{-\lambda_1 \xi_t} X_t^1 \to e^{-\lambda_1 \xi_0} \hat{Z}_\infty$ and this has a one dimensional Gaussian density $\hat{k}(s,x-z)$ if (X^1,ξ) is started at $(X_0^1,\xi_0) = (x,s)$.

We want to show now that $\hat{Z}_\infty$ actually generates $\mathfrak{F}^\infty(X^1,X^2)$, the tail σ-field of the X-process. Let $\Lambda \in \mathfrak{F}^\infty(X^1,X^2)$. It is enough to show that for $0 \le s < \infty$ there exists $\Lambda_s \in \mathfrak{F}^s(X^1)$ such that

$$(2.7) \qquad \Lambda = \Lambda_s \quad P^\mu\text{ - a.s. for all } \mu .$$

We now write $X_\cdot^i$ for the sample function $t \to X_t^i$, $i = 1,2$. Since Λ is X-measurable, it is equivalent to an event of the form $(X_\cdot^1,X_\cdot^2) \in K$, with K a Borel set in the space of functions from $[0,\infty)$ into $R^1 \times R^1$. Also K may be chosen so that $(\eta,\xi) \in K$ implies $(\eta',\xi') \in K$ whenever $\eta(t) = \eta'(t)$ and $\xi(t) = \xi'(t)$ both hold for all sufficiently large t . For η any continuous function from $[0,\infty)$ into R^1 let K_η denote the cross section:

$K_\eta = \{\eta' : (\eta,\eta') \in K\}$. Given that $X^1_\cdot = \eta$ the second component X^2 satisfies

$$(2.8) \qquad dX^2_t = \lambda_2 X^2_t\,dt + dW^2_t + \alpha\eta(t)dt \quad .$$

This gives rise to a one-dimensional diffusion process X^η . If $\eta \equiv 0$ this is the Ornstein-Uhlenbeck process, which is recurrent and so has trivial tail σ-field. Solving (2.8) one sees that in general X^η differs from the Ornstein-Uhlenbeck process only by the addition of a non-random function of t . So X^η has a trivial tail σ-field. Note (1.8) and (1.9) hold for X^η . Now

$$(2.9) \qquad P^\mu[\Lambda | X^1_\cdot = \eta] = P[X^2_\cdot \in K_\eta | X^1 = \eta] = P^\nu[X^\eta \in K_\eta]$$

where ν is the initial distribution X^η inherits from μ , and (2.9) holds for all η outside some set N with $P^\mu[X^1_\cdot \in N] = 0$. By the nature of K , $X^\eta \in K_\eta$ is a tail event for X^η , and so this event has probability 0 or 1 . This shows that P^μ - a.s.

$$P^\mu[\Lambda | \mathcal{F}^0(X^1)] = I_{\Lambda_0}$$

where the right side is the indicator of some $\Lambda_0 \in \mathcal{F}^0(X^1)$, and this implies $\Lambda = \Lambda_0$ P^μ - a.s. Proceeding to consider $s > 0$, we will show

$$(2.10) \qquad P^\mu[\Lambda | \mathcal{F}^0(X^1)] = P^\mu[\Lambda | \mathcal{F}^s(X^1)] \ , \ P^\mu \text{ - a.s.}$$

and that will conclude the proof. To justify (2.10) it suffices to show that the quantities in (2.9) depend on η only through values $\eta(u)$ with $u \geq s$. So let $\eta(u) = \xi(u)$ for $u \geq s$, and show

$$P^\nu[X^\eta \in K_\eta] = P^\nu[K^\xi \in K_\xi] \quad .$$

By the nature of K , $K_\eta = K_\xi$. We claim that

$$P^\nu[X^\eta \in K_\eta | X^\eta_s = x] = P^\nu[X^\xi \in K_\eta | X^\xi_s = x] \quad .$$

Indeed by (1.9) both terms take on the value 0 or 1 , independently of x , and it only remains to argue that both sides take on the same value. This follows immediately from the Markov property and (2.8), which shows, since $\eta(t) = \xi(t)$ for $t \geq s$ that, conditioned on being at x at time s , both X^{η} and X^{ξ} evolve identically; since (1.8) also holds this suffices.

BOUNDED PARABOLIC FUNCTIONS.

Case i). As already noted, there are none, except for constants.

Case ii). In terms of the space-time process $\tilde{X}$ every bounded parabolic function $h(x,s)$ has the representation

$$h(x,s) = E^{(x,s)}H$$

with H invariant for $\tilde{X}$. Since Z_{∞} generates the tail σ-field there is a bounded Borel function g on R^2 such that $H = g(Z_{\infty})$ $P^{(x,s)}$- a.s.; by (1.9) g does not depend on (x,s) . So one obtains the representation

$$h(x,s) = \int_{R^2} k(s,x-z)g(z)dz \tag{2.11}$$

where the "Poisson kernel" $k(s,x-z)$ is simply the Gaussian kernel introduced above. Conversely, each bounded measurable g defines an h by (2.11) . As is easily seen (take Fourier transforms), two functions g_1 and g_2 which are not equal outside a Lebesgue null-set, will give rise to distinct h_1 and h_2 .

Case iii). This is similar to the preceding case. Now for bounded parabolic h one obtains

$$h(x,s) = \int_{R^1} \hat{k}(s,z-x)g(z)dz \tag{2.12}$$

with g a bounded Borel function on R^1 . Again essentially distinct g correspond to distinct h .

THE INVARIANT σ-FIELD.

Case i). The invariant σ-field is trivial.

Case ii). Each point of the unit circle $S^1 = \{x: |x| = 1\}$ corresponds to a unique trajectory of (2.3) and conversely, every trajectory, except the one $x(t) \equiv 0$, crosses S^1 at a unique point. For $x \in R^2 \setminus \{0\}$ let $\theta(x)$ be the point on S^1 which lies on the trajectory through x, that is

$$|\theta(x)| = 1 , \quad x = e^{Bt}\,\theta(x) \quad \text{for some } t .$$

Since for all x and t, $\theta(x) = \theta(e^{Bt}x)$, one has

$$\theta(Z_t) = \theta(e^{-Bt}X_t) = \theta(X_t)$$

and as $t \to \infty$, $\theta(Z_t) \to \theta(Z_\infty)$. Writing $\Theta = \theta(Z_\infty)$, we see that Θ is invariant. For $s \geq 0$ one can apply the shift operator θ_s to Z_t and obtain

$$Z_t \circ \theta_s = e^{-tB}X_t \circ \theta_s = e^{-tB}X_{t+s} \to e^{sB}Z_\infty = Z_\infty \circ \theta_s \quad .$$

Any tail event Λ we know is equivalent to an event $g(Z_\infty)$, with g the indicator function of a set in R^2. Now we see that this event will be invariant if and only if g can be chosen so that $g(z) = g(e^{Bs}z)$, $-\infty < s < \infty$, or equivalently $g(z) = g(\theta(z))$. So the invariant σ-field is generated by Θ. Note $P^x[\Theta \in d\theta]$ equals $k_1(x,\theta)\sigma(d\theta)$, where σ is the uniform measure on S^1 and $k_1(x,\theta) > 0$.

Case iii). Again this is similar to the preceding case. For $z \in R^1 \setminus \{0\}$ let $\hat{\theta}(z)$ equal 1 or -1 according as $z > 0$ or $z < 0$. Deduce as above that $g(\hat{z}_\infty)$ is invariant if and only if it equals $g(\hat{\theta}(z_\infty))$. Let $\hat{\Theta} = \hat{\theta}(\hat{Z}_\infty)$ and conclude that $\hat{\Theta}$ generate the invariant σ-field.

BOUNDED HARMONIC FUNCTIONS.

Case i). All bounded harmonic functions are constants.

Case ii). The bounded harmonic functions are exactly the ones of the form $h(x) = E^x[g(\Theta)]$ with g bounded and measurable on S^1; that is

$$h(x) = \int_{S^1} k_1(x,\theta)g(\theta)d\theta \,. \tag{2.13}$$

Again each g gives an h and distinct g lead to distinct h unless they agree a.e. with respect to surface measure; this uniqueness assertion is a special case of the parabolic uniqueness.

Case iii). The bounded harmonic funstions have the form $h(x) = E^x[f(\hat{\Theta})]$, but since $\hat{\Theta}$ is concentrated on $\{1,-1\}$ this becomes

$$h(x) = g(1)\hat{k}_1(x) + g(-1)[1-\hat{k}_1(x)]$$

where $\hat{k}_1(x) = P^x[\hat{\Theta}=1]$, and $g(1)$ and $g(-1)$ are arbitrary constants.

A CONVERGENCE RESULT.

Assume we are in case ii) and h has the representation (2.13). It follows at once that if $x \to \infty$ so that $\theta(x) \to \theta_0$, and if θ_0 is a continuity point of g, then $h(x) \to g(\theta_0)$. Indeed (2.13) says $h(x) = E_x[g(\theta(Z_\infty)]$. Now Z_∞ is a Gaussian random variable, with covariance not depending on x, and mean x. It follows that for any $\epsilon > 0$, $P^x[|\theta(Z_\infty) - \theta(x)| > \epsilon]$ is very small if only $|x|$ is large enough, and the assertion is justified. So if g is everywhere continuous h(x) will converge as $x \to \infty$ "appropriately", that is along (or near) trajectories of (2.3).

REFERENCES

[1] Blackwell, D. On transient Markov chains with a countable number of states and stationary transition probabilities, Ann. Math. Stat. 26 (1955), 654-658.

[2] Dynkin, E.B. Markov Processes, Springer-Verlag, Berlin, 1965.

[3] Feller, W. Boundaries induced by non-negative matrices. Trans. Amer. Math. Soc. 83, (1956), 19-54.

[4] Freidlin, M.I. The exterior Dirichlet problem in the class of bounded functions, Theory of Probability and Appl. 11 (1966), 407-414.

ON STOCHASTIC BANG-BANG CONTROL

A. V. Balakrishnan*
Systems Science Department
UCLA, Los Angeles
California 90024, USA

Introduction. In this paper we consider a one-dimensional stochastic control problem where the Wiener process model for the observation noise leads to an optimal control which cannot be realized and indicate how the difficulty is circumvented in the (Gauss Measure) white noise model.

1. The Problem

We consider the stochastic regulator problem where the control is to be based on the observed (sensor) data given by:

$$y(t) = S(t) + N(t), \quad 0 < t < T < \infty$$

where N(t) is the stochastic process describing the sensor error and S(t) is the system response. We assume that the system is linear and time-invariant, so that we can write

$$S(t) = \int_0^t H(t-\sigma)u(\sigma)\, d\sigma + N_s(t)$$

where $u(\cdot)$ is the input, $H(\cdot)$ the system weighting Function and $N_s(t)$ is the stochastic process modelling the 'state disturbance' (such as effect of wind-gust in an aircraft flight control system). The control is to be optimised so as to minimise

$$\int_0^T E[||L\ S(t)||^2]\, dt \ , \ 0 < T < \infty \ ,$$

where L is a given rectangular matrix. The control u(t) must depend only on the

* Research supported in part under Grant no. 73-2492, Applied Math Div., AFOSR, USAF.

sensor data available up to time t and furthermore is subject to the constrant:

$$||u(t)|| \leq 1$$

for (almost all) t in [0, T].

To make the problem tractable we now assume that the Laplace transform of H(t) is rational and the $N_s(t)$ is Gauss-Markov, so that we can write:

$$S(t) = Cx(t)$$

$$x(t) = e^{At}x(0) + \int_0^t e^{A(t-\sigma)}Bu(\sigma)\, d\sigma + \int_0^t e^{A(t-\sigma)}F\, d\ \ W(\sigma) \tag{1.2}$$

where A, B, C, F are constant matrices, and W(•) is a Wiener process.

The major question concerns the model to be used for the 'observation' noise N(t). In practice, it is reasonable to assume that it is Gaussian, stationary with spectral density constant accross a frequency band large compared to that of the process x(t). However the precise bandwidth is usually not specifiable, (and the optimization problem is not solvable even if one could specify it precisely) and hence in the older literature (pre 1950) it was allowed to be 'white noise' of infinite bandwith with constant spectral density, and since such a process would have infinite power, various asymptotic interpretations were made which were o.k. so long as the operations or the process y(•) were only linear. However when non-linear operation is inevitable as in the present problem, a 'mathematically more vigorous' model was proposed in the early sixties (see [1, 2]). This is the

2. Wiener Process Model

Here we "integrate" y(•) and obtain

$$Y(t) = \int_0^t S(\sigma)\, d\sigma + \int_0^t N(\sigma)\, d\sigma \tag{2.1}$$

and replace the 'integrated white noise' by a Wiener process. Thus the model becomes:

$$Y(t) = \int_0^t S(\sigma)\, d\sigma + W_o(t) \tag{2.2}$$

where $W_o(t)$ is a Wiener process. Without loss of generality we may choose the Wiener process $W(t)$ such that:

$$W_o(t) = G\, W(t)$$

where

$$GG^* = \text{Identity},$$

and take

$$F\, G^* = 0$$

so that $N_s(t)$ and $N(t)$ are independent processes (assumed thruout).

One immediate advantage is that then the phrase 'control $u(t)$ must depend only on the observation up to time t' can be made quite precise by simply requiring that $u(t)$ is measurable with respect to $B_Y(t)$ where $B_Y(t)$ is the sigma-algebra generated by the process $Y(s)$, $s \le t$. We can then make a more precise statement of the problem also.

With ω denoting, say, the Wiener process $W(\cdot)$ sample paths in the appropriate $C[0, T]$ space let

$$dY(t, \omega) = Cx(t, \omega)\, dt + G\, dW(t)$$

$$dx(t, \omega) = A\, x(t, \omega)\, dt + B\, u(t, \omega)\, dt + FdW(t)$$

$$Y(0, \omega) = 0\ ;\ x(0) \text{ given}$$

and

$u(t, \omega)$ 'adapted to' $Y(t, \omega)$; jointly measurable in t and ω, and

$$|u(t, \omega)| \le 1 \text{ a.s. } [0, T] \times C[0, T]$$

with Lebesgue measure on $[0, T]$ and Wiener measure on $C[0, T]$. We want to choose $u(\cdot, \omega)$ so as to minimise:

$$\int_0^T E[Rx(t, \omega), x(t, \omega) dt$$

where

$$R \text{ is } \geq 0 .$$

No solution to this problem has yet appeared except for the one-dimensional case, (where both the processes $x(t, \omega)$ and $Y(t, \omega)$ are one-dimensional). A complete solution has been obtained for this case by J. Ruzicka in recent papers [3,4]. To explain his solution we need some preliminary preparatory work.

First consider the special choice

$$u(t, \omega) \equiv 0.$$

Then we know that, if we define,

$$Y(t, \omega) - C \hat{\hat{x}}(t,\omega) = \nu(t, \omega)$$

where

$$d\hat{\hat{x}}(t, \omega) = A \hat{\hat{x}}(t, \omega) dt + P(t)C^* d (Y(t, \omega) - C \hat{\hat{x}}(t, \omega))$$

$$\dot{P}(t) = A P(t) + P(t)A^* + FF^* - P(t)C^*C P(t)$$

with

$$P(0) = E ([x(0) - \hat{x}(0)) (x(0) - \hat{x}(0))^*) ,$$

the 'innovation' process $\nu(t, \omega)$ is also a Wiener process with Identity covariance. Ruzicka [4], exploting a result of Liptser-Shiryayev [5], notes that if $u(t, \omega)$ is adapted to $Y(t, \omega)$ then it is also adapted to $\nu(t, \omega)$ and conversely. Moreover

$$\nu(t, \omega) = Y(t, \omega) - C \hat{x}(t, \omega)$$

where

$$d\hat{x}(t, \omega) = A\hat{x}(t, \omega) + B u(t, \omega))$$
$$+ P(t)C^* d Y(t, \omega) - C \hat{x}(t, \omega))$$

where

$$\hat{x}(t, \omega) = E[x(t, \omega) | B_Y(t)]$$

$$= \hat{\hat{x}}(t, \omega) + \int_0^t e^{A(t-s)} B u(s, \omega) ds$$

Moreover

$$\int_0^T E[Rx(t, \omega), x(t, \omega] \, dt$$

$$= \int_0^T E[R\hat{x}(t, \omega), \hat{x}(t, \omega)] \, dt + \int_0^T \text{Tr. } R \, P(t) \, dt$$

Hence ("separation principle") it is enough to consider the problem of choosing $u(t, \omega)$ adapted to $\nu(t, \omega)$ so as to minimise:

$$\int_0^T E[R \, \hat{x}(t, \omega), \hat{x}(t, \omega)] \, dt$$

where

$$d \, \hat{x}(t, \omega) = A \, \hat{x}(t, \omega) \, dt + B \, u(t, \omega) \, dt + P(t)C^* \, d \, \nu(t, \omega) \, .$$

Let us now specialise the latter problem to the one-dimensional version. Without loss of generality (avoiding trivial cases) we may set

$$A = 0$$

$$B = 1$$

$$C = 1$$

$$F \, F^* = GG^* = 1$$

$$R = 1$$

Now

$$\dot{P}(t) = 1 - P(t)^2$$

and for simplicity, let us assume that

$$P(0) = 1$$

$$\hat{x}(0) \text{ given}$$

so that

$$P(t) = 1 \quad 0 \le t \le T$$

Then we have the simplified problem:

$$d\,\hat{x}(t,\omega) = u(t,\omega)\,dt + d\,\nu(t,\omega)$$

$$\hat{x}(0) \text{ given}$$

$$|u(t,\omega)| \leq 1\ ,\ u(t,\omega) \text{ adapted to } \nu(t,\omega)\ ,$$

Minimise

$$\int_0^1 E\,(\hat{x}(t,\omega))^2\,dt\ .$$

Ruzicka shows that the optimal control $u_o(t, \omega)$ is Markovian, and is given by

$$u_o(t,\omega) = -\text{sign}\,\hat{x}(t,\omega)$$

He proves that the Lebesgue measure of the set where $\hat{x}(t, \omega)$ vanishes is zero with probability one and further that the equation:

$$d\,\hat{x}(t,\omega) = -\text{sign}\,\hat{x}(t,\omega)\,dt + d\,\nu(t,\omega)$$

has a strong solution. He gives a special proof for this particular equation; Zwonkin later [6] has proved this in the general case for any bounded measurable function in place of the signum function.

Thus the problem would appear to be solved except for one thing: the stochastic equation characterising the optimal control need not have a (path-wise) solution for sample paths $\nu(t, \omega)$ which are absolutely continuous in [0, T]. Thus let

$$\nu(t,\omega) = \int_0^t \dot{\nu}(\sigma,\omega)\,d\sigma$$

and suppose $\hat{x}(0) = 0$ and

$$|\dot{\nu}(\sigma,\omega)| < 1\ ,\ < \quad 0 < \sigma < T$$

Then the only solution is

$$\hat{x}(\sigma\ \omega) \equiv 0$$

More generally if $\nu(t, \omega)$ is absolutely continuous and

$$-t < \nu(t,\omega) < t$$

in some non-zero neighbourhood of the origin, we do not have a solution. Thus the set of functions $\dot{v}(\cdot, \omega)$ for which the equation does not have a solution is dense in $L_2[0, T]$.

Of course there is no mathematical contradiction here, since on the Wiener process model, absolutely continuous sample paths have zero probability. On the other hand, recalling the way in which this model was arrived at, we note that from the physical point of view, going back to (2.1) (from which the mathematical transition to (2.2) was made) we see that in fact <u>all</u> the sample paths are absolutely continuous in [0, T].

We shall now see how this difficulty can be circumvented in the

3. White Noise Model.

Here we work with the observation noise model in the original (unintegrated) form:

$$y(t) = S(t) + N(t) \tag{2.1}$$

and take $N(t)$ to be white noise in [0, T], in a precise sense, as defined in [7]. Referring to [7] for details, the sample paths of $N(t)$ are now in

$$W = L_2[0, T] \ ; R_n]$$

where n is the dimension of $y(t)$, and the underlying measure is the Gauss measure μ_G on the Hilbert space W, defined by the characteristic function

$$C(h) = \int_W \exp i[h, x] \, d\mu_G(x) \ , \ h \in W$$

$$= \text{Exp} - \frac{1}{2} [h, h]$$

where [,] denotes inner product in W. Since this measure is only a cylinder measure (or weak distribution) not every Borel measurable function can be defined as a random variable. Hence we proceed differently. Let P denote any finite dimensional projection on W. We are interested in functions (random variables) with range in W. Let $f(\cdot)$ be any Borel measurable function mapping W into W, such that $f^{-1}(W) = PW$. Then the function $f(Px)$, $x \varepsilon W$, called a tame function, defines a random-variable, since the probabilities on inverse images of Borel sets are well-difined. We are only interested (in the present application) in L_2-random variables:

$$E\,(||f(Px)||^2) < \infty$$

The class of L_2 functions of the form $f(Px)$ is a linear class and becomes an inner product space under the inner-product

$$[f,\, g] = E([f(Px),\, g(Qx)])$$

where P, Q are finite dimensional projections. Let the completed Hilbert space be denoted $L_2(\mu_G)$. The limit elements are no longer necessarly associable with functions on W. We can identify them by means of an isomorphic map between $L_2(\mu_G)$ and $L_2(\mu_{W_o})$, the L_2-space of random variables defined on $C[0,T;\, R_n]$ with Wiener measure thereon and range in W. Thus let $\{\phi_i\}$ denote any complete orthonormal system in W. Then

$$\zeta_i = \int_0^T [\phi_i(t),\, N(t)]\, dt$$

defines a sequence of zero-one independent Gaussian random variables and passing to the sample space of all real-valued sequences,

$$W_o(t) = \sum_1^{\infty} \phi_i \int_0^t \phi_i(s)\, ds \quad 0 < t < T$$

defines a Wiener process on $C[0,\, T;\, R_n]$. Further we can identify the random variable:

$$\int_0^T [h(t),\, N(t)]\, dt$$

with the variable

$$\int_0^T [h(t), dW_o(t)]$$

In this way we can identify tame functions with corresponding functions on the Wiener process $W_o(\cdot)$. This is clearly a 1:1 isomorphism.

Next let us note that S(t) defines a stochastic process with sample paths in W also. Hence

$$y(t) = S(t) + N(t) \quad 0 < t < T$$

defines a measure μ_y on cylinder sets on W with characteristic function:

$$C_y(h) = \text{Exp}- \frac{1}{2}[h, h].) \quad E[\text{Exp}i \int_0^T [S(t), h(t)]\, dt]$$

Note that this characteristic function is continuous on W and further for any ϕ in W

$$[\phi, y]$$

is a tame function and

$$E\,([\phi, y]^2) = [R\phi, \phi] + [\phi, \phi]$$

where R is non-negative trace-class operator on W into W.

The requirement that the control u(t) depend only on the observation $y(\cdot)$ up to time t requires special consideration. We need a definition which makes sense for each $y(\cdot)$ in W. For this purpose we introduce the notion of a non-linear Volterra operator. Let m denote the dimension of the control u(t) and let

$$W_u = L_2[0, T); R_m]$$

Let $\psi(\cdot)$ denote a function mapping W into W_u. We shall say that $\psi(\cdot)$ is Volterra if the following property is satisfied. There exists a real-valued non-negative function k(t, s) $0 < t, s < T$, square-integrable on $[0, T] \times [0, T]$ such that for any two functions h_1, h_2 in W, letting

$$g_1 = \psi(h_1)$$

$$g_2 = \psi(h_2)$$

we have that

$$|g_1(t) - g_2(t)| \le \int_0^t k(t, s)\ |h_1(s) - h_2(s)||\ ds,$$

$$\text{a.e. } 0 < t < T .$$

Let $\mathcal{V}$ denote the class of such operators. This is clearly a non-empty, linear class.

Let P again denote a finite-dimensional projection on W. Then if $\psi(\cdot) \varepsilon\ \mathcal{V}$ we have:

$$||\psi(Py)|| < M||Py|| + ||\psi(0)||$$

where M is a fixed constant, and hence

$$E\ [||\psi(Py)||^2] < \infty$$

We can define $L_2(\mu_y)$ in a manner similar to our definition of $L_2(\mu_G)$ and note that

$$\psi(Py)\ \varepsilon\ L_2(\mu_y)$$

Let $\{P_n\}$ be any sequence of finite dimensional projections on W converging strongly to the identity. Let $\psi(\cdot)\ \varepsilon\ \mathcal{V}$. Suppose

$$\{\psi(P_n y)\}$$

is Cauchy in $L_2(\mu_y)$ and the limit is independent of the particular approximating sequence. Then we shall call $\psi(\cdot)$ a physical random variable. We note:

<u>Lemma</u> 3.1 suppose $\psi(\cdot)\ \varepsilon\ \mathcal{V}$ and J is any linear Hilbert-Schmidt Volterra operator mapping W into W. Then the composite function $\psi(J.)$ defines a physical random variable.

<u>Proof</u> Is immediate from

$$||\psi(J\ P_n\nu) - \psi(J\ P_m\nu)|| \le M\ ||J(P_n - P_m)\nu||$$

and

$$E(||J(P_n-P_m)\nu||^2 = \mathrm{Tr}\ (P_n-P_m)\ J^*J\ (P_n-P_m)\ .$$

The class of physical random variables will be denoted V_p. This is clearly a linear subspace in $L_2(\mu_y)$. We shall denote the closure by $\mathcal{U}$. Let C denote the subset in W such that

$$C = [u \in W|\ ||u(t)|| \leq 1]$$

Denote by C_u the class of elements in $\mathcal{U}$ such that the range is contained in C. Then C_u is a closed bounded convex set in $\mathcal{U} \subset L_2(\mu_y)$. The controls $u(\cdot)$ will now be constrained to be in C_u.

Consider now the process S(t). We can describe it in white noise terms as in [7]:

$$S(t) = Cx(t)$$

$$\dot{x}(t) = Ax(t) + B\ u(t) + F\ N(t)$$

$$y(t) = S(t) + G\ N(t)$$

where $N(\cdot)$ is white Gaussian noise,

$$GG^* = I$$

$$FG^* = 0$$

Consider now the case $u(t) \equiv 0$:

$$\dot{\tilde{x}}(t) = A\tilde{x}(t) + FN(t)$$

$$\tilde{y}(t) = S(t) + GN(t)$$

Then as in [7] we can show that

$$\eta(t) = \tilde{y}(t) - C\hat{\tilde{x}}(t) \qquad 0 < t < T$$

again defines white noise in W, where

$$\dot{\hat{\tilde{x}}}(t) = A\hat{\tilde{x}}(t) + P(t)C^*(\tilde{y}(t) - C\hat{\tilde{x}}(t))$$

with P(t) defined as before. Let

$$x_u(t) = \int_0^t e^{A(t-\sigma)} B\, u(\sigma)\, d\sigma$$

so that

$$x(t) = x_u(t) + \tilde{x}(t) ,$$

and

$$y(t) = C\, x_u(t) + C\hat{\tilde{x}}(t) + \eta(t)$$

<u>Lemma</u>

Let the control u(t) be defined by

$$u = \psi(J\nu)$$

where

$\psi(\cdot) \in \mathcal{V}$ and J is linear Volterra mapping W into W.

Then we can express $u(\cdot)$ in the same term with respect to $y(\cdot)$:

$$u = \phi(Ly)$$

where $\phi(\cdot) \in \mathcal{V}$ and L is Volterra. And conversely.

<u>Proof</u>

We note that we can write

$$Cx_u = \tilde{\psi}(J\nu) \ , \quad \tilde{\psi}(\cdot) \in \mathcal{V}$$

Hence

$$y = \eta + \tilde{\psi}(J\eta)$$

Hence

$$Jy = J\eta + J\tilde{\psi}(J\eta)$$

This is a non-linear integral equation for $(J\eta)$ and we can readily deduce that

$$J\eta = Jy - H\,(Jy)$$

where $H(\cdot) \in \mathcal{V}$. The converse is proved similarly.

Combining this Lemma with the fact that controls of the form $\phi(Ly)$ are dense in $\mathcal{V}_p$ with respect to strong convergence over W, we see that we may confine ourselves to controls of the form

$$u = \psi(J\nu), \quad \psi \in \mathcal{V}, \; J \sim \text{ linear Volterra.}$$

For controls of this form it follows that the conditional expectation (see [8])

$$\hat{x}(t) = E\,[x(t)|\mathcal{P}(t)Y]$$

where $\mathcal{P}(t)$ is the projection defined by

$$\mathcal{P}(t)f = g \; ; \; f(s) = g(s) \quad 0 < s < t$$
$$= 0 \quad t < s < T$$

is given by

$$\hat{x}(t) = \hat{\hat{x}}(t) + Cx_u(t)$$

Hence the separation principle holds, and it is enough to minimise

$$E||\sqrt{R}\,\hat{x}(\cdot)||^2$$

subject to

$$\dot{\hat{x}}(t) = A\hat{x}(t) + B\,u(t) + P(t)C^*\eta(t) \;\; ; \; \hat{x}(0) \text{ given.}$$

We also note that any control of the 'Markovian' form

$$u(t) = q(\hat{x}(t)) \quad 0 < t < T$$

where $q(\cdot)$ is uniformly Lipschitzian, can also be expressed as $\phi(J\nu)$, $\phi(\cdot) \varepsilon\, \mathcal{V}$, J-linear Volterra.

Let us now go directly to the one-dimensional case, as in section 2. The problem is then to minimise:

$$\int_0^T E[||\hat{x}(\cdot)||^2]$$

where

$$\dot{\hat{x}}(t) = u(t) + \eta(t)$$

$$\eta(t) = y(t) - \hat{x}(t) \;,$$

and the controls u(t) are to be of the form

$$\phi(\cdot) \;\varepsilon\; \mathcal{V}_p$$

and further such that $\phi(\cdot)\ \varepsilon\ C_u$; or,

$$\phi(\eta) \in C \text{ for every } \eta.$$

Define the operator L by

$$L f = g \; ; \; g(t) = \int_0^t f(s)\, ds$$

mapping W into itself. Note that here $W = W_u$. Then it is readily shown that a unique optimal control $u_o(\cdot)$ exists in $\mathcal{U}$ and satisfies

$$x_o = L u_o + L\eta + h \; ; \quad h(t) \equiv \hat{x}(0)$$

$$\operatorname*{Inf}_{u \in C_u} [L^* x_o, u] = [L^* x_o, u_o]$$

where the inner products are taken in $L_2(\mu_G)$. Because of the isomorphism between $L_2(\mu_G)$ and $L_2(\mu_W)$ we note that u_o being unique must be the same as that given in the Wiener process model (for $\eta(\cdot)$). On the other hand u_o cannot be of the form $\psi(J\eta)$ where $\psi(\cdot) \in \mathcal{V}$ and J is linear Volterra.

To see this, let us consider the sequence of admissible controls of the form:

$$u_n(t) = \phi_n(x_n(t))$$

where $\phi_n(x)$ converges to (- sign x) for every $x \neq 0$, and $\phi_n(x)$ is uniformly Lipschitz continuous for every n.

For example we may take:

$$\frac{\pi}{2} \phi_n(x) = -\mathrm{Tan}^{-1} nx$$

For each n, we have that $x_n(t)$ is the solution of

$$\dot{x}_n(t) = \phi_n(x_n(t)) + \eta(t), \quad \text{a.e. } 0 < t < T$$

$$x_n(0) = \hat{x}(0)$$

and is the form:

$$x_n(\cdot) = \psi_n(L\eta)$$

where $\psi_n(\cdot) \in V$, as a Picard iteration procedure readily shows.

Moreover under the isomorphism of $L_2(\mu_G)$ and $L_2(\mu_W)$ this is the same random variable as defined by the strong solution of the Ito equation:

$$d\, x_n(t) = \phi_n(x_n(t))\ \ dt + d\, W(t)$$

Now $\phi_n(x)$ converges to $(-\text{sign } x)$ for every $x \neq 0$, and

$$d\, x_o(t) = -\text{sign } x_o(t)\ \ dt + dW \tag{3.1}$$

has a strong solution. Hence by a theorem of Zwonkin-Krylov [9], $\{x_n(t)\}$ converges in the mean square to $x_o(t)$ for each t. Hence also

$$\lim \int_0^1 E[x_n(t)^2]\ dt = E[||x_o||^2] = \text{Infimum of Cost Functional}$$

Hence we have a minimising sequence of Markovian Controls in V_p. Let us now examine the convergence of

$$\phi_n(\psi_n(L\eta))$$

for each η in W. [This convergence has nothing to do of course with convergence in $L_2(\mu_W)$ since absolutely continuous sample paths have probability zero]. For fixed $\eta(\cdot)$ in W, we have:

$$x_n(t) = \int_0^t \phi_n(x_n(s))\ ds + \int_0^t \eta(s)\ ds + \hat{x}(0)$$

We know that a subsequence of $u_n(t) = q_n(x_n(t)$ converges weakly in W to a function $u(t;\ \eta)$ say. Hence we have that the corresponding $x_n(t)$ converges uniformly to $z(t)$, say, and

$$z(t) = \int_0^t u(s,\ \eta)\ ds + \int_0^t \eta(s)\ ds + \hat{x}(0)\ ,\quad 0 \le t \le T.$$

and

$$\dot{z}(t) = u(t,\ \eta) + \eta(t)\quad \text{a.e. in } [0,\ T]\ .$$

At a point where $z(t) \neq 0$, we can see that we must have

$$u(t,\ \eta) = -\text{sign } z(t)$$

But if $z(t) = 0$, it is not difficult to see that

$$u(t, \eta) = -\eta(t) \quad \text{if} \quad |\eta(t)| < 1$$
$$= -1 \quad \text{if} \quad \eta(t) > 1$$
$$= +1 \quad \text{if} \quad \eta(t) < -1$$

We note in particular that the set of points at which $z(\cdot)$ vanishes need not have Lebesgue measure zero. It is not difficult to see that the whole sequence $x_n(t)$ converges for each $\eta(\cdot)$ to $z(t)$. Thus $z(t)$ is the unique solution of

(3.1) $$\dot{z}(t) = f(z(t), \eta(t))$$
$$z(0) = \hat{x}(0)$$

where

$$f(z, \eta) = -1 + \eta \text{ if } z > 0$$
$$= 1 + \eta \text{ if } z < 0$$
$$= 0 \text{ if } z = 0 \text{ and } |\eta| < 1$$
$$= -1 + \eta \text{ if } z - 0 \text{ and } \eta > 1$$
$$= 1 + \eta \text{ if } z = 0 \text{ and } \eta < -1$$

and the control is given by

(3.2) $$u(t, \eta) = f(z(t), \eta(t)) - \eta(t)$$

of course we have:

$$\dot{z}(t) = -\text{sign } z(t) + \eta(t)$$

for t such that $z(t) \neq 0$.

The control defined by (3.1), (3.2), is not Markovian and not in V_p. We can however obtain a minimising sequence of Markovian controls which are in V_p and which for each sample path $\eta(\cdot)$ converge to a unique limit function, which is then our optimal solution on the white noise model. Obviously, it is an instrumentable solution in contrast to the solution on the Wiener process model. Put another way, more work has to be done before the Wiener-process-model solution is made instrumentable.

Acknowledgment: I am indebted to Professor J. Ruzicka for many stimulating discussions on the topic and in particular for bringing the key Krylov-Zwonkin result to my attention.

REFERENCES

1. W. M. Wonham: Random Differential Equations in Control Theory: in Probabilistic Methods in Applied Mathematics, Vol. 2, Academic Press (1970).

2. R. S. Bucy and P. D. Joseph: Filtering for Stochastic Processes with Application to Guidance, Interscience, N. Y. (1968).

3. J. Ruzicka: On a Class of Stochastic Bang-Bang Control Problems, Lecture Notes in Economics and Mathematical Systems Vol. 107, Springer-Verlag, (1975).

4. J. Ruzicka: On the Separation Principle with Bounded Controls, Journal of Applied Mathematics and Optimization, Vol. 3, (1977).

5. R. Liptzer and A. Shiryayev: Statistics of Random Processes, Vol. 2, Springer-Verlag, 1978.

6. A. K. Zwonkin: "A Drift Annihilating Phase-Space Map for a Diffusion" Math Sbornik, Vol. 93, no. 1, 1974.

7. A. V. Balakrishnan: Applied Functional Analysis, Springer-Verlag, 1976.

8. A. V. Balakrishnan: Likelihood Rations for Signals in Additive White Gaussian Noise, Journal of Applied Math and Optimization, Vol. 3, (1977).

9. A. K. Zwonkin and N. V. Krylov: On Strong Solutions of Stochastic Differential Equations: Proceedings of the Seminar on the Theory of Stochastic Processes, Druskininkai, Lithuania, U.S.S.R. (1974).

STRUCTURE OF MARTINGALES UNDER RANDOM CHANGE OF TIME

H.J. Fischer

1. Definition of transformed family of σ- fields

Let us fix a probability space $(\Omega, \mathcal{F}, P)$, an increasing right-continious family $(\mathcal{G}_s)_{s\geq 0}$ of σ- fields and an adapted to this family right-continious process $(T_s)_{s\geq 0}$, which has the properties:

1. all paths are increasing,
2. $T_0 = 0$, $\lim\limits_{s\to\infty} T_s = \infty$.

Let $\mathbb{H}$ be the class of right-continious increasing families $(\mathcal{H}_t)_{t\geq 0}$ of σ- fields, for which all T_s are stopping times and $\mathcal{H}_{T_s} = \mathcal{G}_s$. If $\mathbb{H} \neq \emptyset$, there is a minimal element $(\mathcal{F}_t)$ in $\mathbb{H}$, defined by $\mathcal{F}_t = \bigcap\limits_{(\mathcal{H}_t)\in\mathbb{H}} \mathcal{H}_t$ for all $t \geq 0$. Obviously, we can characterize properties of $(\mathcal{F}_t)$ (specially, the structure of martingales with respect to $(\mathcal{F}_t)$), in terms of the family $(\mathcal{G}_s)$.

2. Characterization of the family $(\mathcal{F}_t)$ of σ- fields

Let $\overline{\Omega} = \Omega \times (0,\infty]$, $\mathcal{P}(\mathcal{F}) = \sigma\{B\times(t,\infty] : B\in\mathcal{F}_t, t\geq 0\}$ and $\mathcal{P} = \mathcal{P}(\mathcal{G}) = \sigma\{B\times(s,\infty] : B\in\mathcal{G}_s, s\geq 0\}$ be the σ- fields of predictable sets on $\overline{\Omega}$, corresponding to the families $(\mathcal{F}_t)$ and $(\mathcal{G}_s)$. We define $S_t(\omega) = \inf\{s : T_s(\omega) \geq t\}$ for all $t\geq 0, \omega\in\Omega$, $T^*A = \{(\omega,t) : (\omega, S_t(\omega)) \in A\}$ and $T_t^* A = \{\omega : (\omega, S_t(\omega)) \in A\}$ for all $A \subset \overline{\Omega}$ and $t \geq 0$. It is easy to check that T^* and T_t^* have the following properties:

Lemma 1: $T^* \bigcup\limits_k A_k = \bigcup\limits_k T^* A_k$, $T^*A^c = (T^*A)^c$,

$$T_t^* \bigcup_k A_k = \bigcup_k T_t^* A_k, \; T_t^* A^c = (T_t^* A)^c \quad \text{for all}$$

$A, A_k \subset \overline{\Omega}$, $t \geq 0$,

$T^*(B \times (s,\infty]) = (B \times (0,\infty]) \cap [\![T_s,\infty]\!]$,

$T_t^*(B \times (s,\infty]) = B \cap \{T_s < t\}$ for arbitrary $s,t \geq 0$, $B \subset \Omega$.

<u>Theorem 1</u>: $\mathcal{P}(\mathfrak{F}) = \sigma(T^*\mathcal{P} \cup \{(t,\infty]\!], t \geq 0\})$.

Proof: Let us denote by $\mathcal{P}^*$ the σ - field $\sigma(T^*\mathcal{P} \cup \{(t,\infty]\!], t \geq 0\})$. $T^*\mathcal{P} \subset \mathcal{P}(\mathfrak{F})$ since in view of lemma 1 and $\mathfrak{F}_{T_s} = \mathcal{G}_s$ we have $T^*(B_s \times (s,\infty]) \in \mathcal{P}(\mathfrak{F})$ for any $s \geq 0$, $B_s \in \mathcal{G}_s$, and therefore $\mathcal{P}^* \subset \mathcal{P}(\mathfrak{F})$. On the other hand, if $\mathcal{H}_t = \{B \in \mathfrak{F} : B \times (t,\infty] \in \mathcal{P}^*\}$ for any $t \geq 0$, $(\mathcal{H}_t)$ is a right-continious increasing family of σ - fields. The definition of $(\mathcal{H}_t)$ immediatly implies $\mathcal{P}(\mathcal{H}) \in \mathcal{P}^*$. Since in view of lemma 1

$(B_s \cap \{T_s < t\}) \times (t,\infty] = T^*((B_s \cap \{T_s < t\}) \times (s,\infty]) \cap [\![t,\infty]\!] \in \mathcal{P}^*$ for any $s,t \geq 0$, $B_s \in \mathcal{G}_s$, we get $\mathcal{G}_s \subset \mathcal{H}_{T_s}$. Thus we have $(\mathfrak{F}_t \cap \mathcal{H}_t) \in \mathbb{H}$, which implies $\mathcal{P}(\mathfrak{F}) \subset \mathcal{P}(\mathfrak{F} \cap \mathcal{H}) \subset \mathcal{P}(\mathcal{H})$. This relation yields $\mathcal{P}(\mathcal{H}) = \mathcal{P}(\mathfrak{F}) = \mathcal{P}^*$. □

<u>Corrollary 1</u>: $\mathfrak{F}_{t-} = T_t^* \mathcal{P}$ for any $t \geq 0$.

Proof: As we know, $\mathfrak{F}_{t-} = \{\pi(A \cap [\![t]\!]) : A \in \mathcal{P}(\mathfrak{F})\}$, where π is the projection of $\overline{\Omega}$ in Ω. But it is easy to see

$\{\pi(A \cap [\![t]\!]) : A \in \mathcal{P}(\mathfrak{F})\} = \{\pi(A \cap [\![t]\!]) : A \in T^*\mathcal{P}\} =$
$= \{\pi(T^*A \cap [\![t]\!]) : A \in \mathcal{P}\} = T_t^*\mathcal{P}$. □

But by definition $T_t^* A = \pi(A \cap [\![S_t]\!])$ for any $A \subset \overline{\Omega}$, thus in view of well-known properties of predictable sets $T_t^*\mathcal{P} = \mathcal{G}_{S_t-}$. This one can use to get a criterion for $\mathbb{H} \neq \emptyset$:

<u>Corrollary 2</u>: The class $\mathbb{H}$ is non-empty iff the following condition holds: $\mathcal{G}_s = \{A \in \mathfrak{F} : A \cap \{s < S_t\} \in \mathcal{G}_{S_t-}$ for all $t \geq 0\}$ for any $s \geq 0$.

Proof: Let $(\mathfrak{F}'_t)$ be the right-continious increasing family of σ - fields, for which $\mathfrak{F}'_{t-} = \mathcal{G}_{S_t-}$. From right-continiuty of

$(\mathcal{F}'_t)$ and the obvious identity $\{s < S_t\} = \{T_s < t\}$ we get

$$\mathcal{F}'_{T_s} = \{A \in \mathcal{F} : A \cap \{T_s \leq t\} \in \mathcal{F}'_t \text{ for any } t \geq 0\} =$$
$$= \{A \in \mathcal{F} : A \cap \{T_s < t\} \in \mathcal{F}'_{t-} \text{ for any } t \geq 0\} =$$
$$= \{A \in \mathcal{F} : A \cap \{s < S_t\} \in \mathcal{G}_{S_t-} \text{ for any } t \geq 0\}.$$

Thus, the condition is equivalent to $(\mathcal{F}'_t) \in \mathbb{H}$. Therefore the condition implies $\mathbb{H} \neq \emptyset$. On the other hand, the condition is necessary, for in the case $\mathbb{H} \neq \emptyset$ $\mathcal{F}'_{t-} = \mathcal{G}_{S_t-} = \mathcal{F}_{t-}$ for any $t \geq 0$, from which follows $(\mathcal{F}'_t) = (\mathcal{F}_t) \in \mathbb{H}$. □

3. Structure of martingales with respect to $(\mathcal{F}_t)$

Definition: Let $(Y_s)_{s \geq 0}$ be a right-continious adapted to $(\mathcal{G}_s)$ process. We say that (Y_s) is a semi-martingale if there is a finite signed measure P^Y on $(\bar{\Omega}, \mathcal{P})$, such that $P^Y(B_s \times (s, \infty]) = \int_{B_s} Y_s \, dP$, $s \geq 0$, $B_s \in \mathcal{G}_s$. The measure P^Y is unique, if it exists.

We define a measure $P^{1,t}$ by $P^{1,t}(A) = P(T_t^* A)$ for any $t \geq 0$, $A \in \mathcal{P}$.

Lemma 2: For any $A \in \mathcal{P}$ and any $\mathcal{P}$ - measurable random variable Z

$$\int_A Z(\omega, s) \, dP^{1,t} = \int_{T_t^* A} Z(\omega, S_t(\omega)) \, dP .$$

Proof: It is enough to show this for the indicators of elements of $\mathcal{P}$, but in this case the proposition immediatly follows from the definition of $P^{1,t}$. □

Theorem 2: A right-continious process (X_t) is a martingale with respect to $(\mathcal{F}_t)$ iff for any $t \geq 0$

1. $(X_{T_s} \cdot \mathbb{1}_{\{T_s < t\}})$ is a semi-martingale and generates the measure $P^{X,t}$,
2. $P^{X,t} \ll P^{1,t}$,
3. $X_{t-}(\omega) = \dfrac{dP^{X,t}}{dP^{1,t}}(\omega, S_t(\omega))$ a.s.

Proof: "Only if": Suppose (X_t) is a martingale with respect to $(\mathcal{F}_t)$. From martingale property we get the existance of $P^{x,t}$, actually, $P^{x,t}(A) = \int_{T_t^*A} X_t \, dP$ for any $A \in \mathcal{P}$. $P^{x,t} \ll P^{1,t}$ is obvious, and 3. is a conclusion of lemma 2, corrollary 1 and the fact, that right side is $\mathcal{F}_{t-}$ - measurable. "If": Suppose 1. - 3. hold. Let us define $F^t A = A \cap (0, S_t) \cup (T_t^* A \times (0,\infty]) \cap [\![S_t, \infty]\!]$ for any $A \in \mathcal{P}$, $t \geq 0$. It is easy to see that $F^t \bigcup_k A_k = \bigcup_k F^t A$, $F^t A^c = (F^t A)^c$, $F^t(\mathcal{B}_S \times (s,\infty]) = (\mathcal{B}_S \cap \{T_S < t\}) \times (s,\infty]$, $T_{t'}^* F^t A = T_t^* A$ for any $s \geq 0$, $t' \geq t \geq 0$, $\mathcal{B}_S \in \mathcal{G}_S$, $A, A_k \in \mathcal{P}$.

It is enough to show that (X_{t-}) is a martingale with respect to $(\mathcal{F}_{t-})$, which in view of corrollary 1 is equivalent to $\int_{T_t^*A} X_{t-} \, dP = \int_{T_t^*A} X_{t'-} \, dP$ for any $t' \geq t \geq 0$, $A \in \mathcal{P}$. Due to lemma 2 and the properties of F^t this reduces to $P^{x,t}(A) = P^{x,t'}(F^t A)$. But this is true, since right side is a measure and the equation holds for $A = \mathcal{B}_S \times (s,\infty]$, $s \geq 0$, $\mathcal{B}_S \in \mathcal{G}_S$. □

<u>Remark</u>: A criterion for the existance of the measure, generated by a semi-martingale, can be found in the paper /1/, and the structure of relative densities of semi-martingales was investigated in /2/.

<u>Corrollary 3</u>: A right-continious adapted to $(\mathcal{F}_t)$ process (X_t) is a martingale with respect to $(\mathcal{F}_t)$ iff for any $t \geq 0$ $(X_{T_S \wedge t})$ is a uniformly integrable martingale with respect to $(\mathcal{G}_S)$.

Proof: "Only if" is clear. Suppose the condition holds. Let us fix a $t \geq 0$. The process $Y_{t'}^t = E(X_t \mid \mathcal{F}_{t'})$ is a martingale, from uniform integrability of $(X_{T_S \wedge t})$ follows $X_{T_S \wedge t} = E(X_t \mid \mathcal{G}_S) = Y_{T_S}^t$ a.s. Therefore, $Y_{T_S}^t \cdot \mathbb{1}_{\{T_S < t\}} = X_{T_S \wedge t} \cdot \mathbb{1}_{\{T_S < t\}} = X_{T_S} \cdot \mathbb{1}_{\{T_S < t\}}$,

i. e. the defined above measures $P^{X,t}$ and $P^{Y^t,t}$ coincide. For Y^t is a martingale, $\frac{dP^{X,t}}{dP^{1,t}}(\omega, S_t(\omega)) = \frac{dP^{Y^t,t}}{dP^{1,t}}(\omega, S_t(\omega)) = Y^t_{t-}(\omega) = X_t(\omega)$ a. s. From theorem 2 follows, that (X_t) is a martingale with respect to $(\mathcal{F}_t)$. □

4. Example

Let (T_k) be a sequence of random variables, $(\mathcal{G}_k)$ a sequence of σ - fields, such that T_k is $\mathcal{G}_k$ - measurable, $T_k < T_{k+1}$ and $\mathcal{G}_k \subset \mathcal{G}_{k+1} \subset \mathcal{F}$ for any k, $T_0 = 0$, $\lim\limits_{k\to\infty} T_k = \infty$. Put $T_s = T_k$ and $\mathcal{G}_s = \mathcal{G}_k$ for any $s \in [k, k+1)$. Then $\mathcal{F}_t = \{\bigcup_k A_k \cap \{T_k \le t < T_{k+1}\} : A_k \in \mathcal{G}_k\}$ and theorem 2 reduces to the following:

Corrollary 4: The right-continious process (X_t) is a martingale with respect to $(\mathcal{F}_t)$ iff for any $t \ge 0$

1. $\sum_k E|X_{T_k} \cdot \mathbb{1}_{\{T_k < t\}} - E(X_{T_{k+1}} \cdot \mathbb{1}_{\{T_{k+1} < t\}} \mid \mathcal{G}_k)| < \infty$,

2. $\lim\limits_{k\to\infty} E|X_{T_k} \cdot \mathbb{1}_{\{T_k < t\}}| = 0$ and $X_{T_k} \cdot \mathbb{1}_{\{T_k < t\}} = E(X_{T_{k+1}} \cdot \mathbb{1}_{\{T_{k+1} < t\}} \mid \mathcal{G}_k)$ a. s. on $\{P(T_{k+1} \ge t \mid \mathcal{G}_k) = 0\}$,

3. $X_t = \sum_k \frac{X_{T_k} \cdot \mathbb{1}_{\{T_k \le t\}} - E(X_{T_{k+1}} \cdot \mathbb{1}_{\{T_{k+1} \le t\}} \mid \mathcal{G}_k)}{P(T_{k+1} > t \mid \mathcal{G}_k)} \mathbb{1}_{\{T_k \le t < T_{k+1}\}}$ a. s.

Proof: For any $t \ge 0$ the process $(X_{T_s} \mathbb{1}_{\{T_s < t\}})$ is constant on $[k, k+1)$. This implies that measure $P^{X,t}$ is concentrated on $\Omega \times \{1, 2, \dots, \infty\}$, if it exists, and

$$P^{X,t}(A \times \{k+1\}) = \int_A [X_{T_k} \cdot \mathbb{1}_{\{T_k < t\}} - E(X_{T_{k+1}} \cdot \mathbb{1}_{\{T_{k+1} < t\}} \mid \mathcal{G}_k)] \, dP$$

for $k = 0, 1, \dots$, $P^{X,t}(A \times \{\infty\}) = \lim\limits_{k\to\infty} \int_A X_{T_k} \cdot \mathbb{1}_{\{T_k < t\}} \, dP$ . . It's obvious, now, that measure $P^{X,t}$ exists and is finite iff 1. holds and the limit $\lim\limits_{k\to\infty} \int_A X_{T_k} \cdot \mathbb{1}_{\{T_k < t\}} \, dP$ exists for any $A \in \mathcal{F}$. For measure $P^{1,t}$ is defined by

$$P^{1,t}(A \times \{k+1\}) = \int_A P(T_{k+1} \ge t \mid \mathcal{G}_k) \, dP \qquad \text{for } k = 0, 1, \dots,$$

$P^{1,t}(A\times\{\infty\})=0,\quad P^{x,t}\ll P^{1,t}$ is equivalent to condition 2. and

$$\frac{dP^{x,t}}{dP^{1,t}}=\sum_k \frac{X_{T_k}\cdot 1_{\{T_k<t\}}-E(X_{T_{k+1}}\cdot 1_{\{T_{k+1}<t\}}\mid \mathcal{G}_k)}{P(T_{k+1}\geq t\mid \mathcal{G}_k)}\,1_{(k,k+1]}\;.$$

The equality $X_{t-}(\omega)=\frac{dP^{x,t}}{dP^{1,t}}(\omega,S_t(\omega))$ in view of right-continuity of (X_t) is equivalent to 3.□

References:

/1/ Föllmer, H., On the representation of semi-martingales, The Annals of Probability, 1973, Vol. 1, No. 4, 580 - 589

/2/ Airault, H., Föllmer, H., Relative Densities of Semimartingales, Inventions Math. 27, 1974, 299 - 327

ON STOCHASTIC EQUATIONS WITH UNBOUNDED COEFFICIENTS FOR JUMP PROCESSES

Ya.I.Belopolskaya (Kiev)

This paper considers the correctness of a Cauchy problem for a stochastic differential equation for a jump process in the case when so called "drift" coefficient is unbounded. To investigate the problem we use a multiplicative method which is analogous to Dalecky - Trotter method [1], [2], well known in a theory of linear deterministic evolution equations (for non linear case see [3], [4]). This permits us to construct a stochastic process which is prooved to be a unique solution of considered equation under some natural assumptions.

Notice that multiplicative method allows to omit the usual (see [5], [6]) condition of monotonicity of an unbounded coefficient, changing it to the assumption that the evolution family generated by it is quasicontractive.

Put necessary notations. Let H_0 be a real separable Hilbert space with norm $\|\cdot\|_0$, $\mathcal{B}$ - σ-algebra of its Borel sub sets, φ - an unbounded selfadjoint positively defined operator in H_0 with a Hilbert-Schmidt inverse. The domain of φ which is dense in H_0 we denote H_1. It is a Hilert space with a norm $\|\cdot\|_1$.

Let $(\Omega, \mathcal{F}, P)$ be a probability space, $\mathcal{F}_t$ - an icreasing family of σ - subalgebras of $\mathcal{F}$, E_t - conditional expectation with respect to $\mathcal{F}_t$, $\mathcal{H}_k^t$ - Banach spaces of $\mathcal{F}_\tau$ measurable random functions $\xi(t) \in H_k$ with norms $\langle\langle \xi \rangle\rangle_k = \{\sup_{t_0 \le \tau \le t} E\|\xi(\tau)\|_k^2\}^{1/2}$, $k = 0, 1$.

Let $\nu_t(\cdot)$ be a random $\mathcal{F}_t$ - measurable measure on $\mathcal{B}$ with Poisson distribution and a parameter $E\nu_t(A) = t\,\Pi(A)$ (where Π is a measure on $\mathcal{B}$, not necessary bounded). Denote $\nu(\Delta t, A) = \nu_{t+\Delta t}(A) - \nu_t(A)$, $A \in \mathcal{B}$ and $\mu(\Delta t, A) = \nu(\Delta t, A) - E\nu(\Delta t, A)$

Consider a Cauchy problem for a stochastic differential equation

$$d\xi = L(t, \xi(t))dt + a(t, \xi(t))dt + \int_{H_0} f(t, \xi(t), z)\mu(dt, dz) \quad (1)$$

$$\xi(t_0) = \varphi$$

where $a(t,x)$, $f(t,x,z)$, $(t \in [t_0, T]$, $x, z \in H_\kappa)$ are $\mathcal{F}_t$ - measurable functions valued in H_κ. Below we assume that there exists nonrandom constants C_1 and C_2 such that

$$\|a(t,x) - a(t,y)\|_0^2 + \int_{H_0} \|f(t,x,z) - f(t,y,z)\|_0^2 \Pi(dz) \leq$$
$$\leq C_1 \|x-y\|_0^2 \quad (2)$$

$$\|a(t,x)\|_0^2 + \int_{H_0} \|f(t,x,z)\|^2 \Pi(dz) \leq C_2 [1 + \|x\|_0^2] \quad (3)$$

As to a random function $L(t,x) : H_1 \to H_0$ we suppose that it is independent of $\nu_\tau(A)$, $\tau \in [t_0, T]$ and there exists a nonrandom constant C_3 for which

$$\|L(t,x) - L(t,y)\|_0 \leq C_3 \|x-y\|_1 \quad , x, y \in H_1 \quad (4)$$

Moreover we assume there exists a unique solution of the following Cauchy problem

$$\frac{du}{dt} = L(t, u), \quad u(t_0) = \varphi \quad (5)$$

and an evolution family $U(t, t_0) \circ \varphi = U(t, t_0; \varphi)$ generated by it possesses the next properties

1. $\mathcal{U}(t,t_0) : H_\kappa \to H_\kappa \qquad \kappa = 0, 1$

2. There exist nonrandom constants α_κ and β_κ such that

$$\| \mathcal{U}(t,t_0)\circ\varphi - \mathcal{U}(t,t_0)\circ\psi \|_\kappa \le e^{\alpha_\kappa (t-t_0)} \|\varphi-\psi\|_\kappa \quad (6)$$

$$\| \mathcal{U}(t,t_0)\circ\varphi - \varphi \|_\kappa \le \beta_\kappa(\varphi)\, |t-t_0| \quad (7)$$

Consider an auxiliary stochastic equation

$$\xi(t) = \eta(t) + \int_{t_0}^{t} a(s,\xi(s))ds + \int_{t_0}^{t}\int_{H_0} f(s,\xi(s),z)\mu(ds,dz) \quad (8)$$

Below we suppose that $\eta(t)$ is independent of $\nu(\tau)$, $t_0 \le \tau \le T$ and that $\mathcal{F}_t$ is generated by $\eta(t)$, $\nu_\tau - \nu_{t_0}$, $\tau \le t$

Using a technique analogous to [7] one may prove the next facts.

Theorem 1 . Let $\eta(t) \in \mathcal{H}_0^{t_0}$ and estimates (2), (3) are valid. Then there exists a unique (up to a stochastic equivalence) $\mathcal{F}_t$ measurable solution $\xi(t)$ of the equation (8)

If in addition the next estimate

$$\|a(t,x)\|_1^2 + \int_{H_0} \|f(t,x,z)\|_1^2 \Pi(dz) \le C(1+\|x\|_1^2) \quad (9)$$

holds and $\eta(t) \in \mathcal{H}_1^{t_0}$ then $\xi(t) \in \mathcal{H}_1^{t}$ and the next is valid

$$E_{t_0} \|\xi(t) - \eta(t)\|_\kappa^2 \le C_\kappa (t-t_0),\ \kappa = 0, 1 \quad (10)$$

$$E_{t_0} \|\xi(t)\|_\kappa^2 \le e^{C_\kappa(t-t_0)} E_{t_0} \|\eta(t)\|_\kappa^2 \quad (11)$$

Proof. The first part of the theorem is proved in the same way as it has been done in a diffusion case [7] . Let us prove the second part. Consider for this purpose successive approximations $\xi_0(t) = \eta(t)$

$$\xi_i(t) = \eta(t) + \int_{t_0}^{t} a(s,\xi_{i-1}(s))ds + \int_{t_0}^{t}\int_{H_0} f(s,\xi_{i-1}(s),z)\mu(ds,dz)$$

The condition (9) and Granwall inequality lead to an estimate

$$E_{t_o} \|\xi(t)\|_1^2 \leq 3 E_{t_o} \|\eta(t)\|_1^2 e^{\alpha(t-t_o)} + \beta \qquad (11')$$

where α and β depend on C, t_o and T. Further, weak compactness of a sphere in H_1 and estimate $(11')$ involve existence of a limit $y = w\text{-}\lim \varphi \xi_i(t)$ and it follows from the compactness of φ^{-1} that $\varphi^{-1} y = s\text{-}\lim \xi_i(t) = \xi(t)$ and $\xi(t) \in H_1$. Now one may easily check estimates (10) and (11) using Ito's formula and Gronwall inequality.

Denote $T(t, t_o)$ a map $T(t,t_o): \mathcal{H}_K^{t_o} \to \mathcal{H}_K^{t}$ defined by a solution $\xi(t)$ of the equation (8) $T(t, t_o) \circ \eta(t) = \xi(t)$.

<u>Remark.</u> In condition of theorem 1 one may easily prove with a help of Ito's formula that the map $T(t, t_o)$ possesses the next property

$$E_{t_o} \| T(t, t_o) \circ \varphi - T(t, t_o) \circ \psi \|_o^2 \leq e^{c(t-t_o)} \|\varphi - \psi\|_o^2 \qquad (12)$$

Let $q: t_o \leq t_1 \leq \ldots \leq t_n = t$ be a partition of an interval $[t_o, t]$. Construct a multiplicative expression

$$\xi_n(t) = \prod_{K=1}^{n} \circ T_K \circ \mathcal{U}_K \circ \varphi \qquad (13)$$

where $\mathcal{U}_K = \mathcal{U}(t_{K-1}, t_K)$, $T_K = T(t_{K-1}, t_K)$ are generated by solutions of (5) and (8) correspondingly.

We prove some auxiliary facts that will be need below.

<u>Lemma 1.</u> Let conditions $(4), (6), (7)$ are fulfilled as well as conditions of theorem 1. Then the next estimates are valid

$$E_{t_o} \| \prod_{K=1}^{n} \circ T_K \circ \mathcal{U}_K \circ \varphi \|_K^2 \leq e^{c_K(t-t_o)} \{ \|\varphi\|_K^2 + 1 \}, \ \forall \varphi \in \mathcal{H}_K^{t_o} \qquad (14)$$

$$E_{t_0}\|\prod_{\kappa=1}^{n}\circ T_\kappa\circ\mathcal{U}_\kappa\circ\varphi-\prod_{\kappa=1}^{n}\circ T_\kappa\circ\mathcal{U}_\kappa\circ\psi\|_0^2\leq e^{\kappa(t-t_0)}\|\varphi-\psi\|_0^2 \quad (15)$$

$$\Delta=E_{t_0}\|T(t,t_0)\circ\mathcal{U}(t,t_0)\circ\varphi-\prod_{\kappa=1}^{n}\circ T_\kappa\circ\mathcal{U}_\kappa\circ\varphi\|_0^2\leq C_\varphi^n(t-t_0)^3 \quad (16)$$

$$\delta=E_{t_0}\|T(t,s)\circ\mathcal{U}(s,t_0)\circ\varphi-\mathcal{U}(s,t_0)\circ T(t,s)\circ\varphi\|_0^2\leq C_\varphi^1(t-t_0)^3 \quad (17)$$

$\forall\ \varphi,\psi\in\mathcal{H}_0^{t_0}$

Proof. Properties of maps $\mathcal{U}(t,t_0)$ and $T(t,t_0)$ involve immediately (14) and (15). To prove (17) we use equations (5) and (8). Inequalities (4) and (2) lead to estimate

$$\delta=E_{t_0}\|\int_{t_0}^{s}L(\theta,\mathcal{U}(\theta,t_0)\circ\varphi)d\theta+\int_{s}^{t}\int_{H_0}f(\theta,T(\theta,s)\circ\mathcal{U}(s,t_0)\circ\varphi,z)\mu(d\theta,dz)$$
$$+\int_{s}^{t}a(\theta,T(\theta,s)\circ\mathcal{U}(s,t_0)\circ\varphi)d\theta-\int_{s}^{t}a(\theta,T(\theta,s)\circ\varphi)d\theta-$$
$$-\int_{s}^{t}\int_{H_0}f(\theta,T(\theta,s)\circ\varphi,z)\mu(d\theta,dz)-\int_{t_0}^{s}L(\theta,\mathcal{U}(\theta,t_0)\circ T(t,s)\circ\varphi)d\theta\|_0^2$$
$$\leq C(1+|s-t|)\int_{t_0}^{s}E_{t_0}\|T(\theta,s)\circ\mathcal{U}(s,t_0)\circ\varphi-T(\theta,s)\circ\varphi\|_0^2d\theta+$$
$$+(s-t_0)\int_{t_0}^{s}E_{t_0}\|\mathcal{U}(\theta,t_0)\circ\varphi-\mathcal{U}(\theta,t_0)\circ T(t,s)\circ\varphi\|_1^2d\theta$$

and estimate $\delta\leq C_\varphi'(t-s)(s-t_0)^2$ follows from (6), (7), (10), (12).

To prove (16) use triangle inequality in $\mathcal{H}_0^t$

$$\Delta^{1/2}\leq\sum_{i=1}^{n}\{E_{t_0}\|\prod_{\kappa=i}^{n}\circ T_\kappa\circ\mathcal{U}_\kappa\circ T(t_i,t_0)\circ\mathcal{U}(t_i,t_0)\circ\varphi-$$
$$-\prod_{\kappa=i-1}^{n}\circ T_\kappa\circ\mathcal{U}_\kappa\circ T(t_{i-1},t_0)\circ\mathcal{U}(t_{i-1},t_0)\circ\varphi\|_0^2\}^{1/2}$$

that due to (14) leads to

$$\Delta^{1/2}\leq\sum_{i=1}^{n}\{e^{\kappa(t-t_0)}E_{t_0}\|T(t_i,t_0)\circ\mathcal{U}(t_i,t_0)\circ\varphi-$$
$$-T_{i-1}\circ\mathcal{U}_{i-1}\circ T(t_{i-1},t_0)\circ\mathcal{U}(t_{i-1},t_0)\circ\varphi\|_0^2\}^{1/2}$$

and putting (17) into it we have $\Delta\leq C_\varphi^n(t-t_0)^3$

Lemma 2. Let conditions of lemma 1 are fulfilled. Then the sequence (13) is fundamental in the norm of $\mathcal{H}_0^t$.

Proof. First suppose that $\varphi \in \mathcal{H}_1^{t_0}$ and $m = n\ell$ and show that

$$\alpha_{mn} = \Big\{ E_{t_0} \Big\| \prod_{K=1}^{n} \circ T_K \circ \mathcal{U}_K \circ \varphi - \prod_{K=1}^{m} \circ T_K \circ \mathcal{U}_K \circ \varphi \Big\|_0^2 \Big\}^{1/2} \underset{m,n\to\infty}{\to} 0$$

In fact, using triangle inequality we have

$$\alpha_{mn} \le \sum_{K=1}^{n} \Big\{ E_{t_0} \Big\| \prod_{K=i-1}^{n} \circ T_K \circ \mathcal{U}_K \circ \prod_{p=1}^{i-1} \circ S_p \circ \varphi -$$
$$- \prod_{K=i}^{n} \circ T_K \circ \mathcal{U}_K \circ \prod_{p=1}^{i} \circ S_p \circ \varphi \Big\|_0^2 \Big\}^{1/2}$$

where $S_p \circ \varphi = \prod_{s=1}^{\ell} \circ T_{p_s} \circ \mathcal{U}_{p_s} \circ \varphi$ and $T_{p_0} = T_p$, $T_{p_\ell} = T_{p+1}$, $\mathcal{U}_{p_0} = \mathcal{U}_p$, $\mathcal{U}_{p_\ell} = \mathcal{U}_{p+1}$ and estimates (14) and (16) permit to obtain

$$\alpha_{mn} \le c^{\ell}(T-t_0)(t_{K+1}-t_K)^{1/2} \to 0,\ m,n\to\infty,\ \max_K |t_{K+1}-t_K| \to 0$$

To obtain the result for general m, n write

$$\alpha_{mn} \le \Big\{ E_{t_0} \Big\| \prod_{K=1}^{n} \circ T_K \circ \mathcal{U}_K \circ \varphi - \prod_{K=1}^{mn} \circ T_K \circ \mathcal{U}_K \circ \varphi \Big\|_0^2 \Big\}^{1/2} +$$
$$+ \Big\{ E_{t_0} \Big\| \prod_{K=1}^{mn} \circ T_K \circ \mathcal{U}_K \circ \varphi - \prod_{K=1}^{m} \circ T_K \circ \mathcal{U}_K \circ \varphi \Big\|_0^2 \Big\}^{1/2}$$

and apply the previous arguments. At last by denseness for $\varphi \in \mathcal{H}_0^{t_0}$ find $\varphi_s \to \varphi$, $\varphi_s \in \mathcal{H}_1^{t_0}$. Then the inequality

$$\Big\{ E_{t_0} \Big\| \prod_{K=1}^{n} \circ T_K \circ \mathcal{U}_K \circ \varphi - \prod_{K=1}^{m} \circ T_K \circ \mathcal{U}_K \circ \varphi \Big\|_0^2 \Big\}^{1/2} \le e^{K(t-t_0)} \|\varphi_s - \varphi\|_0$$
$$+ \alpha_{mn} + e^{K(t-t_0)} \|\varphi_s - \varphi\|_0$$

proves the theorem

Theorem 2. Let conditions of lemma 1 are fulfilled. Then there exists $\mathcal{F}_t$ measurable random process

$$\xi(t) = P\text{-}\lim \xi_n(t) = V(t,t_0)\circ\varphi \tag{18}$$

and a map $V(t,t_0) : \mathcal{H}_K^{t_0} \to \mathcal{H}_K^{t}$ possesses the next properties: 1) evolution property $V(t,s)\circ V(s,t_0)\circ\varphi = V(t,t_0)\circ\varphi$, $t_0 \le s \le t$

2) the next estimate is valid

$$E_{t_0} \| V(t,t_0)\circ\varphi - \varphi \|_K^2 \le C_\varphi (t-t_0) \qquad (19)$$

Proof. Lemma 2 leads to convergence of $\xi_n(t)$ uniformly in $\mathcal{H}_0^t$. To prove the evolution property choose τ to be a rational point of an interval $[t,t_0]$ such that $\tau = \frac{p(t-t_0)}{m}$,

$$K = \frac{nm}{p+m}, \quad K' = \frac{pn}{p+m}, \quad K+K' = n.$$

Then

$$E_{t_0} \| V(t,\tau)\circ V(\tau,t_0)\circ\varphi - V(t,t_0)\circ\varphi\|_0^2 \le$$

$$\le 3\Big\{E_{t_0} \| V(t,\tau)\circ V(\tau,t_0)\circ\varphi - \prod_{i=1}^{K'} T_i\circ \mathcal{U}_i \circ V(\tau,t_0)\circ\varphi\|_0^2 +$$

$$+ E_{t_0} \| \prod_{i=1}^{K'} T_i\circ \mathcal{U}_i\circ V(\tau,t_0)\circ\varphi - \prod_{i=1}^{K'}\circ T_i\circ\mathcal{U}_i\circ\prod_{j=1}^{K}\circ T_j\circ\mathcal{U}_j\circ\varphi\|_0^2 +$$

$$+ E_{t_0} \| \prod_{i=1}^{K+K'}\circ T_i\circ\mathcal{U}_i\circ\varphi - V(t,t_0)\circ\varphi\|_0^2\Big\} \to 0$$

for $K, K'\to\infty$ due to estimates of lemma 1. Thus the evolution property holds for a dense set of τ and by continuity in t which will be proved below it holds for all $\tau\in[t_0,t]$

To prove the continuity in t thats is an estimate (19) we use estimates (14), (15) of lemma 1.

<u>Theorem 3.</u> Let in conditions of lemma 1 $\varphi\in\mathcal{H}_1^{t_0}$. Then a process $\xi(t)$ of the form (18) possesses a stochastic differential and comes to be a unique solution of an equation (1) If coefficients of the equation (1) are nonrandom, then the process $\xi(t)$ possesses a Markov property.

Proof. To calculate a stochastic differential of the process $\xi(t)$ consider an auxiliary process $\chi(t) = T(t,s)\circ\mathcal{U}(t,s)\circ\xi(s)$ Stochastic differential of this process may be defined by the relation

$$\chi(t+\Delta t) = \xi(t) + \int_t^{t+\Delta t} L(s, U(s,t)\circ\xi(t))ds +$$

$$+ \int_t^{t+\Delta t} a(s, T(s,t)\circ U(t+\Delta t, t)\circ\xi(t))ds +$$

$$+ \int_t^{t+\Delta t}\int_{H_0} f(s, T(s,t)\circ U(t+\Delta t, t)\circ\xi(t), z)\mu(ds, dz)$$

and coincide with right hand of (1) in a limit $\Delta t \to 0$

Let us shaw that processes $\chi(t)$ and $\xi(t) = V(t, t_0)\circ\varphi$ possess the same stochastic differential. To prove this fact it is enough (see [8]) to check two estimates

$$E_t \|\chi(t+\Delta t) - \xi(t+\Delta t)\|^2 = o(\Delta t)$$

$$\|E_t [\chi(t+\Delta t) - \xi(t+\Delta t)]\| = o(\Delta t)$$

One may easily do it with a help of estimates of lemma 1.

Prove the uniqueness of the solution $\xi(t)$, we have obtained. Let $\eta(t)$ be an arbitrary solution of the equation (1) . Consider a function $K(t) = V(s,t)\circ\eta(t), s>t$ and show that it did not depend on t . For this purpose we verify that the stochastic differential of this function is equal to zero. But the last statement is the consequence of estimates

$$E_t \|K(t+\Delta t) - K(t)\|_0^2 \le E_t \|V(s, t+\Delta t)\circ\eta(t+\Delta t) -$$

$$- V(s,t)\circ\eta(t)\|_0^2 \le e^{K(s-t-\Delta t)} E_t \|\eta(t+\Delta t) - V(t+\Delta t, t)\circ\eta(t)\|_0^2 \le$$

$$\le e^{K(s-t-\Delta t)} E_t \|[\eta(t+\Delta t) - \eta(t) - d\xi] + [V(t+\Delta t), t)\circ\eta(t) -$$

$$- \eta(t) - d\xi]\|^2 = o(\Delta t)$$

$$\|E_t [K(t+\Delta t) - K(t)]\| \le \{E \|K(t+\Delta t) - K(t)\|^2\}^{1/2}$$

Thus

$$K(t) = V(s,t) \circ \rho(t) = V(s,t_0) \circ \varphi = K(t_0)$$

and in a limit $s \to t$ we obtain $\rho(t) = V(t,t_0) \circ \varphi$ due to continuosness of $V(s,t_0)$ in s.

Let now coefficients of the equation (1) be nonrandom. It follows from the representation (18) that for a fixed φ the process $\xi(t) = V(t,t_0) \circ \varphi$ does not depend on events of σ-algebra $\mathcal{F}_{t_0}$. Then for any $\mathcal{F}_{t_0}$ – measurable random value λ and any measurable bounded function $g(x)$

$$E \lambda g(\xi(t)) = E E_{t_0} \lambda g(\xi_{t_0, \xi(t_0)}(t)) =$$

$$= E \lambda E g(\xi_\varphi(t)) / \varphi = \xi(t_0))$$

that is $E_{t_0} g(\xi_{t_0,\varphi}(t))$ depend only on $\xi(t_0)$ that involve $P\{\xi(t) \in A / \mathcal{F}_{t_0}\} = P_{t_0,\xi(t_0)}(\xi(s) \in A)$.

References

1. Dalecky Yu.L. Functional integrals connected with operator evolution equations, Russian Math. Surv. 1962,17,5,1-108.
2. Trotter H.F. On the product of semigroups of operators, Proc. Amer. Math. Soc. 1959, 10, 545 - 551.
3. Marsden J. On product formulas for nonlinear semigroups, J. Functional Analysis, 1973, 13, 1, 51 - 72.
4. Dalecky Yu.L., Zaplitnaya A.T. Integrals over a tree space connected with nonlinear parabolic equations. Ukr. mat. journ. 1965, 17, 5, 110 - 114.
5. Krylov N.V., Rosovsky B.L. On a Cauchy problem for linear stochastic equations with partial derivatives, Izv. AN SSSR ser. mat. 1977, 41, 6,

6. Pardoux E. Sur des equations aux derives partielles stochas tic monotones CRAS, ser A, 1972, 275, 2, 101 - 103.

7. Dalecky Yu.L. Infinite dimensional elliptic operators and connected with them parabolic equations, Usp. mat. nauk, 1967, 22, 4, 3 - 54.

8. Gichman I.I. On the theory of differential equations for stochastic processess. Ukr. mat. journal, 1950, 2, 4, 37 - 63.

To the maximum principle theory for problems of control of stochastic differential equations

B. I. Arkin, M.T.Saksonov

I. We shall consider the next problem

$$G(Mx_1, My_1) \to \min \tag{I}$$

$$dy_t = f_0(t, x_t, u_t)dt \quad y_0 = \bar{y} \tag{2}$$

$$dx_t = f(t, x_t, u_t)dt + \sigma(t, x_t)\, dw_t \quad x_0 = \bar{x} \tag{3}$$

$$u_t \in V_t \tag{4}$$

where w_t is one-dimensional normalized Brownian motion on the complete probability space $(\Omega, \mathcal{F}, P)$. The control u_t is taken to be measurable and adapted to $\{\mathcal{Y}_t\}_{t\geq 0}$, $\mathcal{Y}_t \subseteq \mathcal{F}_t$. The solutions (2),(3) are understood as Ito solutions.

2. We will make use of the following notations: ℓ - Lebesque measure on $[0,1]$; $B\mathcal{F}$ and $B\mathcal{Y}$ - σ-algebras, progressively measurable with respect to the flows $\mathcal{F}_t$ and $\mathcal{Y}_t$ respectively; $\overline{B\mathcal{F}}$, $\overline{B\mathcal{Y}}$ are their completions with respect to the measure $\ell \times P$; B^K is the Borel σ-algebra in R^K, $L^n_s[B\mathcal{F}]$ is the space of $\overline{B\mathcal{F}}$ measurable s-integrable functions on $[0,1]\times\Omega$ taking values in R^n, $L^n_s C$ is the space of functions $\varphi_t \in L^n_s[B\mathcal{F}]$ continuous (in t) with probability I such that $M \sup_t |\varphi_t|^s < \infty$. $\mathcal{P}^n$ is the space of $\overline{B\mathcal{F}}$ measurable functions h_t, taking values in R^n and such that $P\{\int_0^1 h_t^2\, dt < \infty\} = 1$; $|x|$ denote the Euclidean norm $x \in R^n$; $\|x\|$ is a norm in functional space. We

will indicate the graph of the point-set correspondence V_t by V, $V=\{(t,\omega,u)\ u\in V_t=V_t(\omega)\}$. The unessential constant arising in the proof we will mark $R_1 \dots R_n$

3. Formulation of the problem

$$G: R^n\times R^m \to R^1 \ ; \qquad f_0 : [0,1]\times \Omega\times R^n\times R^K \to R^m$$

$$f\ [0,1]\times\Omega\times R^n\times R^K \to R^n; \quad \sigma: [0,1]\times\Omega\times R^n \to R^n$$

V_t is the point-set correspondence from $[0,1]\times\Omega$ in , We fix $p\geq 2, \tau\geq 1$. A function $u_t(\omega)$, such that $u_t\in L^K_\tau[B\mathcal{F}]$, $u_t\in$ $\in V_t$ $\ell\times P$ a.s. will be called a control. We will call the pair (y^u_t, x^u_t, u_t) admissible if $x^u_t\in L^n_p[B\mathcal{F}]$, u_t is a control, and (2),(3) holds $\ell\times P$ a.s. An admissible pair (y^*_t, x^*_t, u^*_t) will be called optimal if it minimizes the functional (I) on the set of admissible pairs.

We suppose that the following conditions are held.

II$_1$) The graph V is $\overline{B\mathcal{G}}\times B^K$ measurable, $\mathcal{G}_t, \mathcal{F}_t$ - are two increasing right continuous sequences of σ-algebras, $\mathcal{G}_t\subset\mathcal{F}_t$ and $\mathcal{G}_0$ contains all P-null sets.

II$_2$) The functions f_0, f, σ are measurable in (t,ω,u) with respect to the σ-algebra $\overline{B\mathcal{F}}\times B^K$ and are continuously differentiable in x, $\bar{x}, \bar{y}$ - are two $\mathcal{F}_0$ measurable random variables $M|\bar{x}|^p<\infty$, $M|\bar{y}|<\infty$

II$_3$) $\exists K>0$ such that $|f'_x(t,\omega,x,u)|\leq K$

$|\sigma'_x(t,\omega,x)|\leq K$ uniformly in (t,ω,x,u)

II$_4$) $\exists q, \xi_t$ the constant $N>0$ such that $1\leq q\leq p$

$\xi_t\in L'_{p'}[B\mathcal{F}]$ $\quad p'^{-1}+p^{-1}q=1$

$$|f_0(t,x,u)|\leq N(\xi_t|x|^q+|u|^\tau+1),\ |f'_{0x}(t,x,u)|\leq N(\xi_t|x|^{q-1}+1)$$

We shall adduce a number of preliminary result from the stochastic differential equations theory before than the main result will be formulated.

Let A_t, B_t be bounded, measurable and adapted with $\{\mathcal{F}_t\}_{t\geq 0}$ $n\times n$ matrix-valued functions.

Lemma I ([I],[2]). The stochastic equation

$$d\Phi_t = A_t \Phi_t dt + B_t \Phi_t dW_t \qquad \Phi_0 = I \qquad t\in[0,1]$$

where Φ_t is $n\times n$ random matrix, has unique solution Φ_t, $\Phi_t \in L_s^{n^2}C\ \forall s\geq 1$, $\exists \Phi_t^{-1}\ \forall t$ and $\Phi_t^{-1}\in L_s^{n^2}C\ \forall s\geq 1$

Lemma 2 ([3]). Let the control u_t and the function $\varphi_t \equiv \overline{x}$ be such that $\int_0^t f(\tau,\varphi_\tau,u_\tau)d\tau + \int_0^t \sigma(\tau,\varphi_\tau)dW_\tau \in L_2^n C$, then there exists a unique x_t^u such that (3) hold.

Definition ([4]). Let $\varphi(t,\omega,u)$ be a $\overline{B^1\times\mathcal{F}}\times B^K$ measurable function. The $\overline{B\mathcal{G}}\times B^K$ measurable function $\overline{\varphi}_{\{V\mathcal{G}\}}(t,\omega,u)$ such that for any $\overline{B\mathcal{G}}$ measurable function u_t, $u_t\in V_t$ ($\ell\times P$ a.s.), $M\int_0^1|\varphi(t,\omega,u_t)|dt<\infty$ there hold

$$\overline{\varphi}_{\{V\mathcal{G}\}}(t,\omega,u_t) = M[\varphi(t,\omega,u_t)/\mathcal{F}_t] \qquad \ell\times P\text{ - a.s.}$$

we call the Regular Conditional Expectation of φ on the set V with respect to the flow $\mathcal{G}_t$

Theorem. Under the conditions II_1)-II_4), let (y_t^*, x_t^*, u_t^*) be an optimal pair in the problem (I)-(4), and the following conditions be fulfilled

II_5) $\int_0^1 \|f(\tau,x_\tau^*,u_\tau^*)\| \in L_2^1 C \quad \sigma(t,x_t^*)\in L_2^n[\overline{B\mathcal{F}}]$

The continuous function $G(x,y)$ is continuously differentable in the point (Mx_1^*, My_1^*)

II_6) There exists an increasing sequence point-set correspondences V_t^j with the graph V^j, and sequence $\overline{B\mathcal{F}}$ measurable functions γ_t^j such that V^j is $\overline{B\mathcal{G}}\times B^K$ measurable

$u_t^*\in V_t^1$, $\bigcup_{j=1}^{\infty} V_t^j = V_t \qquad \ell\times P$ - a.s. $M\int_0^1|\gamma_t^j|^s dt<\infty\ \forall s\geq 1$

and $\sup_{u \in V_t^j} |f(t,x_t^*,u) - f(t,x_t^*,u_t^*)| \leq \gamma_t^j$

Then for $\mathcal{H}(t,\omega,u) = (p_t, f(t,x_t^*,u) - f(t,x_t^*,u_t^*)) - (\lambda_0^*, f_0(t,x_t^*,u) - f_0(t,x_t^*,u_t^*))$

where $p_t = -M[\Phi_t^{*-1}\{\Phi_t^* \lambda_1^* + \int_t^1 \Phi_\tau^* f_{0x}'^*(\tau,x_\tau^*,u_\tau^*)d\tau \cdot \lambda_0^*\}/\mathcal{F}_t]$ (6)

$$\lambda_1 = G_x'(Mx_1^*, My_1^*) \quad , \quad \lambda_0 = G_y'(Mx_1^*, My_1^*) \qquad (7)$$

and the matrix Φ_t is a solution of the system

$$d\Phi_t = f_x'(t,x_t^*,u_t^*)\Phi_t dt + \sigma_x'(t,x_t^*)\Phi_t dW_t \quad \Phi_0 = I \qquad (8)$$

the following equation holds

$$\sup_{u \in V_t} \overline{\mathcal{H}}_{\{V,\mathcal{Y}\}}(t,\omega,u) = 0 \quad {}^{*)} \qquad (9)$$

Under other assumptions the necessary conditions for such problems are considered in [I], [5].

4. Proof of the theorem. It is easy to prove the next technical lemma which we are need of

Lemma 3. Let $\tau_1, \tau_2 \ldots \tau_n, s$ be a constant such that $\tau_i, s \geq 1$ $\sum_{i=1}^n \frac{1}{\tau_i} = \frac{1}{s}$, $\varphi_j^i(\omega)$ is a functions with $M|\varphi_j^i(\omega)|^{\tau_i} < \infty$, $i = 1 \ldots n$ $j = 0,1\ldots$, then $M|\prod_{i=1}^n \varphi_j^i|^s < \infty \ \forall j$, and if $M|\varphi_j^i - \varphi_0^i|^{\tau_i} \xrightarrow[j\to\infty]{} 0$ then $M|\prod_{i=1}^n \varphi_j^i - \prod_{i=1}^n \varphi_0^i|^s \xrightarrow[j\to\infty]{} 0$

Lemma 4. Let a_t, b_t be such that $P\{\int_0^1 |a_t| dt < \infty\} = 1$ $P\{\int_0^1 b_t^2 dt < \infty\} = 1$ and $dx_t = (A_t x_t + a_t)dt + (B_t x_t + b_t)dW_t$ $x_0 = \bar{x}$ then $x_t = \Phi_t\{\bar{x} + \int_0^t \Phi_s^{-1}[a_s - B_s b_s]ds + \int_0^t \Phi_s^{-1} b_s dW_s$

The proof follows immediatly from Ito's formula.

---

*) The existence of the function $\overline{\mathcal{H}}_{\{V,\mathcal{Y}\}}(t,\omega,u)$ is assured by the [4, предложение 4]

At first we prove the theorem under the addition assumptions

II_7) There exists function γ_t and constant N_1 such that $M\int_0^1 |\gamma_t|^s dt < \infty \quad \forall s \geq 1$

$$\sup_{u\in V_t} | f(t,x_t^*,u_t^*) - f(t,x_t^*,u)| \leq \gamma_t \quad \sup_{u\in V_t} |u-u_t^*| \leq N_1$$

<u>Lemma 5</u>. Under the conditions II_I)-II_7) for any control there exists unique solution x_t^u of (3), $x_t^u \in L_p^n[B\mathcal{F}]$ and

$$M \sup_t |x_t^u - x_t^*|^s < \infty \qquad \forall s \geq 1$$

<u>Proof</u>. Let $\varphi_t \equiv \overline{x}$, then by the II_3), II_5), II_7)

$$\int_0^t f(\tau,\varphi_\tau,u_\tau)d\tau \in L_2^1 C, \quad \sigma(t,\varphi_t) \in L_2^n[B\mathcal{F}]$$

By the lemma 2 there exists a unique $x_t^u \in L_2^n C$. Let $q_t = x_t^u - x_t^*$, then $q_t = \int_0^t f_x'(s,x_s^*+\theta_s q_s,u_s)ds + \int_0^t f(s,x_s^*,u_s) -$

$$- f(s,x_s^*,u_s^*)ds + \int_0^t \sigma_x'(s,x_s^*+\tilde{\theta}_s q_s)\, q_s\, dw_s \tag{I0}$$

So q_t is a solution of linear equation (I0) (it is easy to prove that $\theta_s, \tilde{\theta}_s$ can be chosen such that coefficients of this equation will be $\overline{B\mathcal{F}}$ measurable) and the proof of the lemma follows from II_3) and II_7).

We fix any control u_t . Let

$$F_0(t,x) = f_0(t,x,u_t) - f_0(t,x,u_t^*), \quad \Psi_0(t,x) = f_0(t,x,u_t^*)$$

$$F(t,x) = f(t,x,u_t) - f(t,x,u_t^*), \quad \Psi(t,x) = f(t,x,u_t^*)$$

and consider an auxiliary problem

$$G(Mx_1, My_1) \to min \tag{II}$$

$$dy_t = [\alpha_t F_0(t,x_t) + \Psi_0(t,x_t)]dt \qquad y_0 = \overline{y} \tag{I2}$$

$$dx_t = [\alpha_t F(t,x_t) + \Psi(t,x_t)]dt + \sigma(t,x_t)dw_t \quad x_0 = \overline{x} \tag{I3}$$

$$0 \leq \alpha_t \leq 1 , \quad \alpha_t \in L_\infty^1[B\mathcal{Y}] \tag{I4}$$

Lemma 6. Under the conditions $II_I)$-$II_7)$ tne pair (y_t^*, x_t^* $\alpha_t^* \equiv 0$) is optimal in the problem (II)-(I4).

Lemma was proved in [4] for $\sigma \equiv 0$. But that proof is suitable for our case.

Lemma 7. Let $II_I)$-$II_7)$ be held. Then for any $\alpha_t \in L_\infty^1[B\mathcal{Y}]$

$$M\int_0^1 \alpha_t\{\lambda_0 F_0(t,x_t^*) + [\lambda_1 \Phi_1 + \lambda_0 \int_t^1 f'_{0x}(\tau, x_\tau^*, u_\tau^*)\cdot \Phi_\tau d\tau]\cdot \Phi_t^{-1} F(t,x_t^*)dt \geq 0$$

here $\lambda_0, \lambda_1, \Phi_t$ is defined by (7), (8)

Proof. Let $\alpha_t \in L_\infty^1[B\mathcal{Y}]$, $0 \leq \alpha_t \leq 1$. We denote $y_t^\varepsilon, x_t^\varepsilon$ the solution of the system (I2),(I3) with the control $\varepsilon\alpha_t$, $y_t^0, x_t^0 = y_t^*, x_t^*$. If there exists tne sequence $\{\varepsilon_j\}$, the constant $\mathcal{J}$ such that $\lim\limits_{\varepsilon_j \to 0} \dfrac{G(Mx_1^{\varepsilon_j}, My_1^{\varepsilon_j}) - G(Mx_1^*, My_1^*)}{\varepsilon_j} = \mathcal{J}$

then by lemma 6 $\mathcal{J} \geq 0$. We shall show the existence of such sequence and we shall compute $\mathcal{J}$. It will be proof of the lemma 7. Using the Gronwall's lemma and the well known unequality for stochastic integrals, it is easy to see that

$$\overline{\lim_{\varepsilon \to 0}} \frac{1}{\varepsilon} \| x_t^\varepsilon - x_t^* \|_{L_s^n[B\mathcal{F}]} < \infty, \quad \overline{\lim_{\varepsilon \to 0}} \frac{1}{\varepsilon} \sqrt[2s]{M|x_1^\varepsilon - x_1^*|^{2s}} < \infty \tag{I5}$$

we shall prove that $\overline{\lim\limits_{\varepsilon \to 0}} \dfrac{M|y_1^\varepsilon - y_1^*|}{\varepsilon} < \infty$ (I6)

Indeed $\overline{\lim\limits_{\varepsilon \to 0}} \dfrac{M|y_1^\varepsilon - y_1^*|}{\varepsilon} = \overline{\lim\limits_{\varepsilon \to 0}} \dfrac{1}{\varepsilon} M|\int_0^1 \varepsilon\alpha_t \cdot F_0(t,x_t^\varepsilon) + \Psi_0(t,x_t^\varepsilon) - \Psi_0(t,x_t^*) dt|$

$$\leq \overline{\lim_{\varepsilon \to 0}} M\int_0^1 |F_0(t,x_t^\varepsilon)| dt + \overline{\lim_{\varepsilon \to 0}} \frac{1}{\varepsilon} M\int_0^1 |\Psi_0(t,x_t^\varepsilon) - \Psi_0(t,x_t^*)| dt$$

but $M\int_0^1 |F_0(t,x_t^\varepsilon)| dt \leq R_1 \cdot M\int_0^1 (\xi_t |x_t^\varepsilon|^q + |u_t|^r + |u_t^*|^r + 1) dt \leq$

$$\leq (R_2 M\int_0^1 |\xi_t|^{\frac{p}{p-q}} dt)^{\frac{p-q}{p}} (R_2 M\int_0^1 |x_t^\varepsilon|^p dt)^{q/p} + 2R_1 M\int_0^1 (|u_t|^r + |u_t^*|^r) dt + R_1$$

$$\frac{1}{\varepsilon}M\int_0^1|\Psi_0(t,x_t^\varepsilon)-\Psi_0(t,x_t^*)|dt\le\frac{1}{\varepsilon}\|x_t^\varepsilon-x_t^*\|_{L_p^n}\cdot\|\Psi'_{0x}(t,x_t^*+\theta(x_t^\varepsilon-$$

$$-x_t^*))\|_{L^{n\times n}_{p(p-1)^{-1}}}\le R_3\left(M\int_0^1|\xi_t|^{p(p-1)^{-1}}(|x_t^*|+|x_t^*-x_t^\varepsilon|)^{\frac{p(q-1)}{p-1}}dt\right)^{\frac{p-1}{p}}$$

(I6) is proved.

$$\Delta x_t^\varepsilon=x_t^\varepsilon-x_t^*=\int_0^t[\varepsilon\alpha_t\cdot F_x'(\tau,x_\tau^*+\theta_1\Delta x_\tau^\varepsilon)\Delta x_\tau^\varepsilon+\Psi_x'(\tau,x_\tau^*+\theta_2\Delta x_\tau^\varepsilon)\cdot$$

$$\cdot\Delta x_\tau^\varepsilon+\varepsilon\alpha_\tau\cdot F(\tau,x_\tau^*)]d\tau+\int_0^t\sigma_x'(\tau,x_\tau^*+\theta_3\Delta x_\tau^\varepsilon)\,\Delta x_\tau^\varepsilon\,dW_\tau \tag{I7}$$

where θ_1, θ_2, θ_3 is a $n\times n\times n\times n$ matrix random functions, and $F_x'(\tau,x_\tau^*+\theta_1\Delta x_\tau^\varepsilon)$, $\Psi_x'(\tau,x_\tau^*+\theta_2\Delta x_\tau^\varepsilon)$, $\sigma_x'(\tau,x_\tau^*+\theta_3\Delta x_\tau^\varepsilon)$ are $\overline{B\mathcal{F}}$ measurable function. So Δx_t^ε is the solution of linear equation (I7) with the bounded coefficients. $\Phi_t(\varepsilon)$ is a solution of equation

$$d\Phi_t(\varepsilon)=[\varepsilon\alpha_t\cdot F_x'(t,x_t^*+\theta_1\Delta x_t^\varepsilon)+\Psi_x'(t,x_t^*+\theta_2\Delta x_\tau^\varepsilon)]\Phi_t(\varepsilon)dt+$$

$$+\sigma_x'(t,x_t^*+\theta_3\Delta x_t^\varepsilon)\cdot\Phi_t(\varepsilon)\,dW_t \qquad \Phi_0(\varepsilon)=I$$

Let us compute $\mathcal{J}$.

$$\frac{1}{\varepsilon}[G(Mx_1^\varepsilon,My_1^\varepsilon)-G(Mx_1^*,My_1^*)]=\frac{1}{\varepsilon}G_x'(Mx_1^*,My_1^*)\cdot M\Delta x_1^\varepsilon+$$

$$+\frac{1}{\varepsilon}G_y'(Mx_1^*,My_1^*)\,M\Delta y_1^*+\frac{1}{\varepsilon}o(|M\Delta x_1^\varepsilon|,|M\Delta y_1^\varepsilon|)=\frac{1}{\varepsilon}\Delta G_1(\varepsilon)+\frac{1}{\varepsilon}\Delta G_2(\varepsilon)+$$

$+\frac{1}{\varepsilon}\Delta G_3(\varepsilon)$ By (I5),(I6) $\lim_{\varepsilon\to0}\frac{1}{\varepsilon}\Delta G_3(\varepsilon)=0$

$$\frac{1}{\varepsilon}\Delta G_1(\varepsilon)=\frac{1}{\varepsilon}M\Delta x_1^\varepsilon=\frac{1}{\varepsilon}\lambda_1M\int_0^1\Phi_1(\varepsilon)\cdot\Phi_t^{-1}(\varepsilon)\cdot\varepsilon\alpha_t\cdot F(t,x_t^*)dt$$

$$\frac{1}{\varepsilon}\Delta G_2(\varepsilon)=\frac{1}{\varepsilon}\lambda_0M\int_0^1[\varepsilon\alpha_t F_0(t,x_t^\varepsilon)+\Psi_0(t,x_t^\varepsilon)-\Psi_0(t,x_t^*)]dt \tag{I8}$$

Let $\{\varepsilon_j\}$ be such sequence that $x_t^{\varepsilon_j}\longrightarrow x_t^*$ $\ell\times P$ a.s. It is not difficult to show that $\forall\ s\ge1$

$$M|\Phi_1(\varepsilon_j)-\Phi_1|^{2s}\xrightarrow[j\to\infty]{}0 \qquad \|\Phi_t^{-1}(\varepsilon_j)-\Phi_t^{-1}\|_{L^{n^2}_{2s}[B\mathcal{F}]}\xrightarrow[j\to\infty]{}0 \tag{I9}$$

$$\lim_{\varepsilon_j \to 0} M \int_0^1 \Big| \int_0^1 (\Phi_\tau^{-1}(\varepsilon) - \Phi_\tau^{-1}) \alpha_\tau \cdot F(\tau, x_\tau^*) d\tau \Big|^{2s} dt = 0 \qquad (20)$$

$$\lim_{\varepsilon_j \to 0} M \int_0^1 | \Phi_t(\varepsilon) - \Phi_t |^{2s} dt = 0 \qquad (2I)$$

From (I9), (2I) and lemma 3

$$\lim_{\varepsilon_j \to 0} \frac{1}{\varepsilon_j} \Delta G(\varepsilon_j) = \lambda_1 \cdot M \int_0^1 \Phi_1 \cdot \alpha_t \cdot \Phi_t^{-1} \cdot F(t, x_t^*) dt \qquad (22)$$

Will count the limits of each item in (I8)

By II_4), $\{ F_0(t, x_t^{\varepsilon_j}) \}$ is uniformly integrable and $\alpha_t \cdot F_0(t, x_t^{\varepsilon_j}) \to \alpha_t F_0(t, x_t^*)$ $\ell \times P$- a.s. and we obtain that

$$\lim_{\varepsilon_j \to 0} \frac{1}{\varepsilon_j} \lambda_0 \cdot M \int_0^1 \varepsilon_j \alpha_t \cdot F_0(t, x_t^{\varepsilon_j}) dt = \lambda_0 \cdot M \int_0^1 \alpha_t \cdot F_0(t, x_t^*) dt \qquad (23)$$

Making use (20), (2I) and the fact that

$$M \int_0^1 | \Psi'_{0x}(t, x_t^* + \Theta \Delta x_t^{\varepsilon_j}) - \Psi'_{0x}(t, x_t^*) |^{p(p-1)^{-1}} dt \underset{j \to \infty}{\longrightarrow} 0$$

and lemma 3, we obtain

$$\lim_{\varepsilon_j \to 0} \frac{1}{\varepsilon_j} \lambda_0 M \int_0^1 [\Psi_0(t, x_t^{\varepsilon_j}) - \Psi_0(t, x_t^*)] dt = \lim_{\varepsilon_j \to 0} \lambda_0 M \int_0^1 \frac{1}{\varepsilon_j} \Psi'_{0x}(t, x_t^* + \Theta \Delta x_t^{\varepsilon})$$
$$\Delta x_t^{\varepsilon_j} dt = \lim_{\varepsilon_j \to 0} M \int_0^1 \Psi'_{0x}(t, x_t^* + \Theta \Delta x_t^{\varepsilon_j}) \cdot \Phi_t(\varepsilon_j) \int_0^t \Phi_\tau^{-1}(\varepsilon_j) \alpha_\tau F(\tau, x_\tau^*) d\tau dt =$$
$$= M \int_0^1 \alpha_t \lambda_0 [\int_t^1 \Psi'_{0x}(\tau, x_\tau^*) \cdot \Phi_\tau d\tau] \Phi_t^{-1} \cdot F(t, x_t^*) dt \qquad (24)$$

It follows from (22), (23), (24) $J = M \int_0^1 \alpha_t \{ \lambda_0 \cdot F_0(t, x_t^*) +$

$$+ (\lambda_1 \Phi_1 + \lambda_0 \cdot \int_t^1 \Psi'_{0x}(\tau, x_\tau^*) \cdot \Phi_\tau d\tau) \cdot \Phi_t^{-1} \cdot F(t, x_t^*) dt \geqslant 0$$

The lemma 7 is proved.

Let $\widetilde{V}_t^j = \{u : u \in V_t^j,\ |u - u_t^*| \leq j\}$. By the lemma 7 for any $\overline{B\mathcal{Y}}$ measurable $u_t \in \widetilde{V}_t^j$ $\ell \times P$ - a.s. there holds $M\int_0^1 \mathcal{H}(t,\omega,u_t(\omega))dt \leq 0$ $\forall \alpha_t$. It follows that $\overline{\mathcal{H}}_{\{V,\mathcal{Y}\}}(t,\omega,u_t(\omega)) \leq 0$ $\ell \times P$- a.s. By [4. предложение I] for any $j \geq 1$

$$\sup_{u \in \widetilde{V}_t^j} \overline{\mathcal{H}}_{\{V,\mathcal{Y}\}}(t,\omega,u) = 0 \qquad \ell \times P\text{- a.s.} \qquad (25)$$

and since $\bigcup_{j=1}^{\infty} \widetilde{V}_t^j = V_t$ $\ell \times P$ - a,s. we obtain

$$\sup_{u \in V_t} \overline{\mathcal{H}}_{\{V,\mathcal{Y}\}}(t,\omega,u) = 0 \qquad \ell \times P\text{- a.s.}$$

The theorem is proved.

<u>Remark</u>. Let $\mathcal{Y}_t = \mathcal{F}_t$. Then II_6) automatically holds. It sufficient to set $\gamma_t^j = j$ and $V_t^j = \{u : u \in V_t;\ |f(t,x_t^*,u) - f(t,x_t^*,u_t^*)| \leq j\}$

References

I. I.M.Bismut, Conjugate convex functions in optimal stochastic control, J. of Math. Anal. Appl., 44, p. 384 (I973).

2. I.M.Bismut, Linear quadratic optimal stochastic control with random coefficients, SIAM, J. Control, I4, p. 4I9 (I976).

3. И.И.Гихман, А.В.Скороход. Управляемые случайные процессы. Киев, "Наукова Думка", I977.

4. В.И.Аркин, М.Т.Саксонов. Вероятностные процессы и управление экономическими системами. Москва, ЦЭМИ, I978.

5. H.J.Kushner, Necessary conditions for continuous parameter stochastic optimization problems, SIAM, J. Control,, II, p. 587 (I972).

DIFFUSION PROCESSES WITH SINGULAR CHARACTERISTICS

S.V.Anulova

The last papers by N.I.Portenko are devoted to the study of diffusions in Euclidean space, which have a singular drift, localized on a hypersurface S (cf.[1]). A.N.Kolmogorov has suggested a natural extension of the class of processes considered by Portenko. It covers, in particular, processes whose diffusion as well as drift has a singularity of the δ-function type concentrated on S. Just those very processes and their modifications are considered in this communication. The structure of such processes has much in common with that of diffusionswith Ventzell's boundary conditions in the region bounded by S; however, it differs from the structure of the latter by the penetrability of the boundary S.

Let R^n be the n-dimensional Euclidean space, $n \geqslant 2$,
$\Lambda_+ = \{x \in R^n; x^1 > 0\}, \Lambda_- = \{x \in R^n, x^1 < 0\}, \Lambda = \Lambda_+ \cup \Lambda_-, \Gamma = R^n \setminus \Lambda$.

Let a function $f = f(t,x): [0,\infty) \times R^n \to R^1$ with a compact support be bounded and continuous together with its derivatives $\frac{\partial f}{\partial t}, \frac{\partial f}{\partial x^i}, \frac{\partial^2 f}{\partial x^i \partial x^j}$, $i,j = 2,\dots,n$. Let the derivatives $\frac{\partial f}{\partial x^1}$, $\frac{\partial^2 f}{\partial x^1 \partial x^i}$, $i=1,\dots,n$, be bounded and continuous on each of the sets $\Lambda_+ \cup \Gamma, \Lambda_- \cup \Gamma$ (on Γ one-sided derivatives $\frac{\partial f^{\pm}}{\partial x^1}$, $\frac{\partial^2 f}{\partial x^1 \partial x^i}$, $i=1,\dots,n$, are meant). The set of all such f is denoted by $\hat{C}_0^{1,2}$.

Let the following borel functions bounded by a constant C (each in its own norm) be given:

$$a = (a^{ij}(t,x))_{i,j=1}^{n} \quad : [0,\infty)\times\Lambda \to S^n,$$

$$b = (b^{i}(t,x))_{i=1}^{n} \quad : [0,\infty)\times\Lambda \to R^n,$$

$$g = (g^{ij}(t,x))_{i,j=2}^{n} \quad : [0,\infty)\times\Gamma \to S^{n-1},$$

$$c = (c^{i}(t,x))_{i=2}^{n} \quad : [0,\infty)\times\Gamma \to R^{n-1},$$

$$c_{\pm} = c_{\pm}(t,x) : [0,\infty)\times\Gamma \to [0,\infty),$$

$$\rho = \rho(t,x) : [0,\infty)\times\Gamma \to [0,\infty)$$

(S^k is the set of all symmetric nonnegative definite k x k-matrices). Define operators A and B acting on $f \in \hat{C}_0^{1,2}$ according to the formulas

$$Af(t,x) = \sum_{i,j=1}^{n} a^{ij}(t,x)\frac{\partial^2 f}{\partial x^i \partial x^j}(t,x) + \sum_{i=1}^{n} b^i(t,x)\frac{\partial f}{\partial x^i}(t,x),$$
$$(t,x)\in[0,\infty)\times\Lambda,$$

$$Bf(t,x) = \sum_{i,j=2}^{n} g^{ij}(t,x)\frac{\partial^2 f}{\partial x^i \partial x^j}(t,x) + \sum_{i=2}^{n} c^i(t,x)\frac{\partial f}{\partial x^i}(t,x) +$$

$$+ c_+(t,x)\frac{\partial f^+}{\partial x^1}(t,x) - c_-(t,x)\frac{\partial f^-}{\partial x^1}(t,x), \quad (t,x)\in[0,\infty)\times\Gamma.$$

An operator is called continuous if its coeffitients are continuous.

Denote Ω the set of all continuous functions $x = x_t$:

$[0,\infty) \to R^n$. For $s,t \in [0,\infty)$, $s \le t$ set

$$\mathcal{M}^s = \sigma\{x_u, u \in [s,\infty)\}, \quad \mathcal{M}^s_t = \sigma\{x_u, u \in [s,t]\}.$$

Fix $(s,x) \in [0,\infty) \times R^n$.

Definition 1. A measure P on $(\Omega, \mathcal{M}^s)$ is called a solution of the (A,B,ρ) -submartingale problem starting from (s,x) if $P(x_s = x) = 1$ and for any $f \in \hat{C}^{1,2}_0$ satisfying the condition $(\rho \frac{\partial}{\partial t} + B) f \ge 0$ the process

$$f(t,x_t) - \int_s^t I_\Lambda \left(\frac{\partial}{\partial t} + A\right) f(u,x_u)\, du, \quad t \in [s,\infty),$$

is an $(\mathcal{M}^s_t)$ -submartingale.

Definition 2. We say that a solution of the (A,B,ρ) -submartingale problem possesses local time, if there exists a continuous nondecreasing $(\mathcal{M}^s_t)$ -adapted process $\varphi = \varphi_t : [s,\infty) \times \Omega \to [0,\infty)$ such that $\varphi_s = 0$, $M\varphi_t < \infty$, $\varphi_t = \int_s^t I_\Gamma(x_u)\, d\varphi_u$ for any $t \in [s,\infty)$ and for any $f \in \hat{C}^{1,2}_0$ the process

$$f(t,x_t) - \int_s^t I_\Lambda \left(\frac{\partial}{\partial u} + A\right) f(u,x_u)\, du - \int_s^t I_\Gamma \left(\rho \frac{\partial}{\partial u} + B\right) f(u,x_u)\, d\varphi_u$$

is an $(\mathcal{M}^s_t)$ -martingale.

The following result analogous to theorems 2.4[3] and 3.2[4] is easily obtained:

Lemma 1. If B, ρ are continuous, $c_+ + c_- \ge \varepsilon > 0$ then any solution of the (A,B,ρ) -submartingale problem possesses local time.

The assertions of the following lemma are proved in the same way as the correspondent ones in [2].

Lemma 2. Suppose a solution of the (A,B,ρ) -submartingale problem possesses local time φ. If $\rho+c_++c_-\geqslant\varepsilon>0$, then:

1) for any $t,u\in[s,\infty)$, $t\leqslant u$

$$M(\varphi_u-\varphi_t)^3\leqslant N(C,\varepsilon,n)(u-t+\sqrt{u-t})^3;$$

2) for any $t\geqslant s$, $p\geqslant n+1$, $f\in L_p([s,t]\times\Lambda)$

$$M\Big|\int_s^t I_\Lambda(x_u)\sqrt[n+1]{\det a(u,x_u)}\,f(u,x_u)\,du\Big|\leqslant N(C,\varepsilon,t-s,n)\|f\|_{L_p};$$

3) for any $t\geqslant s$, $p\geqslant n$, $\psi\in L_p(\Gamma)$

$$M\Big|\int_s^t\sqrt[n]{\det g(u,x_u)}\,\psi(x_u)\,d\varphi_u\Big|\leqslant N(C,\varepsilon,t-s,n)\|\psi\|_{L_p};$$

under an additional assumption $\rho\geqslant\varepsilon>0$ the estimate holds for any $\psi\in L_p([s,t]\times\Gamma)$.

The proof of the following lemma is obtained through reproducing the argument of sections 3[3] and 5[4].

Lemma 3. If A,B,ρ are continuous, $c_++c_-\geqslant\varepsilon>0$, then there exists a solution of the (A,B,ρ) -submartingale problem starting from any point.

The main result is formulated in the following theorem.

Theorem. Suppose

1) the operator $A|_{\Lambda_+}$ (resp. $A|_{\Lambda_-}$) is either continuous or uniformly elliptic;

2) $a^{11}\geqslant\varepsilon>0$;

3) $c_++c_-+\rho\geqslant\varepsilon>0$;

4) one of the following conditions a)-d) is satisfied:

a) B,ρ are continuous;

b) A, B, ρ are time independent, B is either continuous or uniformly elliptic;

c) B, ρ are time-independent, B is uniformly elliptic;

d) B is uniformly elliptic, $\rho \geq \varepsilon > 0$.

Then there exists a function mapping every point $(s,x) \in [0,\infty) \times R^n$ into a measure $P_{s,x}$ on $(\Omega, \mathcal{M}^s)$ which is a solution of the (A,B,ρ) - submartingale problem starting from (s,x) such that $(x_t, \mathcal{M}_t^s, P_{s,x})$ is a strong Markov process.

Proof. All the omitted details of the proof can be found in [2]. Let us take a sequence A_n, B_n, ρ_n , n=1,2,..., of elements satisfying the conditions of Lemma 3, tending to A, B, ρ in a suitable sense. Fix $(s,x) \in [0,\infty) \times R^n$ and denote P_n the solution of the (A_n, B_n, ρ_n) -submartingale problem starting from (s,x) , n=1,2,... Assertion 1) of Lemma 2 provides the relative compactness of the family $\{P_n\}$. One can show by the standard argument that every limit point of $\{P_n\}$ is a solution of the (A,B,ρ) -submartingale problem starting from (s,x) (in the discontinuous case the L_ρ- estimates of Lemma 2 are of vital importance).The existence of the corresponding strong Markov process is proved by a method due to N.V.Krylov.

1 Portenko N.I., On stochastic differential equations with generalized trend vector, II Vilnius conference on probability theory, abstracts of communications, vol.1, Vilnius,

1977.

2 Anulova S.V., On stochastic differential equations with boundary conditions in a halfspace, Izv. AN SSSR, ser. math., 1979.

3 Stroock D.W., Varadhan S.R.S., Diffusion processes with boundary conditions, Comm. Pure Appl. Math., 1971, vol. XXIV, 2, 147-225.

4 Anderson R.F., Diffusions with second order boundary conditions, Indiana Univ. Math. J., vol. 25, 4, 367-395.

CONSTRUCTION AND PROPERTIES OF A CLASS OF STOCHASTIC INTEGRALS

J.M.Stoyanov* and O.B.Enchev**

1. Introduction

Let on the complete probability space $(\Omega, \mathcal{F}, \mathbb{P})$ the random processes $f=(f(t,\omega), t\in T)$ and $X=(X_t(\omega), t\in T)$ be given where $T=[0,1]$. It is possible under some conditions for f and X to construct a random element of the type $\int_0^t f(s,\omega)dX_s(\omega),\ t\in T$ which is called a stochastic integral.

The development of the theory of the random processes and their applications shows that the stochastic integrals are very important.

The first results concerning the construction and the properties of the stochastic integrals belong to the Japan mathematician K.Ito and the Soviet mathematician I.Gihman. Their papers published about 30 years ago are well known. The exact coordinates of these papers can be found in [1] , [2] , [9] and [10] .

Now we have a well developed theory of the stochastic integrals and the corresponding stochastic differential equations when X is either the Wiener process $W=(W_t, t\geq 0)$ or the (centred) Poisson process $\mathcal{T}=(\pi_t, t\geq 0)$ and the integrand f is a nonanticipating functional of X (see [1] , [2] , [4]).

Many general results have been obtained during the last 10 years about the stochastic integration when $X=M=(M_t, t\geq 0)$ is either a square integrable martingale, or a local martingale, or a weak martingale and the random function f is again adapted with respect to the family of σ-algebras generated by M . Important results in this direction belong to H.Kunita, S.Watanabe, P.-A. Meyer,

*Inst.Mathematics,Bulg.Acad.Sci., 1090 Sofia,Box 373,Bulgaria
**Dept.Mathematics,VIMMESS Institute, 7000 Russe, Bulgaria.

C.Doleans-Dade, C.Dellacherie, N.Kazamaki, P.Protter, M.Metivier, J.Jacod, and other. (see [3], [9] [10]).

Some new possibilities on the stochastic integration can be found in Daleckii-Paramonova [6] , [7] and Kabanov-Skorohod [8].

In 1975 J.Yeh published his paper [11] containing construction of a stochastic integral of the type

(1) $$\int_0^1 f(t)\,dX_t(\omega)$$

where $f=(f(t), t\in T)$ is a nonrandom function and $X=(X_t(\omega), t\in T)$ is a Gaussian process. We would like to note especially that all assumptions in [11] concern only the covariance function of the process X plus a natural condition of an integrability of f . Properties of a martingale type or of a nonanticipateness are not required.

The present paper is devoted to the construction and the properties of the stochastic integral (1) when $f=(f(t,\omega), t\in T)$ is a random. We show in two cases the possibility to integrate the random function f with respect to the random process X under conditions concerning only the covariance functions of f and X .

2. Stochastic integral for independent processes

Let $L_2=L_2(\Omega,\mathcal{F},\mathbb{P})$ be the space of the real one-dimensional random variables with a finite second moment. The inner product of the elements $\xi_1, \xi_2 \in L_2$ will be denoted by $(\xi_1,\xi_2)=\mathbb{E}\{\xi_1\xi_2\}$ and the norm of $\xi\in L_2$ by $\|\xi\|=\sqrt{(\xi,\xi)}$. Let $\mathcal{L}_2=\mathcal{L}_2(T,\Omega,\mathcal{F},\mathbb{P})$ be the space of L_2-continuous (t,ω)-measurable functions $f=(f(t,\omega), t\in T, \omega\in\Omega)$ with $\mathbb{E}\{\int_T \|f(t,\omega)\|^2 dt\}<\infty$. The Lebesgue measures in $\mathcal{R}^1$ and $\mathcal{R}^2$ will be denoted by μ_1 and μ_2 , respectively.

Definition 1. The random process $X=(X_t, t\in T)$ belongs to the class $\mathcal{K}$ if the following conditions hold: 1. $X_t\in L_2$ $\forall t\in T$; 2. The covariance function $K(s,t)=\mathbb{E}\{X_sX_t\}$ is continuous on the square $D=T\times T$; 3. The derivatives $\frac{\partial^2 K}{\partial s^2}$ and $\frac{\partial^2 K}{\partial s\,\partial t}$ exist and are bounded on the open triangles $D_1=\{(s,t)\in D, s<t\}$ and $D_2=\{(s,t)\in D, s>t\}$; 4.The function $\frac{\partial^2 K}{\partial s \partial t}$ is continuous on $D_1 \cup D_2$ possibly with an exception of a μ_2-null set.

Let us note that the class $\mathcal{K}$ cantains many processes, for example such important processes as the Wiener process W, the Poisson process π and other ones defined by their covariance functions.

Let $\mathcal{P}_n=\{0=a_0<a_1<\dots<a_n=1\}$ be a partition of the interval $T=[0,1]$, $|\mathcal{P}_n|=\max_k |a_k-a_{k-1}|$ be its diameter and $\mathcal{A}_n=\{\alpha_{kn}\in[a_{k-1},a_k], k=\overline{1,n}\}$. The quantity

$$S_n = S_n(f, X; \mathcal{P}_n, \mathcal{A}_n) = \sum_{k=1}^{n} f(\alpha_{kn})(X_{a_k} - X_{a_{k-1}})$$

is said to be an integral sum of f about X.

Definition 2. If at $n\to\infty$ and $|\mathcal{P}_n|\to 0$ the sequence $\{S_n\}$ is fundamental in the space L_2 and the limit does not depend on $\mathcal{P}_n$ and $\mathcal{A}_n$, then this limit is called a stochastic integral of the random function f with respect to the random process X. The following notations will be used:

$$J(f) = J^X(f) = \int_T f(t,\omega)\,dX_t(\omega) = \int_T f(t)\,dX_t .$$

Theorem 1. Let the random function $f\in\mathcal{L}_2$ and the random process $X\in\mathcal{K}$. If f and X are independent the stochastic integral $J(f)$ exists and it is an element of the space L_2.

It is interesting to consider the stochastic integral $J(f)$ as a function of its upper limit. Put $J_t(f)=J(f\cdot\chi_{[0,t]})$ where $\chi_{[0,t]}$ is an indicator of the subinterval $[0,t]$, $t\in T$. Then $J(f)=J_1(f)$.

The next theorem contains some properties of $J_t(f)$, $t\in T$.

Theorem 2. Let f and g be elements of the space $\mathcal{L}_2$ and any one of them does not depend on the process X where $X\in\mathcal{K}$. Then:

a) for arbitrary real numbers c_1 and c_2

$$J_t(c_1 f + c_2 g) = c_1 J_t(f) + c_2 J_t(g), \quad t\in T;$$

b) if $\mathbb{E}X_s=0$ $\forall s\in T$ then $\mathbb{E}\{J_t(f)\}=0$ $\forall t\in T$;

c) $(J_t(f), J_t(g)) = \int_{[0,t]} (f(s), g(s))\,\gamma(s)\,d\mu_1 + \int_{[0,t]\times[0,t]} (f(s), g(u))\,\frac{\partial^2 K}{\partial u\,\partial s}(s,u)\,d\mu_2$

where $\gamma(s) = \lim_{u \downarrow s} \frac{1}{u-s}\left(K(u,s) - K(s,s)\right) - \lim_{u \uparrow s} \frac{1}{u-s}\left(K(u,s) - K(s,s)\right);$

d) $\|J_t(f)\|^2 = \int_{[0,t]} \|f(s)\|^2 \gamma(s)\,d\mu_1 + \int_{[0,t]\times[0,t]} (f(s), f(u))\,\frac{\partial^2 K}{\partial u\,\partial s}(s,u)\,d\mu_2\,;$

e) if $C(s,t)$ is the covariance function of f then

$$\Gamma(s,t) = \mathbb{E}\{J_s(f) J_t(f)\} = \int_0^{s\wedge t} C(u,u)\,\gamma(u)\,d\mu_1 + \int_{[0,s]\times[0,t]} C(u,v)\,\frac{\partial^2 K}{\partial v\,\partial u}(u,v)\,d\mu_2\,.$$

Example 1. Let $W = (W_t, t \in T)$ and $W^* = (W_t^*, t \in T)$ be two independent Wiener processes, $\pi = (\pi_t, t \in T)$ and $\pi^* = (\pi_t^*, t \in T)$ be two independent Poisson processes with parameters λ and λ^*, respectively, where W and π are independent. Then:

$$\|J_t^W(W^*)\|^2 = \mathbb{E}\left\{\left(\int_0^t W_s^* \, dW_s\right)^2\right\} = \frac{1}{2} t^2,$$

$$\|J_t^W(\pi)\|^2 = \mathbb{E}\left\{\left(\int_0^t \pi_s \, dW_s\right)^2\right\} = \frac{\lambda^2}{2} t^3 + \frac{\lambda}{2} t^2,$$

$$\|J_t^{\pi}(W)\|^2 = \mathbb{E}\left\{\left(\int_0^t W_s \, d\pi_s\right)^2\right\} = \frac{\lambda^2}{2} t^3 + \frac{\lambda}{2} t^2,$$

$$\|J_t^{\pi}(\pi^*)\|^2 = \mathbb{E}\left\{\left(\int_0^t \pi_s^* \, d\pi_s\right)^2\right\} = \frac{\lambda^2 \lambda^{*2}}{4} t^4 + \frac{\lambda\lambda^*(\lambda + \lambda^*)}{3} t^3 + \frac{\lambda\lambda^*}{2} t^2.$$

Example 2. Let $f \in \mathcal{L}_2$ and W be the Wiener process (f and W are independent). Then:

$$\Gamma(s,t) = (J_s^W(f), J_t^W(f)) = \int_0^{s\wedge t} C(u,u)\,d\mu_1$$

where $C(u,v)$ is the covariance function of f.

3. Stochastic integral for processes satisfying property (G)

Let us consider another situation without the restriction for the random processes f and X to be independent. Assume for convenience that f and X have zero mean, i.e. $\mathbb{E} f(t) = \mathbb{E} X_t = 0 \quad \forall t \in T$.

Definition 3. Suppose f and X are arbitrary random processes. We shall say that the pair (f,X) satisfies the property (G) if for each choise of the time-moments t_1, t_2, t_3 and t_4 from T the 4-dimensional random vector $(f(t_1), f(t_2), X_{t_3}, X_{t_4})$ has a Gaussian distribution.

Theorem 3. Let the L_2-continuous random function $f \in \mathcal{L}_2$ and the process $X \in \mathcal{K}$ be such that $\mathbb{E}\{f^2(s) X_t^2\} < \infty$, $\forall s,t \in T$ and mutual covariance function $M(s,t) = \mathbb{E}\{X_s f(t)\}$ have bounded derivatives $\frac{\partial^2 M}{\partial s^2}$ and $\frac{\partial^2 M}{\partial s \partial t}$ on the square $(0,1)\times[0,1]$ with an exception may be of a μ_2-null set. If the pair (f,X) satisfies the property (G) then the stochastic integral $J(f)$ exists and it is an element of the space L_2.

Now let $X \in \mathcal{K}$, $f,g \in \mathcal{L}_2$, $K(s,t)$ be the covariance function of X and $M_f(s,t)$ and $M_g(s,t)$ be the mutual covariance functions of (X,f) and (X,g) respectively.

Theorem 4. Suppose every one of the pairs (f,X) and (g,X) satisfies the conditions of theorem 3. Then:

a) $$(J(f), J(g)) = \int_T (f(t), g(t))\,\gamma(t)\,d\mu_1 + \int_D (f(s), g(t)) \frac{\partial^2 K}{\partial t \partial s}(s,t)\,d\mu_2$$
$$+ \int_D \frac{\partial M_f}{\partial s}(s,t) \frac{\partial M_g}{\partial t}(t,s)\,d\mu_2 + \Big(\int_T \frac{\partial M_f}{\partial s}(s,s)\,d\mu_1\Big)\Big(\int_T \frac{\partial M_g}{\partial s}(s,s)\,d\mu_1\Big);$$

b) $$\|J(f)\|^2 = \int_T \|f(t)\|^2 \gamma(t)\,d\mu_1 + \int_D (f(s), f(t)) \frac{\partial^2 K}{\partial t \partial s}(s,t)\,d\mu_2$$
$$+ \int_D \frac{\partial M_f}{\partial s}(s,t) \frac{\partial M_f}{\partial t}(t,s)\,d\mu_2 + \Big(\int_T \frac{\partial M_f}{\partial s}(s,s)\,d\mu_1\Big)^2.$$

4. Some remarks

1) The proofs of our results are based on special estimates for the covariance functions of X, f, g, ect. One can prove that under additional assumptions the integral $J_t(f)$, $t \in T$ has other interesting properties.

2) Theorems 1 and 2 given above are an extension of theorems 1 and 4 from the paper of J.Yeh [11].

References

[1] I.Gihman and A.Skorohod: Stochastic Differential Equations. Naukova Dumka,Kiev,1968. Engl.transl. Springer-Verlag,1972.

[2] I.Gihman and A.Skorohod: Theory of Random Processes.Volume 3. Nauka,Moscow,1975. Engl.transl. Springer-Verlag,1977.

[3] A.Kussmaul: Stochastic Integration and Generalized Martingales Pitman,London,1977.

[4] R.Liptser and A.Shiryayev: Statistics of Random Processes. Nauka,Moscow,1974. Engl.transl. Springer-Verlag,Volume 1,1977; Volume 2,1978.

[5] E.Begovatov: Stochastic Integral for a Gaussian Process. Theory Probab.Appl.,14 (1969),370-372.

[6] Yu.Daleckii and S.Paramonova: Stochastic Integrals with Respect to a Normal Distributed Additive Set Functions. Dokl.Akad.Nauk SSSR, 208 (1973),512-515.

[7] Yu.Daleckii and S.Paramonova: On a Formula from the Theory of Gaussian Measures and on Estimation of Stochastic Integrals. Theory Probab.Appl.,19 (1974),844-849.

[8] Yu.Kabanov and A.Skorohod: Extended Stochastic Integrals. In: Proc.School-Seminar on the Theory of Random Processes, Druskininkai 1974,Part 1,Vilnius,1975,121-167.

[9] M.Metivier and J.Pellaumail: A Basic Course on General Stochastic Integration. In: Séminaire de Probabilités, Univ.de Rennes,1977.

[10] P.-A.Meyer: Un cours sur les Intégrales Stochastiques. In: Lecture Notes in Mathem., 511 (1976),245-400.

[11] J.Yeh: Stochastic Integral of L_2-Function with Respect to Gaussian Processes. Tohoku Mathem.J.,27 (1975),175-186.

THE ASYMPTOTIC STATISTICAL PROBLEMS FOR FIELDS OF DIFFUSION TYPE

A.F.Taraskin (Kuibyshev)

§1. Introduction

Let we have the continuous random field $X = \{X_{s,t}, s \geqslant 0, t \geqslant 0\}$ which satisfies an stochastic differential equation of the form [1,2]

$$dX_{s,t} = a_{s,t}(X;\theta)\,ds\,dt + \sigma_{s,t}(X)\,dW_{s,t}$$

and it is required to get the statistical inference of the value θ having one realization of this field at some part of the positive quadrant.

The similar problems for random processes, which are resolutions of the Ito stochastic differential equations, are considered by many authors. Particularity there are some asymptotic problems of an estimating of unknown drift parameters and testing of hypotheses for diffusion processes basing on the maximum likelihood (ML) method be considered in [3,4].

In this paper we extend the asymptotic ML theory for diffusion random fields [5]. There is a likelihood ratio for the considering fields in [5]. This likelihood ratio is used for the ML method and includs the stochastic integral in Wiener field. Then we give some necessary information about of the stochastic integral.

§3 contains some sufficient conditions for the asymptotic normality of stochastic integral; the asymptotic likelihood results, on which our interest centres are state, are gotten in §4. These results follow easily from the limit theorem in §3, which is the central limit theorem for the considering problems.

§2. Notation and preliminaries

Let R_+^2 be the positive quadrant of the plane. Thre is a partial ordering $\prec$ in R_+^2. Let $z=(s,t)$, $s\geq 0$, $t\geq 0$ denote points on R_+^2. $z_1 \prec z_2$ will denote $s_1 \leq s_2$ and $t_1 \leq t_2$. Let $z_1=(s_1,t_1)$, $z_2=(s_2,t_2)$, the condition $s_1 < s_2$, $t_1 < t_2$ will be denoted by $z_1 \prec\prec z_2$. If $z_1 \prec\prec z_2$, $(z_1,z_2]$ will denote the rectangle $(s_1,s_2]\times(t_1,t_2]$ and if Y_z is a function, $Y(z_1,z_2]$ will denote $Y_{s_2,t_2}-Y_{s_2,t_1}-Y_{s_1,t_2}+$ $+Y_{s_1,t_1}$. R_z will denote the rectangle $\{\zeta : 0\prec\zeta\prec z\}$.

Let $(\Omega,\mathcal{F},P)$ be a probability space and let $\{\mathcal{F}_z, z\in R_+^2\}$ be a family of sub-σ-fields of $\mathcal{F}$ satisfying [6] :

(F1) if $z_1 \prec z_2$ then $\mathcal{F}_{z_1}\subset\mathcal{F}_{z_2}$;

(F2) $\mathcal{F}_0$ contains all null sets of $\mathcal{F}$;

(F3) for each z $\mathcal{F}_z = \bigcap \mathcal{F}_\zeta$, $z\prec\prec\zeta$;

(F4) for each z $\mathcal{F}_z^1$ and $\mathcal{F}_z^2$ are conditionally independent given $\mathcal{F}_z$, where $\mathcal{F}_z^1=\mathcal{F}_{s,\infty}$ and $\mathcal{F}_z^2=\mathcal{F}_{\infty,t}$.

Let $W=\{W_z, \mathcal{F}_z, z\in R_+^2\}$ be the standard random Wiener field and $\mathcal{L}_W^2$ be the class of all $\mathcal{F}_z$-adapted measurable random fields $\varphi=\{\varphi_z, z\in R_+^2\}$ for which

$$E\left\{\int_{R_z}\varphi_\zeta^2\,d\zeta\right\} < \infty$$

for all $z\in R_+^2$. For each field $\varphi\in\mathcal{L}_W^2$ exists a stochastic integral of the form [6,7]

$$J_z(\varphi)=\int_{R_z}\varphi_\zeta\,dW_\zeta \qquad (1)$$

Properties of the integrals (1) are in [6,7]. Wecan say, that the field $J_z(\varphi)$ will be a strong square-integrable martingale with the characteristic

(2) $$\langle \mathcal{J}(\varphi)\rangle_z = \int_{R_z} \varphi_\zeta^2 \, d\zeta ,$$

and for each positive ε and N to take place [7]

(3) $$P\{\sup_{0\prec\zeta\prec z} |\mathcal{J}_\zeta(\varphi)| > \varepsilon\} \leq \frac{N}{\varepsilon^2} + P\{\langle \mathcal{J}(\varphi)\rangle_z > N\}.$$

§3. Asymptotic normality of the stochastic integral

Definition. The nonrandom nonnegative function $b = \{ b_z , z \in R_+^2\}$ is a normalization function along the line Γ on R_+^2, if

(i) for any $z_1 \prec z_2$ $b(z_1, z_2] \geqslant 0$;

(ii) for any $z \in R_+^2$

(4) $$b_z = \int_{R_z} \rho_\zeta^2 \, d\zeta ,$$

where $\rho = \{\rho_z , z \in R_+^2\}$ is a measurable function.

(iii) the line Γ has a parametric representation $\{z : z = \gamma(\tau), \tau \geqslant 0\}$, where $\gamma(\tau_1) \prec \gamma(\tau_2)$ for any τ_1, τ_2 ($\tau_1 \leqslant \tau_2$) and $|\gamma(\tau)| \longrightarrow \infty$ when $\tau \longrightarrow \infty$;

(iiii) $b_z \longrightarrow \infty$ when $|z| \longrightarrow \infty$ and $z \in \Gamma$.

Theorem 1. Let $\varphi \in \mathcal{L}_W^2$, b be a normalization function along the line Γ and

(A) $$b_z^{-1} \int_{R_z} (\varphi_\zeta - \rho_\zeta)^2 d\zeta \xrightarrow{P} 0$$

be satisfied when $|z| \longrightarrow \infty$ and $z \in \Gamma$. Then the distribution of random variable

$$b_z^{-\frac{1}{2}} \int_{R_z} \varphi_\zeta \, dW_\zeta \tag{5}$$

converges to the standard normal $\mathcal{N}(0,1)$ distribution.

Proof. Note, that the stochastic integral

$$\int_{R_z} \rho_\zeta \, dW_\zeta$$

has a normal distribution with zero mean and dispersion

$$\int_{R_z} \rho_\zeta^2 \, d\zeta .$$

Thus for proving this theorem it is enough to identify, that

$$b_z^{-\frac{1}{2}} \int_{R_z} \varphi_\zeta \, dW_\zeta - b_z^{-\frac{1}{2}} \int_{R_z} \rho_\zeta \, dW_\zeta \xrightarrow{P} 0, \tag{6}$$

when $|z| \longrightarrow \infty$, $z \in \Gamma$. According to the inequality (3) for a stochastic integral we have

$$P\left\{ \left| \int_{R_z} (\varphi_\zeta - \rho_\zeta) \, dW_\zeta \right| > \varepsilon b_z^{\frac{1}{2}} \right\} \leq$$

$$\leq \frac{16N}{\varepsilon^2} + P\left\{ \int_{R_z} (\varphi_\zeta - \rho_\zeta)^2 d\zeta > N b_z \right\}$$

for any positive ε and N. Due to arbitrarily ε, N and the condition (A) the relation (6) follows from this inequality.

Thus the theorem is proved.

Remark 1. It is easy to prove that the condition (A) is equivalent to

$$(A_1) \qquad b_z^{-1} \int_{R_z} \varphi_\zeta^2 \, d\zeta \xrightarrow{P} 1$$

and

$$(A_2) \qquad b_z^{-1} \int_{R_z} \varphi_\zeta \, \rho_\zeta \, d\zeta \xrightarrow{P} 1$$

when $|z| \to \infty$, $z \in \Gamma$.

§4. Asymptotic likelihood theory for diffusion fields

Let $(C, \mathcal{L})$ be a measurable space of the continuous functions $x = \{x_z , z \in R_+^2\}$ turning into zero on the axes and with σ-algebra $\mathcal{L} = \sigma\{x_z , z \in R_+^2\}$. Suppose $\mathcal{L}_z = \sigma\{x_\zeta , \zeta \prec z\}$. The continuous $\mathcal{F}_z$-adapted random field $X = \{X_z, z \in R_+^2\}$ is called the diffusion field if the functionals $a_z(x)$ and $\sigma_z(x)$, $x \in C$ exist such as

(i) $a_z(x)$ and $\sigma_z(x)$ are measurable functions of their arguments and $\mathcal{L}_z$-measurable for any $z \in R_+^2$;

(ii) for any $z \in R_+^2$

$$P\left\{\int_{R_z} |a_\zeta(X)| d\zeta < \infty\right\} = P\left\{\int_{R_z} \sigma_\zeta^2(X) d\zeta < \infty\right\} = 1;$$

and when

(7) $$X_z = \int_{R_z} a_\zeta(X)\,d\zeta + \int_{R_z} \sigma_\zeta(X)\,dW_\zeta .$$

The equation (7) can be considered as a stochastic equation which has a unique solution [2] under certain condition of the functionals $a_z(x)$ and $\sigma_z(x)$.

The functional $\sigma_z(x)$ in the form (7) is completely known, but $a_z(x)$ is dependent on the unknown parameter θ having values in the open set $K \subset R^n$. We denote the probability measure in the space $(C, \mathcal{L})$ which is in accordance with the solution of the equation (7) in the rectangle R_z, by μ^z. We make the following assumptions.

B$_1$. $a_z(x) = a_z(x;\theta) = \alpha_z(x) + (\theta, \varphi_z(x))$, $\varphi_z(x) = (\varphi_{1z}(x), \dots, \varphi_{nz}(x))$, where $\varphi_{kz}(x)$, $k = \overline{1,n}$ and $\alpha_z(x)$ are given $\mathcal{L}_z$-measurable functionals on C. The symbol $(\cdot,\cdot)$ means a scalar production in R^n.

B$_2$. The measures $\mu^z_{\theta_i}$, $i = 1,2$ corresponding to the solutions of equation (7) with $a_z(x) = a_z(x;\theta)$ are mutual equivalent for all $\theta_1, \theta_2 \in K$ and each $z \in R^2_+$.

Let $\theta_0 = (\theta_{01}, \dots, \theta_{0n})$ is a given vector in K. Let $\mathcal{L}_z(x;\theta) = \log((d\mu^z_\theta / d\mu^z_{\theta_0})(x))$. Using formula for density from paper [5], we obtain

(8) $$\mathcal{L}_z(X;\theta) = \int_{R_z} (\theta - \theta_0, \varphi_\zeta(X))\,\sigma_\zeta^{-2}(X)\,[dX_\zeta - \alpha_\zeta(X)\,d\zeta] - \frac{1}{2}\int_{R_z} \left[(\theta, \varphi_\zeta(X))^2 - (\theta_0, \varphi_\zeta(X))^2\right]\sigma_\zeta^{-2}(X)\,d\zeta .$$

Let us consider ML estimator of the unknown parameter

$\theta=(\theta_1,\dots,\theta_n)$, using the observation of the field X in the rectangle R_z, and its asymptotic behaviour when the rectangle unlimitedly expand.

Theorem 2. Let be satisfied the assumptions B_1, B_2 and exist the normalization function along the line Γ such that there are fulfilled the conditions by $|z|\longrightarrow\infty$, $z\in\Gamma$:

$$(C_1) \qquad b_z^{-1}\int_{R_z}\varphi_{k\zeta}(X)\varphi_{j\zeta}(X)\sigma_\zeta^{-2}(X)d\zeta \xrightarrow{P} b_{kj}(\theta)$$

where $B=\|b_{kj}(\theta)\|$, $k,j=\overline{1,n}$ is positively definite matrix and $b_{kj}(\theta)<\infty$;

(C_2) for each $\lambda\in R^n$

$$b_z^{-1}\int_{R_z}(\lambda,\varphi_\zeta(X))\sigma_\zeta^{-1}(X)\rho_\zeta\, d\zeta \xrightarrow{P} (B\lambda,\lambda)^{\frac{1}{2}}.$$

Then ML estimator $\hat\theta_z$ is consistent. A vector $b_z^{\frac{1}{2}}(\hat\theta_z-\theta)$ is asymptoticly normal with zero mean and the covariance matrix B^{-1} (B^{-1} is matrix which is reverce to B).

Proof. If θ is true value of the parameter, X is the solution of equation (7) at $a_z(x)=\alpha_z(x)+(\theta,\varphi_z(x))$ and

$$B_z=\left\|\int_{R_z}\varphi_{k\zeta}(X)\varphi_{j\zeta}(X)\sigma_\zeta^{-2}(X)d\zeta\right\|, \quad k,j=\overline{1,n}$$

is the non-singular matrix, then the ML estimator has the representation

$$(9) \qquad \hat\theta_z = \theta + B_z^{-1}\int_{R_z}\varphi_\zeta(X)\sigma_\zeta^{-1}(X)dW_\zeta .$$

If the matrix $B_{\mathfrak{z}}$ is singular, then the estimator $\hat{\theta}_{\mathfrak{z}}$ are supposed to be equal to a certain vector in K. Let there is (9) for $\omega \in \Omega_{\mathfrak{z}} \subset \Omega$. If there is ($C_1$) that $P\{\Omega_{\mathfrak{z}}\} \to 1$ and $b_{\mathfrak{z}} B_{\mathfrak{z}}^{-1} \chi_{\Omega_{\mathfrak{z}}} \xrightarrow{P} B^{-1}$ (χ_A is a characteristic function of the set A). Besides, using the inequality (3), we obtain

$$b_{\mathfrak{z}}^{-1} \int_{R_{\mathfrak{z}}} \varphi_{\kappa\zeta}(X) \sigma_{\zeta}^{-1}(X)\, dW_{\zeta} \xrightarrow{P} 0.$$

Thus, the second item in the right hand of (9) converge on the probability to the zero vector, that means there is a consistency of the estimator $\hat{\theta}_{\mathfrak{z}}$. For proof of the second assertion of the theorem it is enough to show that the vector

$$b_{\mathfrak{z}}^{-\frac{1}{2}} \int_{R_{\mathfrak{z}}} \varphi_{\zeta}(X) \sigma_{\zeta}^{-1}(X)\, dW_{\zeta}$$

is asymptoticaly normal with the zero mean and the covariance matrix B. For this it is enough to notice, that for each $\lambda \in R^n$ the value

$$b_{\mathfrak{z}}^{-\frac{1}{2}} \int_{R_{\mathfrak{z}}} (\lambda, \varphi_{\zeta}(X)) \sigma_{\zeta}^{-1}(X)\, dW_{\zeta}$$

is asymptoticaly normal $\mathcal{N}(0, (B\lambda, \lambda))$ on the strength of the conditions (C_1), (C_2) and of the remark 1.

Let us consider a special case. Let us together wich the assumptions B_1 , B_2 there is valid assumption

B_3. $\varphi_{\mathfrak{z}}(x) \sigma_{\mathfrak{z}}^{-1}(x) = (\beta_1(\mathfrak{z}), \dots, \beta_n(\mathfrak{z}))$ where $\beta_K(\mathfrak{z})$, $K = \overline{1,n}$ are the random fields independent from $W_{\mathfrak{z}}$, for which

$$(10) \qquad b_{\kappa j}(z) = \int_{R_z} \beta_\kappa(\zeta)\beta_j(\zeta)\,d\zeta < \infty, \quad \kappa, j = \overline{1,n}.$$

<u>Theorem 3.</u> Let the assumptions $B_1 - B_3$ are fulfilled and the normalization function b_z along the line Γ to exists such that by $|z| \longrightarrow \infty$, $z \in \Gamma$ is fulfilled the condition

$$(C_1') \qquad b_z^{-1} \int_{R_z} \beta_\kappa(\zeta)\beta_j(\zeta)\,d\zeta \xrightarrow{P} b_{\kappa j},$$

where $B = \| b_{\kappa j} \|$, $\kappa, j = \overline{1,n}$ is positively definite matrix. Then the ML estimator $\hat{\theta}_z$ is consistence and the vector $b_z^{\frac{1}{2}}(\hat{\theta}_z - \theta)$ is asymptoticaly normal with zero mean and covariance matrix B^{-1}.

The proof of this theorem is standard. For that reason we omit it.

<u>Remark 2.</u> If $\beta_\kappa(z)$, $\kappa = \overline{1,n}$ are the nonrandom functions on R_+^2 and matrix $B_z = \| b_{\kappa j}(z) \|$, $\kappa, j = \overline{1,n}$ is the non-singular, where $b_{\kappa j}(z)$ are defined by (10), then the ML estimator $\hat{\theta}_z$ of the parameter θ is an effective and has normal distribution $\mathcal{N}(0, B^{-1})$.

Let consider asymptotic properties of the likelihood ratio test for verification of the parametric hypothesis. Let the observable field has the representation (7) and the supposition B_1 and B_2 are fulfilled and the K set includes a zero vector. Suppose that we wish to test the $H_0 : \theta = (0, \dots, 0)$ hypothesis against a general alternative imposing no restriction other than $\theta \in K$. We shall solve this problem using the ML ratio. As a critical region we take the following set of the functions

$$C_z = \{X : \mathcal{L}_z(X;\hat{\theta}_z) > c_\gamma\} \tag{11}$$

when $\mathcal{L}_z(x;\theta)$ is defined (8), in which $\theta_0 = (0,\dots,0)$, and c_γ is constant for which the following inequality has to be satisfied

$$P(\{X_\zeta,\ \zeta \prec z\} \in C_z) \leq \gamma$$

if the hypothesis H_0 is true and γ is a significance level. The limit distribution of the statistic $\mathcal{L}_z(X;\hat{\theta}_z)$, on which the considered test is based, is defined by the following theorem.

Theorem 4. Let the conditions of the theorem 2 are fulfilled and $\hat{\theta}_z$ is a ML estimator of the θ parameter. Then the value $2\mathcal{L}_z(X;\hat{\theta}_z)$ has the limiting distribution χ^2_n (when $|z| \to \infty$, $z \in \Gamma$) if the hypotesis H_0 is true.

Proof. From (9) and (8) we can obtain

$$2\mathcal{L}_z(X;\hat{\theta}_z) = (B_z\hat{\theta}_z, \hat{\theta}_z)$$

And now the assertion of the theorem follows from the condition (C_1) and that the value $b_z^{\frac{1}{2}}\hat{\theta}_z$ is asymptotic normal with zero mean and covariance matrix B^{-1}.

Remark3. If the conditions of the theorem 3 is fulfilled, $\beta_K(z)$, $K=\overline{1,n}$ are non-random functions on R^2_+ and B_z is non-singular, then the value $2\mathcal{L}_z(X;\hat{\theta}_z)$ has χ^2_n distribution for fixed z.

Remark 4. If is easy to prove that under the conditions of the theorem 2 the test (11) is consistency.

REFERENCES

[1]. Cairoli R. Sur une èquation différentielle stochastique. C. R. Acad. Sc. Paris, t. 274, 1972, A1739-A1742.

[2]. Tsarenko T.I. A stochastic analog of an equation of the hyperbolic type is introduced for random fields with the help of a stochastic integral in Wiener measure. "Cybernetics" N4, 1972, p. 115-117 (in Russian).

[3]. Taraskin A.F. On the asymptotic normality of vector-valued stochastic integrals and estimates of drift parameters of a multidimensional diffusion proceses. Theor. Probability and Math. Statist., N2, 1974, p. 209-224 (translation from Russian).

[4]. Taraskin A.F. Statistical problems for a certain class of stochastic differential equations. "Mathematical Physics" N10, Naukova Dumka, Kiev, 1971, p. 91-99. (in Russian).

[5]. Knopov P.S., Statland E.S. On the absolute continuous of measures corresponding to solutions of some stochastic differential equations. Materialy vsesojusnogo simposiuma po statistike slushainich processov. Kiev, 1973, p. 93-96 (in Russian).

[6]. Cairoli R., Walsh J.B. Stochastic integrals in the plane. Acta Mathematica, v. 134, N1-2, 1975, p. 111-183.

[7]. Gikhman I.I., Pyasetskaya T.E. Two types of stochastic integrals from the martingale measures on the plane. "Doklady AN Ukrainskoj SSR", A N11, 1975, p. 963-966 (in Russian).

A NOTE ON STRONG SOLUTIONS OF STOCHASTIC DIFFERENTIAL EQUATIONS WITH RANDOM COEFFICIENTS

B.L.ROZOVSKY

It is well known that in the classical Itô theorem concerning the existence of the strong solution of the equation

$$dx(t)= a(x(t),t,\omega)dt + b(x(t),t,\omega)dw(t), \qquad (1)$$
$$x(0) = x_0$$

local Lipschitz condition is required (see for example [1], [2]).

The aim of the present report is to show that this assumption is a little bit too demanding.

Let $(\Omega, \mathcal{F}, \mathcal{F}_t, P)$ be a given complete probability space together with an increasing family of σ-algebras $\mathcal{F}_t$, $t \in [0,T]$, such that $\mathcal{F}_t \subseteq \mathcal{F}$ and let $a(x,t,\omega)$, $b(x,t,\omega)$ be m-vector and $m \times n$-matrix functions respectively, both are defined on $R^m \times]0,T] \times \Omega$. We shall assume them throughout as Borel measurable with respect to x ($\forall t, \omega$) and as progressively measurable with respect to ω ($\forall x$). We denote $w(t)$ n-vector standard Wiener process with respect to the family $\mathcal{F}_t$.

<u>Theorem</u>. Suppose that the following conditions are satisfied:

A) $x \longrightarrow a(x,t,\omega)$ is continuous function on R^m

(a.s.);

B) for every $\tau < \infty$ there exists such a constant N_τ that

$$(a(x) - a(y), x - y) + \frac{1}{2}|b(x) - b(y)|^2 \leq N_\tau |x - y|^2$$

$$\forall x, y \in \{z : z \in R^m, |z| < \tau\};$$ [1)]

C) there exists such a constant C that

$$|a(x)| \leq C(1 + |x|), \quad |b(x)| \leq C(1 + |x|) \quad \text{(a.s.)}$$

$$\forall x \in R^m;$$

D) $E x_0^2 < \infty$.

Then there exists a unique continuous progressively measurable strong solution x of the equation (1) and

$$E \sup_t |x(t)|^2 < \infty . \qquad (2)$$

---

1) We denote $(\cdot, \cdot)$ scalar product in R^m, $|\cdot|$ norm in Euclidian space; $|b|^2 = \sum_{ij} |b_{ij}|^2$ where b_{ij} are elements of the matrix b. In cases where there is no danger of confusion, we shall omit arguments t, ω

We shall call assumption (B) "monotony property". It is clear that monotony and continuity of the coefficients follows from Lipschitz conditions, but not the other way round.

<u>Example</u>. Let $m = n = 1$,

$$dx(t) = -|x(t)|^{p-1}(\operatorname{sign} x(t))\, v(t,\omega)\, dt + \tag{3}$$
$$|x(t)|^{p/2}\, m(t,\omega)\, dw(t), \qquad x(0) = x_0 ,$$

where $p \in \,]1, 2[$, v and m are progressively measurable processes, $\operatorname*{ess\,sup}_{t,\omega} v(t,\omega) < \infty$ and

$$m^2(t,\omega) \leq 8(p-1)/p^2 \; v(t,\omega) \qquad (a.s.) \; (4)$$

Evidently neither drift nor diffusion coefficients of (3) are Lipschitz continuous but monotony property does hold. Really by using Hadamard's formula we have

$$-2v(|x|^{p-1}\operatorname{sign} x - |y|^{p-1}\operatorname{sign} y)(x-y) +$$

$$m^2(|x|^{p/2} - |y|^{p/2})^2 =$$

$$[-2v\int_0^1 (p-1)|x + \tau(y-x)|^{p-2}\, d\tau +$$

$$\frac{m^2}{2}\Big(\int_0^1 p\,|x + \tau(y-x)|^{p/2-1}\operatorname{sign}(x + \tau(y-x))\,d\tau\Big)^2] \times$$

$$\times (x-y)^2 \leq$$

$$[-2(p-1)v + \frac{p^2}{4}m^2] \int_0^1 |x+\tau(y-x)|^{p-2} d\tau \times$$

$$\times (x-y)^2 \leq 0 .$$

It should be worth while to note that if coefficients do not depend on ω then the assertion of the theorem is a simple corollary of known results. Really the existence of a weak solution follows from the results of A.V.Skorohod [3] because the coefficients are continuous. From the other hand strong uniqueness is a trivial consequence of the assumption (B). Therefore, applying results of T.Yamada and S.Watanabe [4] we see that this solution is a strong one. Unfortunately in our situation this method is inapplicable.

The approach which is developed here is related to Minty-Browder method of monotony operators in nonlinear analysis (see for example [5]). This method has been vastly used and enriched by many authors. In the theory of stochastic differential equations it has been used firstly by A.Bensoussan and R.Temam [6] ($b \equiv 1$) and then by E.Pardoux (b is localy Lipschitz)[7].

Proof of the theorem. As it has been mentioned above uniqueness is obvious and we can restrict ourselves to the existence problem.

Without loss of generality we can suppose that N_t does not depend on t . Extansion onto the general ("local") case could be done by the standard method.

Let $\mathcal{J}_\varepsilon$ be Friedrichs' averaging operator

$$\mathcal{J}_\varepsilon f(x) = \varepsilon^{-m} \int_{R^m} f(y) j\left(\frac{|x-y|}{\varepsilon}\right) dy ,$$

where $$j(y) = \begin{cases} \exp\left\{\frac{|y|^2}{|y|^2-1}\right\} & , \text{ if } |y|<1 \\ 0 & , \text{ if } |y|\geq 1. \end{cases}$$

From the properties of $\mathcal{J}$ (see for example [8]) it follows easily that:

I) $a_\varepsilon(x,t,\omega) = \mathcal{J}_\varepsilon a(x,t,\omega)$ and $b_\varepsilon(x,t,\omega) = \mathcal{J}_\varepsilon b(x,t,\omega)$ satisfy conditions of the classical Itô theorem cited above;

II) a_ε and b_ε satisfy conditions (A), (B), (C) of our theorem uniformly with respect to ε ;

III) $a_\varepsilon(x,t,\omega) \to a(x,t,\omega)$ and $b_\varepsilon(x,t,\omega) \to b(x,t,\omega)$ strongly in $L_2(]0,T]\times\Omega)$ as $\varepsilon \to 0$.

From (I), (II) we see that there exists a unique continuous progressively measurable strong solution x^ε of the equation

$$dx^\varepsilon(t) = a_\varepsilon(x^\varepsilon(t),t,\omega)\,dt + b_\varepsilon(x^\varepsilon(t),t,\omega)\,dw(t), \quad x^\varepsilon(0) = x_0 \tag{4}$$

and

$$\sup_{\varepsilon<1} E \sup_t |x^\varepsilon(t)|^2 < \infty . \tag{5}$$

From the inequality (5) it follows immediately that there

exists such a subsequence $\{\nu\}$ and such a progressively measurable elements of $L_2(]0,T]\times\Omega)$ x, a_∞, b_∞ and $x_\infty \in L_2(\Omega;\mathcal{F}_T)$ that

$$x^\nu \longrightarrow x \ , \quad a_\nu \longrightarrow a_\infty \ , \quad b_\nu \longrightarrow b_\infty$$

weakly in $L_2(]0,T]\times\Omega)$ (6)

and

$$x^\nu(T) \longrightarrow x_\infty \qquad \text{weakly in } L_2(\Omega) \tag{7}$$

as $\nu \to 0$.

Therefore, going to the limit in (5) we see that there exists such a continuous progressively measurable modification of x [1] that :

$$x(t) = x_0 + \int_0^t a_\infty(s)\,ds + \int_0^t b_\infty(s)\,dw(s) \tag{8}$$

$$\forall\, t,\omega \in [0,T]\times\Omega' \quad (P(\Omega')=1),$$

$$x_\infty = x(T) \quad (a.s.) . \tag{9}$$

We have to clarify convergence in the stochastic integral only: since stochastic integral is a strongly continuous operator in $L_2(]0,T]\times\Omega)$ it is also weakly continuous in this space.

To complete our proof we need only to verify equalities

$$a_\infty(t) = a(x(t)) \quad \text{(a.s.)}, \tag{10}$$

$$b_\infty(t) = b(x(t)) \quad \text{(a.s.)}. \tag{11}$$

Let y be some element of $L_2(]0,T]\times\Omega)$ adopted to the family $\{\mathcal{F}_t\}$. We shall denote

---

1) We shall identify them throughout.

$$\mathcal{O}_\nu = E\int_0^T e^{-Nt}\{2(a_\nu(x^\nu(t)) - a_\nu(y(t)), x^\nu(t) -$$

$$y(t)) - N|x^\nu(t) - y(t)|^2 +$$

$$|b_\nu(x^\nu(t)) - b_\nu(y(t))|^2\}\,dt.$$

It follows from (II) that a suitable choice of the constant N garantees inequality

$$\mathcal{O}_\nu \leq 0 \tag{12}$$

Let us also denote

$$\mathcal{O}_\nu^1 = E\int_0^T e^{-Nt}\{2(a_\nu(x^\nu(t)), x^\nu(t)) -$$

$$N|x^\nu(t)|^2 + |b_\nu(x^\nu(t))|^2\}\,dt$$

Applying Itô's formula to $|x^\nu(t)|^2$ we get

$$\mathcal{O}_\nu^1 = Ee^{-NT}|x^\nu(T)|^2 - E|x_0|^2.$$

Hence we obtain by (7), (9)

$$\liminf_{\nu\to 0} \mathcal{O}_\nu^1 \geq Ee^{-NT}|x(T)|^2 - E|x_0|^2.$$

From the other hand applying the Itô formula to $|x(t)|^2$ we obtain from (8)

$$Ee^{-NT}|x(T)|^2 - E|x_0|^2 =$$

$$E\int_0^T e^{-Nt}\{2(a_\infty(t), x(t)) - N|x(t)|^2 + |b_\infty(t)|^2\}dt.$$

Therefore,

$$\liminf_{\nu\to 0} \mathcal{O}_\nu^1 \geq E\int_0^T e^{-Nt}\{2(a_\infty(t), x(t)) - N|x(t)|^2 + |b_\infty(t)|^2\}dt \tag{13}$$

By (III), (6), (12), (13) we see that

$$0 \geq \liminf_{\nu\to 0} \mathcal{O}_\nu \geq E\int_0^T e^{-Nt}\{2(a_\infty(t)-a(y(t)), x(t)-y(t)) - N|x(t)-y(t)|^2+|b_\infty(t) - b(y(t))|^2\}dt \tag{14}$$

Then letting in this inequality $y = x$ we get (11)

Now let $y = x + \lambda z$ where z is an arbitrary element of $L_2([0,T]\times\Omega)$ and $\lambda \in R^1_+$, then it follows from (14) that

$$E\int_0^T e^{-Nt}\{2(a_\infty(t) - a(x(t)+\lambda z(t)), z(t)) -$$

$$N\lambda |z(t)|^2 \} dt \leq 0.$$

Letting that λ tends to zero we get from this inequality, according to the Lebesgue theorem,

$$E \int_0^T (a_\infty(t) - a(x(t)), z(t))dt \leq 0.$$

Hence we see that (10) is true. The proof is complete.

By using methods analogous to those developed by Pardoux this result can be extended to the case of equations in Banach space with unbounded operator coefficients, in particular to the case of partial stochastic differential equations.

The present note was initiated in connection with the requirements of this theory, for it is in this theory that the monotony property occurs much more often than the Lipschitz condition.

REFERENCES

[1] Itô K., On stochastic differential equations. Mem.Math. Amer.Soc., 1951.

[2] Гихман И.И., Скороход А.В., Стохастические дифференциальные уравнения, Киев, I968.

[3] Скороход А.В., Исследования по теории случайных процессов, Киев I96I.

[4] Yamada T., Watanabe S., On the uniqueness of solutions of stochastic differential equations, I J.Math.Kyoto Univ., vol.11, N°1 (1971) 155-167.

[5] Nirenberg L., Topics in nonlinear functional analysis, N.-Y., 1974.

[6] Bensoussan A., Temam R., Equations aux dérivées partieles stochastiques non linéaires. Isr.J. of Math., 11, (1972) p.95-129.

[7] Pardoux E. These, L'Université Paris Sud, 1975.

[8] Bers L., John F., Shechter M., Partial differential equations. N.-Y., 1964.

NON-EQUILIBRIUM SOLUTIONS OF AN INFINITE SYSTEM OF STOCHASTIC DIFFERENTIAL EQUATIONS

Hermann Rost
Heidelberg

1. Introduction

We deal in this paper with the following infinite system of stochastic differential equations

$$(1) \qquad X_i(t) = x_i + \int_0^t c_i(X(s))ds + W_i(t) \quad , \; i = 1,2,...$$

where $W_1, W_2,..$ are independent standard Wiener processes on R, the c_i are functions of the argument $x = (x_1,x_2,...)$ with $x_j \in R$, and $X = (X_1,X_2,..)$ is the process to be constructed.

One thinks of (1) as the Einstein-Smoluchowski description of a system of interacting particles in a viscous medium; in particular, one is interested in the case where c_i is of the form

$$(2) \qquad c_i(x) = - \sum_{j \neq i} \Phi'(x_i - x_j)$$

and Φ is a "nice" (smooth, finite range) potential on R.

This model has first been studied by Lang ([1]), who was able to show the existence and uniqueness of solutions in the <u>equilibrium case</u>, i. e. if one restricts oneself to those solutions X(t), for which the induced point process (= the sequence $X_i(t)$, i = 1,2,..., without labelling of particles) is stationary in time. Here we show that within the class of tempered solutions for suitable initial values x there is exactly one solution of (1). Many of the ideas used in this paper go back to Lanford ([2]) and Dobrushin and Fritz ([3]), who studied an analogous system of equations in the case of deterministic Newtonian dynamics. Our restriction to the dimension one is due to the fact that the main a priori estimate (lemma 1) can only be proven in that case.

We divide the paper into two parts : an existence theorem for (1), based on a compactness argument, and a strong uniqueness statement, for which we need some Lipschitz condition for the c_i .

2. Existence

Notations and assumptions. If $x = (x_1, x_2, ..)$ is a point configuration and $u \in R$ we shall denote by $N(x,u,\rho)$ the number of indices i such that $|x_i - u| \leq \rho$.

The functions c_i are assumed to satisfy

(A 1) there is a constant L (range of interaction) and a constant C such that

$$|c_i(x)| \leq C \cdot N(x, x_i, L) \quad \text{for all} \quad i \geq 1 .$$

The density fluctuations of a configuration are measured by a function g, for which we assume the following :

(A 2) g is defined on R_+, positive and concave, $g(0) = L$ and $g(v) = \log v$ for v large.

(This last assumption might be replaced by the condition that $\int^{\infty} (v \cdot g(v))^{-1} dv = \infty$. It is clear, however, that only those g's are of interest for which $\overline{N}(X)$ is finite a.s. with respect to any Gibbs state for Φ, if the c_i are defined by (2); but this implies, generally, that $g(v)$ may not grow slower than $\log v$. So there is not too much freedom to choose g in view of these two requirements.)

For a configuration x we consider as a measure of density

$$\overline{N}(x) := \sup_{\rho \geq g(|u|)} \sup_{u \in R} \rho^{-1} \cdot N(x,u,\rho) \; ;$$

we denote by $\overline{X}$ the set of all x with $\overline{N}(x)$ finite. A solution $X(t)$, $t \geq 0$, of (1) or a similar equation is called tempered, if a.s.

$$\sup_{t \leq T} \overline{N}(X(t)) < \infty \quad \text{for all } T .$$

(In particular, for a tempered solution the right hand side in (1) is well defined.)

Theorem 1. *Let (A 1) and (A 2) be satisfied. Then for any* $x \in \overline{X}$ *there exists a tempered solution to (1) on a suitable probability space with suitably defined independent Wiener processes* W_i.

We give the proof by a series of lemmas.

Lemma 1. Consider the deterministic equation

$$(3) \qquad X_i(t) = x_i + \int_0^t c_i(X(s))ds \quad , \quad i = 1,2,... .$$

Then for any $\overline{n}$ there exists a function $u(t)$, depending only on $\overline{n}$ (and, of course, on the specific form of g and the constants in

(A 1)), such that for any tempered solution of (3)

$$\sup_{t'\leq t} \overline{N}(X(t')) \leq u(t) \quad \text{for all} \quad t \geq 0, \text{ whenever } \quad \overline{N}(x) \leq \overline{n} .$$

Proof of lemma 1. For all $u \in R$ and $\rho \geq g(|u|)$ one has the estimate

$$N(X(t),u,\rho) \leq N(x,u,r(0)) ,$$

where $r(s)$, $0 \leq s \leq t$, satisfies the integral equation

$$r(s) = \rho + C \cdot \int_s^t \overline{N}(X(s')) \cdot g(|u| + r(s'))ds' ,$$

i.e. $r(s)$ is the maximal possible distance from u at time s of a particle which at time t has a distance less than ρ from u . Put

$$I(t) = \int_0^t \overline{N}(X(s))ds ;$$

then the ratio $q = q(t) = r(t)/\rho$ satisfies for all t

(4) $$q \leq 1 + CI \cdot g(|u| + \rho q)/\rho .$$

We look - at the moment for t fixed - for the maximal possible value q^* of all those q's , where the maximum is taken over all u, ρ with $g(|u|) \leq \rho$. For fixed $|u| = v$ the right side in (4) is decreasing in ρ ; hence we take the supremum over all v of solutions q of

(5) $$q \leq 1 + CI \cdot g(v + g(v) \cdot q)/g(v) .$$

Again, by concavity and positivity of g , the right side in (5) is decreasing in v , and q^* turns out to be a solution of

(6) $$q^* = 1 + CI \cdot g(Lq^*)/L .$$

For convenience we introduce $r^* = Lq^*$ as new variable; it satisfies

(6') $$r^* = L + CI \cdot g(r^*)$$

and is related to $\overline{N}(X(t))$ by

(7) $$\overline{N}(X(t)) \leq \overline{n} \cdot r^*(t)/L .$$

To study the dependence of r^* on I we put $a := CI$ and call, by some abuse of language, $r^*(a)$ the solution of

(8) $$r^* = L + a \cdot g(r^*) .$$

If we differentiate (8) we obtain

(9) $$\frac{dr^*}{da} = g(r^*) \cdot (1 - a \cdot g'(r^*))^{-1} .$$

Because of (8) a/r^* and hence $a \cdot g'(r^*)$ goes to zero if a or r^* tend to infinity; so for a , r^* large one has

(10) $$\frac{dr^*}{da} \leq 2g(r^*) .$$

This last estimate gives together with (7) for large t

(11) $$L^{-1}\cdot\bar{n}\cdot r^{*}(a(t)) \geqslant \bar{N}(X(t)) = C^{-1}\cdot\frac{da}{dt} = C^{-1}\cdot\frac{da}{dr}\cdot\frac{dr^{*}(a(t))}{dt} \geqslant$$
$$\geqslant (2Cg(r^{*}))^{-1}\cdot\frac{dr^{*}(a(t))}{dt} .$$

So , for large t, $r^{*}(a(t))$ is bounded by a suitable solution of the differential equation

$$L^{-1}\cdot\bar{n}\cdot u(t) = (2Cg(u))^{-1}\cdot\frac{du}{dt} ,$$

which is finite for every t since the integral $\int^{\infty}(u\cdot g(u))^{-1}$ diverges. This together with (7) proves the assertion of the lemma.

Lemma 2. Let x be in $\bar{X}$ (it suffices to assume : $\liminf |x_i/i| > 0$) and $W_1, W_2, \ldots$ be independent standard Wiener processes; then for any fixed T the quantity

(12) $$\sup_i \sup_{t\leq T} |W_i(t)|/g(x_i)$$

is finite almost surely.

Proof. By a Borel-Cantelli argument.

We will call an equation of the form

(1′) $$X_i(t) = x_i + \int_0^t \tilde{c}_i(X(s))ds + \tilde{W}_i(t) , \quad i = 1,2,\ldots$$

with $\tilde{c}_i = c_i\cdot 1_{\{i\in I\}}$, $\tilde{W}_i = W_i\cdot 1_{\{i\in I\}}$ and I some finite subset of the natural numbers a finite subsystem of (1).

Lemma 3. For $\varepsilon>0$, T and $\bar{n}$ in R_+ there exists a finite number K such that for any tempered solution of any finite subsystem of (1) one has

$$P(\bar{N}(X(t)) > K \text{ for some } t\leq T) \leq \varepsilon ,$$

whenever $\bar{N}(x)\leq\bar{n}$.

Proof. Choose $\bar{W}$ so that the quantity in (12) exceeds $\bar{W}$ with a probability less than ε. Then replace $CI(t)$ in Lemma 1 by $CI(t) + \bar{W}$, and take as K the expression $\sup_{t\leq T}(L^{-1}\cdot\bar{n}\cdot u(t))$.

The theorem now follows from lemma 3 by the usual compactness argument : take the set $E = R\times R\times R\ldots$ with its product topology; take the measures on $\mathcal{C}([0,\infty),E)$ which are induced by solutions of finite subsystems of (1); the class of these measures is tight and any limit point of it is a soltuion of (1). The temperedness constant K in lemma 3 gives again a bound for the limiting process.

3. Uniqueness

In this section we introduce a new assumption, dealing with the Lipschitz continuity of the drift functions c_i :

(A 3) there is a constant $K < \infty$ such that c_i is Lipschitz continuous in the argument x_j , for all $j \neq i$, with constant K, and is Lipschitz continuous in the argument x_i on the set $\{x : N(x, x_i, L) = n\}$ with constant $n \cdot K$ for all $n \geq 0$.

(This hypothesis is satisfied if (2) holds with Φ symmetric on R, twice differentiable, $\Phi(u) = 0$ for $|u| > L$.)

Theorem 2. *Let independent Wiener processes W_i , $i \geq 1$, on some probability space be given; let x be in $\overline{X}$. Then any two tempered solutions of (1) coincide for all time points almost surely. The unique solution may be obtained as (a.s. existing) pointwise limit of a sequence of solutions of finite subsystems.*

Proof. We call X^n the solution of the subsystem

$$(13) \qquad \begin{aligned} X_i^n(t) &= x_i + \int_0^t c_i(X^n(s))ds + W_i(t) && \text{for } i \in M_n , \\ X_i^n(t) &= x_i && \text{for } i \notin M_n , \end{aligned}$$

where $M_n = \{i : |x_i| \leq r(n)\}$. The numbers $r(n)$ will be chosen later in a convenient way.

For a process $X(t)$, $t \in T$, starting at x , we define two new concepts : its logarithmic fluctuation

$$\sup_i \sup_{t \leq T} |X_i(t) - x_i| / g(|x_i|) ,$$

and the (random) set G_i , or $G_i(X)$, of all indices j for which $\inf\{|X_i(t) - X_j(t)| : t \leq T\}$ is less than L .

One remarks the following fact : if on some point of the probability space the logarithmic fluctuation is bounded by a number D , then the cardinality $|G_i|$ of G_i on that point can be estimated by $A \cdot g(|x_i|)$, where A depends only on D (and the density $\overline{N}(x)$ of the initial configuration, which is considered as fixed here).

We choose now the constants D and A in such a way that for any tempered solution X of (1) or a finite subsystem the logarithmic fluctuation is bounded by D and $|G_i| \leq A \cdot g(|x_i|)$ for all i , on a set of probability greater than $1 - \varepsilon$. This is possible according to lemma 3.

If X and Y are two solutions of (1) one has

$$(14)\quad |X_i(t) - Y_i(t)| \leq \int_0^t |c_i(X(s)) - c_i(Y(s))|\,ds \leq$$

$$\leq 2K \cdot \int_0^t \sum_{j \in G_i(X) \cup G_i(Y)} |X_j(s) - Y_j(s)|\,ds$$

Suppose now that the two solutions X and Y are tempered; if we choose the sequence $r(n)$, $n \geq 0$, in such a way that

$$(15)\quad r(n+k) - r(n) \geq D \cdot (g(r(n)) + g(r(n+k))) + L \quad \text{for } n \geq 0,\ k \geq 1 ,$$

we deduce from (14) and the property of A that with probability at least $1-2\varepsilon$ the following inequality holds :

$$(16)\quad \sup_{i \in M_n} |X_i(t) - Y_i(t)| \leq 2KAg(r(n)) \cdot \int_0^t \sup_{M_{n+1}} |X_i(s) - Y_i(s)|\,ds$$

hence, by induction,

$$(17)\quad \sup_{t \leq T} \sup_{M_0} |X_i(t) - Y_i(t)| \leq \sup_{t \leq T} \sup_{M_n} |X_i(t) - Y_i(t)| \cdot \frac{1}{n!} \cdot (2KA)^n \prod_{k<n} g(r(k)) \leq$$

$$\leq \frac{1}{n!} \cdot 2D \cdot (2KA)^n \prod_{k \leq n} g(r(k)) \quad =: b_n \quad .$$

It is easy to see that condition (15) can be satisfied; moreover, one can chose the $r(n)$ in such a way that $g(r(n)) = o(n)$ (take e.g. $r(n) = a + bn^2$ with suitable a and b). But then the sequence b_n converges to zero and it follows that $X_i(t) = Y_i(t)$ for $t \leq T$ on the set in consideration. Since ε, T and $r(0)$ are arbitrary the uniqueness result is proven.

For the proof of the approximation statement we observe two facts : first, $g(r(n)) = o(n)$ implies $\sum b_n < \infty$; second, for finite subsystems of the form (13) we have

$$\sup_{t \leq T} \sup_{M_0} |X^n(t) - X^{n-1}(t)| \leq b_n ,$$

since the "Wiener term" in both processes is the same for the indices i in M_{n-1} . So we get converge of $X_i^n(t)$ as n increases, uniformly in $t \leq T$ and $i \in M_0$, with probability at least $1 - 2\varepsilon$.

Remark. The proof of theorem 2 gives at the same time the continuous dependence of $X(t)$ on the initial condition x , as x lies in the set $\{\bar{N}(x) \leq a\}$ for some finite a . So one can define by the solution of (1) a Markovian evolution on $\bar{X}$.

The construction given here seems to be simpler than the one used in ([1]), if one wants to prove strong uniqueness. In order to show, however, that Gibbs states are invariant under the dynamics induced by the above mentioned Markovian evolution on the space of non-labelled point configurations one needs an additional argument : an a priori estimate for the logarithmic fluctuation of solutions of finite subsystems with reflecting boundary conditions. In the present context we did not care about a suitable modification of lemma 3 in that direction.

Literature.

[1] R. Lang : Unendlich-dimensionale Wienerprozesse mit Wechselwirkung. Z.Wahrscheinlichkeitstheorie verw. Geb. 38, 55-72 and 39, 277-299(1977).

[2] O.E. Lanford : The classical mechanics of one-dimensional systems of infinitely many particles. Commun. math. Phys. 9,161-191 and 11,257-292(1969).

[3] R.L. Dobrushin and J. Fritz : Non-equilibrium dynamics of one-dimensional infinite particle systems with a hard core interaction. Commun. math. Phys. 55,275-292(1977).

[4] J.Fritz : Stochastic dynamics of two-dimensional infinite particle systems. Preprint. Budapest 1978.

ON CONDITIONS FOR UNIFORM INTEGRABILITY FOR CONTINUOUS EXPONENTIAL MARTINGALES

A.A.Novikov

1. Introduction. Let $(\mathcal{M}_t)$, $t \in [0, \infty)$, be a continuous local martingale, $\mathcal{M}_0 = 0$, on a complete probability space $(\Omega, \mathcal{F}, P)$ with a non-decreasing right continuous family $(\mathcal{F}_t)$ of sub - σ - fields $\mathcal{F}$. Let $<\mathcal{M}>_t$ be the increasing process of $(\mathcal{M}_t)$ such that $\lim_{t \to \infty} <\mathcal{M}>_t = <\mathcal{M}>_\infty < \infty$ a.s. and set

$$\mathcal{Z}_t = exp(\mathcal{M}_t - 1/2 <\mathcal{M}>_t).$$

In this paper the question about sufficient conditions for the uniform integrability of $(\mathcal{Z}_t)$ is considered. This question is arised in some problems of the theory of stochastic differential equations [1] (in these problems $\mathcal{Z}_t$ plays the role of a density of one probability measure with respect to another). It is well known that $(\mathcal{Z}_t)$ is a local martingale and that $(\mathcal{Z}_t)$ is uniform integrable iff $E\mathcal{Z}_\infty = 1$, where $\mathcal{Z}_\infty = \lim_{t \to \infty} \mathcal{Z}_t$ a.s., but the direct verification of this equality is usually hard to carry out.

In the case when $\mathcal{M}_t$ is a classical Ito stochastic integral the following result was proved in [2] (see ibid the references on the preceding works of Girsanov, Gihman and Skorokhod, Liptzer and Shiryaev):

$$E\, exp(1/2 <\mathcal{M}>_\infty) < \infty \Rightarrow E\mathcal{Z}_\infty = 1. \tag{1}$$

Recently Lepingle and Memin [3] have proved this assertion in the common case and Kazamaki [4] has showed that if $\mathcal{M}_t$ is a continuous martingale then for each $t>0$

$$E\exp(1/2\langle \mathcal{M}\rangle_t)<\infty \Rightarrow E\exp(1/2\mathcal{M}_t)<\infty \Rightarrow E\mathcal{Z}_t=1. \tag{2}$$

It should be noted that as it is shown in [2] the assertion (1) doesn't hold with the constant $1/2-\varepsilon$, $(\varepsilon>0)$ instead $1/2$.

2. <u>Main result</u>. The next theorem contains the result which is an improvement both (1) and (2).

<u>Theorem 1.</u> *Let $\mathcal{M}$ be a class of all stopping times with respect to $(\mathcal{F}_t)$. Then*

$$E\exp(1/2\langle \mathcal{M}\rangle_\infty - C\langle \mathcal{M}\rangle_\infty^{1/2})<\infty \Rightarrow$$

$$\Rightarrow \sup_{\tau\in\mathcal{M}} E\exp\{1/2(\mathcal{M}_\tau - C\langle \mathcal{M}\rangle_\tau^{1/2})\}<\infty \Rightarrow E\mathcal{Z}_\infty=1,$$

where C is an arbitary non-negative constant.

<u>Proof.</u> At the beginning we shall prove the first implication. By the Schwarz inequality

$$E\exp\{1/2(\mathcal{M}_\tau - C\langle \mathcal{M}\rangle_\tau^{1/2})\} = E\mathcal{Z}_\tau^{1/2}\exp\{1/4\langle \mathcal{M}\rangle_\tau - C/2\langle \mathcal{M}\rangle_\tau^{1/2}\}\le$$
$$\le (E\mathcal{Z}_\tau)^{1/2}(E\exp\{1/2\langle \mathcal{M}\rangle_\tau - C\langle \mathcal{M}\rangle_\tau^{1/2}\})^{1/2}, \quad \tau\in\mathcal{M}. \tag{3}$$

It is clear that $\mathcal{Z}_t$ is a positive supermartingale with $\mathcal{Z}_0=1$ and hence $E\mathcal{Z}_\tau\le 1$ for any $\tau\in\mathcal{M}$. Now applying the elementary inequality $1/2x - Cx^{1/2}\le 1/2y - Cy^{1/2} + C^2/2$, $(y\ge x\ge 0)$ we get from (3)

$$\sup_{\tau\in\mathcal{M}} E\exp\{1/2(\mathcal{M}_\tau - C\langle \mathcal{M}\rangle_\tau^{1/2})\}\le (E\exp\{1/2\langle \mathcal{M}\rangle_\infty - C\langle \mathcal{M}\rangle_\infty^{1/2} + C^2/2\})^{1/2}<\infty.$$

In order to prove the second implication we introduce the stopping time

$$\tau_a=\inf\{t\ge 0: \mathcal{M}_t\le \langle \mathcal{M}\rangle_t - C\langle \mathcal{M}\rangle_t^{1/2} - a\}, \quad 0<a<\infty,$$

where $\inf\{\emptyset\}=\infty$, and set

$$\mathcal{Z}_t(\lambda)=\mathcal{Z}_t^{\lambda}\exp\left(\frac{\lambda-\lambda^2}{2}<\mu>_t\right),\quad 0<\lambda<1.$$

It is obvious that $\mathcal{Z}_t(\lambda)=\exp(\lambda\mu_t-\lambda^2/2<\mu>_t)$ is a positive local martingale. Choose the number p such that

$1<p<1/\lambda$ and $\tau\equiv(\lambda-\lambda^2)p/(p-1)<1$, $(f.e.\ p=[\lambda+\frac{(1-\lambda)^2}{2}]^{-1})$.

Then by the Holder inequality

$$E\sup_{t\geq 0}\mathcal{Z}_{t\wedge\tau_a}(\lambda)\leq\left(E\sup_{t\geq 0}\mathcal{Z}_t^{p\lambda}\right)^{1/p}\left(E\exp\left(\frac{\tau}{2}<\mu>_{\tau_a}\right)\right)^{1-1/p}. \tag{4}$$

Since $\mathcal{Z}_t$ is a positive supermartingale then by the Doob inequality $P\{\sup_{t\geq 0}\mathcal{Z}_t\geq x\}\leq E\mathcal{Z}_0/x=1/x$, $x>0$, and hence $E\sup_{t\geq 0}\mathcal{Z}_t^{p\lambda}<\infty$, $(0<p\lambda<1)$. Now show that the second multiplier in (4) is also finite and therefore the stopped martingale $\mathcal{Z}_{t\wedge\tau_a}(\lambda)$ is uniform integrable. By the definision of τ_a and the continuity of μ_t

$$\mu_{\tau_a}\geq<\mu>_{\tau_a}-C<\mu>_{\tau_a}^{1/2}-a. \tag{5}$$

Applying the elementary inequality $\tau/2\,x\leq 1/2\,x-Cx^{1/2}+$ $+\frac{C^2}{2(1-\tau)}$, $(x\geq 0,\ 0<\tau<1)$, we get

$$E\exp\left\{\frac{\tau}{2}<\mu>_{\tau_a}\right\}\leq E\exp\left\{1/2\left(\mu_{\tau_a}-C<\mu>_{\tau_a}^{1/2}\right)+\frac{C^2}{2(1-\tau)}+\frac{a}{2}\right\}<\infty.$$

The uniform integrability of $\mathcal{Z}_{t\wedge\tau_a}(\lambda)$ implies that

$$E\mathcal{Z}_{\tau_a}(\lambda)=1,\quad(0<\lambda<1). \tag{6}$$

Now show that we can pass to the limit as $\lambda\to 1$ under the sing of the expectation. Indeed, by (5) we get

$$\mathcal{Z}_{\tau_a}(\lambda) = \mathcal{Z}_{\tau_a} \exp\{(\lambda-1)\mathcal{M}_{\tau_a} + \frac{1-\lambda^2}{2} <\mathcal{M}>_{\tau_a}\} \le$$

$$\le \mathcal{Z}_{\tau_a} \exp\{a + \frac{C^2}{2} - \frac{1}{2}[(1-\lambda)<\mathcal{M}>^{1/2}_{\tau_a} - C]^2\} \le \mathcal{Z}_{\tau_a} \exp\{a + \frac{C^2}{2}\}.$$

Since $\mathcal{Z}_{\tau_a}(\lambda) \to \mathcal{Z}_{\tau_a}$ as $\lambda \to 1$ we have by the dominated convergence theorem from (6)

$$E\mathcal{Z}_{\tau_a} = 1 . \qquad (7)$$

The last step of proving of theorem 1 consists in the passage to the limit as $a \to \infty$. Since $\mathcal{M}_{\tau_a} = <\mathcal{M}>_{\tau_a} - C<\mathcal{M}>^{1/2}_{\tau_a} - a$ on the set $\{\tau_a < \infty\}$ then

$$EI\{\tau_a<\infty\}\mathcal{Z}_{\tau_a} = EI\{\tau_a<\infty\}\exp\{1/2(\mathcal{M}_{\tau_a} - C<\mathcal{M}>^{1/2}_{\tau_a}) - \frac{a}{2}\} \le$$

$$\le \sup_{\tau\in\mathfrak{M}} E\exp\{1/2(\mathcal{M}_\tau - C<\mathcal{M}>^{1/2}_\tau)\}\exp(-a/2) \to 0, \quad a\to\infty.$$

From (7) we get that $\lim_{a\to\infty} EI\{\tau_a=\infty\}\mathcal{Z}_\infty = 1$.
Since $E\mathcal{Z}_\infty \le 1$ as a result we recieve that $E\mathcal{Z}_\infty = 1$.

3. <u>Remarks.</u> a) The Kazamaki's result (2) follows from the theorem 1 with $C=0$. Indeed, if $\mathcal{M}_t$ is a martingale and $E\exp\{1/2\mathcal{M}_t\}<\infty$, then $\exp\{1/2\mathcal{M}_{t\wedge\tau}\}$ is a positive submartingale, $(t\wedge\tau = \min(t,\tau))$. Hence by the Fatou's lemma

$$E\exp\{1/2\mathcal{M}_\tau\} \le \lim_{t\to\infty} E\exp\{1/2\mathcal{M}_{t\wedge\tau}\} \le \sup_{t\ge 0} E\exp\{1/2\mathcal{M}_t\}.$$

Therefore

$$\sup_{\tau\in\mathfrak{M}} E\exp\{1/2\mathcal{M}_\tau\} = \sup_{t\ge0} E\exp\{1/2\mathcal{M}_t\} ,$$

which with theorem 1 gives in particular the statement (2).

b) The next example shows that the condition of the second implication can fulfil ever if $E\langle\mathcal{M}\rangle_\infty^\nu = \infty$ at some $\nu > 0$. Let W_t be a standart wiener process, $\mathcal{M}_t = W_{t\wedge\gamma}$, where $\gamma = \inf\{t \geq 0 : W_t \geq b\sqrt{t} + a\}$, $b > 0$, $a > 0$. Then $\langle\mathcal{M}\rangle_t = t\wedge\gamma$, $\langle\mathcal{M}\rangle_\infty = \gamma < \infty$ a.s. and evidently on one hand

$$\sup_{\tau\in\mathcal{M}} E\exp\{1/2(\mathcal{M}_\tau - b\langle\mathcal{M}\rangle_\tau^{1/2})\} \leq \exp(a/2) < \infty .$$

On the other hand it is known[5] that $E\gamma^{\nu(b)} = \infty$, where the function $\nu(b)$ is positive continuous such that $\nu(b)\downarrow 0$, $b\to\infty$.

c) In the case when $\mathcal{M}_t$ is a stopped wiener process the result of theorem 1 can be slightly improved. Let $\mathcal{N}$ be a class of continuous monotonic positive functions such that $g(t)/\sqrt{t}\uparrow$, $g(t)/t\downarrow 0$, $t\to\infty$ and

$$\int^\infty \frac{g(t)}{t^{3/2}} \exp\left\{-\frac{g^2(t)}{2t}\right\} dt = \infty$$

(f.e. $g(t) = (2t\ln\ln t)^{1/2}$, $t\to\infty$) . According to the Kolmogorov's criterion the functions $g(t)$ from the class $\mathcal{N}$ are lower for a wiener process, that is $P\{W_t < g(t), t\to\infty\} = 0$.

Theorem 2. *Let* $\mathcal{M}_t = W_{t\wedge\sigma}$, $\sigma\in\mathcal{M}$ *and* $\sigma < \infty$ *a.s. Then for each* $g(t)\in\mathcal{N}$

$$E\exp\{1/2\,\sigma - g(\sigma)\} < \infty \Rightarrow$$

$$\Rightarrow \sup_{\tau\in\mathcal{M}} E\exp\{1/2(W_{\tau\wedge\sigma} - g(\tau\wedge\sigma)\} < \infty \Rightarrow E\mathcal{Z}_\infty = 1.$$

Sketch of proof. The first implication proves by the same manner as the first implication of theorem 1. Further set

$$\tau_a = \inf\{t \geq 0 : w_t \leq t - g(t) - a\}, \quad 0 < a < \infty.$$

It is clear that $\tau_a < \infty$ a.s. According to one result of Shepp [6] (theorem 4) for any continuous function $g(t)$

$$E \exp\{W_{\tau_a} - 1/2\, \tau_a\} = P\{\tau'_a < \infty\}, \tag{8}$$

where $\tau'_a = \inf\{t \geq 0 : w_t \leq -g(t) - a\}$. Since $g(t)$ is a lower function then $P\{\tau'_a < \infty\} = 1$. It follows from (8) that $E Z_{\tau_a} = 1$, where $Z_{\tau_a} = \exp\{W_{\tau_a \wedge \sigma} - 1/2\, \tau \wedge \sigma\}$.

The remainder part of the proof repeats in fact the last step of proving of theorem 1 and so is omitted.

REFERENCES

1. Липцер Р.Ш., Ширяев А.Н., Статистика случайных процессов, Москва, 1974.

2. Новиков А.А., Об одном тождестве для стохастических интегралов, Теория вероятностей и ее применение, ХУII, № 4 (1972), 761-765.

3. Lepingle D., Memin J., Sur l'integrabilite uniforme des martingales exponentielles, Z.Wahrscheinlichkeitstheorie verw Gebiete, 42, № 3 (1978), 175-203.

4. Kazamaki N., On a problem of Girsanov, Tôhoky Math. J., 29, № 4 (1977), 35-45.

5. Новиков А.А., О моментах остановки винеровского процесса, Теория вероят. и ее примен., XVI, № 3 (1971), 458-465.

6. Shepp L.A., Explicit solutions to some problems of optimal stopping, Ann. Math. Stat., 40, № 3 (1969), 993-1010.

ON WEAK COMPACTNESS OF THE SETS OF MULTIPARAMETER STOCHASTIC PROCESSES

R.Morkvėnas

The theory of weak convergence of one-parameter stochastic processes having no discontinuities of the second kind is well developed by many authors. Criteria for conditional compactness or tightness of the sets of probability measures on the Skorohod space $\mathcal{D}$ [0, 1] play an important role in this theory. For many applications very convenient are the criteria, expressed in terms of conditional distributions of increments of stochastic processes in the small time intervals, i.e. by the conditions of the Kinney - Dynkin type. The most general results of this kind are proved by B.Grigelionis [1] and V.Mackevičius [2].

There are much less results, concerning the weak convergence of discontinuous multiparameter stochastic processes. The multiparameter Skorohod space was investigated by G.Neuhaus [3] and M.L.Straf [4]. They proved the tightness criteria in terms of moduli of "continuity", which, unfortunately, are not always convenient for applications. N.N.Čentsov [5] and P.Bickel and M.J.Wichura [6] derived tightness criteria expressed in the so called Čentsov type conditions, generalizing the well known results by N.N.Čentsov and P.Billingslley (see [7]) to the multiparameter case. In this paper we give a Kinney - Dynkin type criterion for tightness, generalizing those of [1] and [2]. For simplicity we consider here the two-parameter case, but analogical results are also true for d-parameter ($d > 2$) stochastic processes.

Let T denote the unit square $[0,1]^2$ and let $S(T)$ be

the space of all simple functions $x : T \to R^1$ i.e. such that x is a linear combination of indicators $I_{E_1 \times E_2}(s,t)$, where each E_i is either a left closed right open subinterval of $[0, 1]$ or the singleton $\{1\}$, and $x(0,t) = x(s,0) = 0$ for all s, t in $[0, 1]$. Let $\mathcal{D}(T)$ be the uniform closure of $S(T)$ in the space of all bounded functions from T to R^1. One can indroduce a metric topology on $\mathcal{D}(T)$ (see [3], [4], [6]), which is analogous to the Skorohod J_1 topology on $\mathcal{D}[0, 1]$ With respect to this topology $\mathcal{D}(T)$ is separable and complete.

Let X_I denote a family $\{X^{(\alpha)}(s,t) ; (s,t) \in T, \alpha \in I\}$ of stochastic processes on tha probability space $(\Omega, \mathcal{F}, P)$ (which may depend on α ; I - arbitrary set), adapted to the family $\{\mathcal{F}(s,t), (s,t) \in T\}$ of sub-σ-algebras of $\mathcal{F}$ such that $\mathcal{F}(s,t) \subset \mathcal{F}(s',t')$, if $s \le s'$, $t \le t'$. We assume that the paths of $X^{(\alpha)}$, $\alpha \in I$ are in $\mathcal{D}(T)$ with probability 1. We say that family X_I is weakly compact, if the set of the corresponding probability measures on $\mathcal{D}(T)$ is conditionally compact in the weak convergence topology. For $0 \le s' \le s'' \le 1$, $0 \le t' \le t'' \le 1$ denote:

$$\mathcal{F}(s', s''; t', t'') = \sigma\{\mathcal{F}(s', t''), \mathcal{F}(s'', t')\},$$

$$\Delta X(s', s''; t', t'') = X(s',t') - X(s',t'') - X(s'',t') + X(s'',t'').$$

<u>Theorem 1</u>. X_I is weakly compact if

(i) the sets of distributions in R^k of collections $\{(X^{(\alpha)}(s_1,t_1), \dots, X^{(\alpha)}(s_k,t_k)), \alpha \in I\}$ are weakly compact for all $(s_1,t_1), \dots, (s_k,t_k)$ in T and $k \ge 1$,

(ii) there exists a nonrandom function $\beta^{(\alpha)}(h, \varepsilon)$, $h > 0$, $\varepsilon > 0$ such that a.s.

$$P\{|\Delta X^{(\alpha)}(s',s'';t',t'')|\geq\varepsilon\mid\mathcal{F}(s',s'';t',t'')\}\leq\beta^{(\alpha)}(h,\varepsilon)$$

for all $0\leq s'\leq s''\leq 1$, $0\leq t'\leq t''\leq 1$ with $(s''-s')\wedge(t''-t')\leq h$, and

$$\lim_{h\to 0}\ \sup_{\alpha\in I}\ \beta^{(\alpha)}(h,\varepsilon)=0\ ,\ \varepsilon>0.$$

The proof of this theorem is based on the following lemmas.

Lemma 1. X_I is weakly compact if condition (i) of Theorem 1 is satisfied and there exist nonrandom functions $\beta_i^{(\alpha)}(h,\varepsilon)$, $h>0$, $\varepsilon>0$, $i=1,2$ such that a.s.

$$P\{\|X^{(\alpha)}(s'',\cdot)-X^{(\alpha)}(s',\cdot)\|\geq\varepsilon\mid\mathcal{F}(s',1)\}\leq\beta_1^{(\alpha)}(h,\varepsilon)$$

for all $0\leq s'\leq s''\leq s'+h\leq 1$,

$$P\{\|X^{(\alpha)}(\cdot,t'')-X^{(\alpha)}(\cdot,t')\|\geq\varepsilon\mid\mathcal{F}(1,t')\}\leq\beta_2^{(\alpha)}(h,\varepsilon)$$

for all $0\leq t'\leq t''\leq t'+h\leq 1$, and

$$\lim_{h\to 0}\ \sup_{\alpha\in I}\ \beta_i^{(\alpha)}(h,\varepsilon)=0\ ,\ \varepsilon>0\ ,\ i=1,2.$$

(Here $\|x(s'',\cdot)-x(s',\cdot)\| = \sup_{0\leq t\leq 1}|x(s'',t)-x(s',t)|$).

The proof of this lemma follows from the Corollary to Theorem 2 [6] repeating almost exactly the proof of Theorem 2 [1].

Lemma 2. Let conditions of Theorem 1 be satisfied. Then a.s.

$$P\{\sup_{t\leq t'\leq 1}|\Delta X^{(\alpha)}(s',s'';t,t')|\geq\varepsilon\mid\mathcal{F}(s',s'',t,1\}\leq\frac{\beta^{(\alpha)}(h,\frac{\varepsilon}{2})}{1-\beta^{(\alpha)}(h,\frac{\varepsilon}{2})}$$

for all $0\leq t\leq 1$, $0\leq s'\leq s''\leq s'+h$.

P r o o f. Suppose $s'\leq s''\leq s'+h$ to be fixed. It is enough to show, that for all $n\geq 1$, $t=t_0<t_1<\ldots<t_n=1$ a.s.

$$P\{\max_{1\le k\le n} |\Delta X^{(\alpha)}(s',s'';t_0,t_k)|\geq \varepsilon \,|\, \mathcal{F}(s',s'';t,1)\} \leq \frac{\beta^{(\alpha)}(h,\frac{\varepsilon}{2})}{1-\beta^{(\alpha)}(h,\frac{\varepsilon}{2})} . \quad (1)$$

Let us denote

$$A_j = \{ |\Delta X^{(\alpha)}(s',s'',t_0,t_k)| < \varepsilon , \; k=1,\dots,j-1, |\Delta X^{(\alpha)}(s',s'';t_0,t_j)|\geq\varepsilon \} ,$$

$$B_j = \{ |\Delta X^{(\alpha)}(s',s'';t_j,t_n)| \geq \frac{\varepsilon}{2} \} .$$

It is easy to see that

$$B_0 = \bigcup_{j=1}^{n} (A_j \cap B_j^c) .$$

Let $A \in \mathcal{F}(s',s'';t,1)$. Then

$$P\{B_0 \cap A\} \geq \sum_{j=1}^{n} P\{A_j \cap B_j^c \cap A\} =$$

$$= \sum_{j=1}^{n} \int_{A\cap A_j} [1 - P\{B_j \,|\, \mathcal{F}(s',s'';t_j,1)\}]\, dP \geq$$

$$\geq (1-\beta^{(\alpha)}(h,\tfrac{\varepsilon}{2}))\, P\{A \cap [\max_{1\le k\le n} |\Delta X^{(\alpha)}(s',s'';t_0,t_k)| \geq \tfrac{\varepsilon}{2}]\} .$$

This completes the proof of (1) and the lemma.

Lemma 3. Under the conditions of Theorem 1 we have, that a.s.

$$P\{\sup_{0\le t'\le t''\le t'''\le 1} MX^{(\alpha)}(s',s'';t',t'',t''')\geq 2\varepsilon \,|\, \mathcal{F}(s',1)\} \leq$$

$$\leq \left(\frac{\beta^{(\alpha)}(h,\frac{\varepsilon}{2})}{1-\beta^{(\alpha)}(h,\frac{\varepsilon}{2})}\right)^2$$

for all $0 \leq s' \leq s'' \leq s'+h$, where

$$MX^{(\alpha)}(s',s'';t',t'',t''') =$$

$$= \min\left(|\Delta X^{(\alpha)}(s',s'';t',t'')|,\ |\Delta X^{(\alpha)}(s',s'';t'',t''')|\right).$$

P r o o f. Let $s' \le s'' \le s'+h$ be fixed. It is enough to show, that for all $n \ge 1$, $0 = t_0 < t_1 < \dots < t_n = 1$ a.s.

$$P\{\max_{1 \le i < j < k \le n} MX^{(\alpha)}(s',s'';t_i,t_j,t_k) \ge 2\varepsilon \mid \mathcal{F}(s',1)\} \le$$

$$\le \left(\frac{\beta^{(\alpha)}(h,\frac{\varepsilon}{2})}{1-\beta^{(\alpha)}(h,\frac{\varepsilon}{2})}\right)^2. \qquad (2)$$

Denote

$$C_j = \{\max_{j<k\le n} |\Delta X^{(\alpha)}(s',s'';t_j,t_k)| \ge \varepsilon\}.$$

Obviously,

$$\{\max_{1\le i<j<k\le n} MX^{(\alpha)}(s',s'';t_i,t_j,t_k) \ge 2\varepsilon\} \subset \bigcup_{j=1}^{n}(A_j \cap C_j). \qquad (3)$$

(We use notations of Lemma 2, with $t_0 = 0$). It follows from Lemma 2, that

$$P\{C_j \mid \mathcal{F}(s',s'';t_j,t_n)\} \le \frac{\beta^{(\alpha)}(h,\frac{\varepsilon}{2})}{1-\beta^{(\alpha)}(h,\frac{\varepsilon}{2})} \qquad (4)$$

for all $j = 0, 1, \dots, n-1$. Let $A \in \mathcal{F}(s',1)$. Then using (3) and (4) we have that

$$P(\{\max_{1\le i<j<k\le n} MX^{(\alpha)}(s',s'';t_i,t_j,t_k) \ge 2\varepsilon\} \cap A) \le \sum_{j=1}^{n} P\{A_j \cap C_j \cap A\} =$$

$$= \sum_{j=1}^{n} \int_{A_j \cap A} P\{C_j \mid \mathcal{F}(s',s'';t_j,t_n)\}\, dP \le$$

$$\le \frac{\beta^{(\alpha)}(h,\frac{\varepsilon}{2})}{1-\beta^{(\alpha)}(h,\frac{\varepsilon}{2})} \; P\{G_0 \cap A\} . \tag{5}$$

Now (2) follows from (5) and (4). Lemma is proved.

P r o o f o f T h e o r e m 1.

It is easy to see, that

$$\| X^{(\alpha)}(s'',\cdot) - X^{(\alpha)}(s',\cdot)\| \le | X^{(\alpha)}(s'',1) - X^{(\alpha)}(s',1)| +$$

$$+ \sup_{0\le t'\le t''\le t'''\le 1} M X^{(\alpha)}(s',s'';t',t'',t''') . \tag{6}$$

It follows from the condition (ii) of the theorem, that

$$P\{ | X^{(\alpha)}(s'',1) - X^{(\alpha)}(s',1) \ge \frac{\varepsilon}{2} \,|\, \mathcal{F}(s',1)\} \le \beta^{(\alpha)}(h,\frac{\varepsilon}{2}) \tag{7}$$

for all $0 \le s' \le s'' \le s'+h \le 1$ **(This is because** $X^{(\alpha)}(s'',1) - X^{(\alpha)}(s',1) = \Delta X^{(\alpha)}(s',s'';0,1)$, $\mathcal{F}(s',1) = \mathcal{F}(s',s'',0,1)$ **).**

Thanks to Lemma 3 and (7), it follows from (6), that

$$P\{ \| X^{(\alpha)}(s'',\cdot) - X^{(\alpha)}(s',\cdot)\| \ge \varepsilon \,|\, \mathcal{F}(s',1)\} \le \beta^{(\alpha)}(h,\frac{\varepsilon}{2}) + \left(\frac{\beta^{(\alpha)}(h,\frac{\varepsilon}{8})}{1-\beta^{(\alpha)}(h,\frac{\varepsilon}{8})}\right)^2 \tag{8}$$

for all $0 \le s' \le s'' \le s'+h \le 1$.

Symmetrically,

$$P\{ \| X^{(\alpha)}(\cdot,t'') - X^{(\alpha)}(\cdot,t')\| \ge \varepsilon \,|\, \mathcal{F}(1,t')\} \le \beta^{(\alpha)}(h,\frac{\varepsilon}{2}) + \left(\frac{\beta^{(\alpha)}(h,\frac{\varepsilon}{8})}{1-\beta^{(\alpha)}(h,\frac{\varepsilon}{8})}\right)^2 \tag{9}$$

for all $0 \le t' \le t'' \le t'+h \le 1$.

It follows from (8) and (9) that the conditions of Lemma 1 are satisfied. This completes the proof of the theorem.

Using the idea of the proof of Theorem 1 [2] one can get the following generalisation of Theorem 1.

<u>Theorem 2</u>. X_I is weakly compact if condition (i) of Theorem 1 is satisfied and for every $\delta > 0$, $\alpha \in I$ there exist a set $A^{(\alpha)} = A^{(\alpha)}(\delta) \in \mathfrak{F}(1,1)$ and nonrandom function $\beta^{(\alpha)}_{A^{(\alpha)}}(h,\varepsilon)$, $h>0$, $\varepsilon>0$ such that

$$P(A^{(\alpha)}) \geqslant 1-\delta$$

$$P\{|\Delta X^{(\alpha)}(s',s'';t',t'')| \geq \varepsilon \,|\, \mathfrak{F}(s',s'';t',t'')\} \leq \beta^{(\alpha)}_{A^{(\alpha)}}(h,\varepsilon)$$

a.s. on $A^{(\alpha)}$ for all $0 \leq s' \leq s'' \leq 1$, $0 \leq t' \leq t'' \leq 1$ with $(s''-s') \wedge (t''-t') \leq h$ and

$$\lim_{h \to 0} \sup_{\alpha \in I} \beta^{(\alpha)}_{A^{(\alpha)}}(h,\varepsilon) = 0, \quad \varepsilon > 0.$$

It also appeares that Kinney - Dynkin type conditions are sufficient for the paths of a stochastic multiparameter process to belong to the space $\mathfrak{D}(T)$ with probability 1.

<u>Theorem 3</u>. Let $X(s,t)$, $(s,t) \in T$ be a separable stochastic process, vanishing on the axes $s=0$ and $t=0$. If there exists a nonrandom function $\beta(h,\varepsilon)$, $h>0$, $\varepsilon>0$ such that a.s.

$$P\{|\Delta X(s',s'';t',t'')| \geq \varepsilon \,|\, \mathfrak{F}(s',s'';t',t'')\} \leq \beta(h,\varepsilon)$$

for all $0 \leq s' \leq s'' \leq 1$, $0 \leq t' \leq t'' \leq 1$ with $(s''-s') \wedge (t''-t') \leq h$, and

$$\lim_{h \to 0} \beta(h,\varepsilon) = 0, \quad \varepsilon > 0$$

then the paths of $X(s,t)$ are in $\mathfrak{D}(T)$ with probability 1.

P r o o f. Thanks to monotonicity with respect to h of the modulus of "continuity" in $\mathfrak{D}(T)$, $w''_h(x)$ (see Theorem 2 of [6]) one can get in the same way as in the proof of Theorem

1, that

$$\lim_{h\to 0} w''_h(X) = 0 .$$

According to [4] and [6], the last fact ensures that X is in $\mathfrak{D}(T)$ with probability 1.

Remark. Stochastically continuous (respectively, stochastically continuous uniformly in α) processes with independent increments satisfy the Kinney - Dynkin type condition of Therem 3 (respectively of Theorem 1).

REFERENCES

1. Grigelionis B., On conditional compactness of the sets of probability measures on $\mathfrak{D}_{[0,\infty)}(\mathcal{X})$. - "Liet.matem.rink.", 1973, v. 13, N 4, p. 83-97 (In Russian).
2. Mackevičius V., On the weak convergence of stochastic processes on the spaces $\mathfrak{D}_{[0,\infty)}(\mathcal{X})$. - "Liet.matem.rink.", 1974, v.14, N 4, p.117-122 (In Russian).
3. Neuhaus G., On weak convergence of stochastic processes with multidimensional time parameter. - "Ann.Math.Statist.", 1971, v. 42, N 4, p.1285-1295.
4. Straf M.L., Weak convergence of stochastic processes with several parameters. - "Proc. 6th Berkeley Symp. Math.Statist. Probab.", 1972, v. 2, p.187-221.
5. Čentsov N.N., Limit theorems for certain classes of random functions. - Trudy vsesojuzn. sovešč. po teorii ver. i matem. statist., Erevan, 1960, p.280-285 (In Russian).
6. Bickel P.J., Wichura M.J., Convergence criteria for multiparameter stochastic processes and some applications.- "Ann.

Math. Statist.", 1971, v.42, N 5, p.1656-1670.

7. Billingsley P., Convergence of probability measures. J. Wiley, 1968.

LIMIT THEOREMS FOR STOCHASTIC EQUATIONS WITH PARTIAL DERIVATIVES

S. Ja. Mahno

Let $(\Omega, \mathcal{F}, P)$ be a probability space with increasing family $\mathcal{F}_t \subset \mathcal{F}$. Let we consider Cauchy problem

$$d_t \xi(t,x) = [L_{t,x}\xi(t,x) + A(t,x,\xi(t,x))]\,dt + B(t,x,\xi(t,x))\,dw(t), \qquad (I)$$

$$\xi(0,x) = \varphi(x),$$

where $L_{t,x} = \frac{1}{2}\sum_{i,j=1}^{n} a_{ij}(t,x)\frac{\partial^2}{\partial x_i \partial x_j} + \sum_{i=1}^{n} b_i(t,x)\frac{\partial}{\partial x_i}$, $x \in R_n$, $t \in [0,T]$, $w(t)$ is n-dimensional $\mathcal{F}_t$-measurable Wiener process.

<u>Definition</u>. A random field $\xi(t,x)$ which is $\mathcal{F}_t$-measurable for any $t \in [0,T]$ and twicely differentiable in x in the mean square, is called a solution of the problem (I) if

$$\xi(t,x) = \varphi(x) + \int_0^t [L_{s,x}\xi(s,x) + A(s,x,\xi(s,x))]\,ds + \int_0^t B(s,x,\xi(s,x))\,dw(s). \qquad \text{a.s.}$$

We'll suppose that the following conditions are satisfied:

I) the functions $a_{ij}(t,x)$ and $b_i(t,x)$ are continuous in t , each of them has two derivatives bounded in x and

$$Sp\,\|a_{ij}(t,x)\|_{i,j=\overline{1,n}} + |b(t,x)|^2 \leq K(1+|x|^2);$$

2) the functions $\varphi(x)$, $A(t,x,y)$, $B(t,x,y)$ are continuous in t , each of them has two derivatives bounded in x , y and

$$|\varphi(x)|^2 \leq K(1+|x|^2),$$

$$|A(t,x,y)|^2 + |B(t,x,y)|^2 \leq K(1+|x|^2+|y|^2).$$

According to [1] equation (I) has a unique solution,

$$\sup_{t,x} \frac{M|\xi(t,x)|^2}{1+|x|^2} < \infty \qquad (2)$$

and the problem (I) is equivalent to the following integral equation

$$\xi(t,x) = u(t,x) + \int_0^t\int A(s,y,\xi(s,y))\,\mu_{T-t,x}(T-s,dy)\,ds + $$

$$+ \int_0^t\int B(s,y,\xi(s,y))\,\mu_{T-t,x}(T-s,dy)\,dw(s), \qquad (3)$$

where $u(t,x)$ is a solution to Cauchy problem

$$\frac{\partial u}{\partial t} = L_{t,x}u,$$

$$u(0,x) = \varphi(x),$$

$\mu_{T-t,x}(T-s,A) = P\{\zeta_{T-t,x}(T-s) \in A\}$ and stochastic process $\zeta_{t,x}(s)$ is a solution to equation

$$\zeta_{t,x}(s) = x + \int_t^s b(T-v,\zeta_{t,x}(v))\,dv + \int_t^s \sigma(T-v,\zeta_{t,x}(v))\,dw,$$

$\sigma(t,x)\cdot\sigma^*(t,x) = a(t,x)$.

Henceforth we will use the following analogue of Gronool-Bellman lemma. The proof of this lemma is based on the same principle as the lemma mentioned above [2 p.4I].

<u>Lemma I.</u> Let $m_{t,x}(s,A)$ be a measure in A such that $m_{t,x}(s,R_n)=1$ and $\int |y|^2 m_{t,x}(s,dy) \le C(1+|x|^2)$. If for some functions $\alpha(t) \ge 0$, $0 \le z(t,x) \le C(1+|x|^2)$ with a constant $K>0$

$$z(t,x) \le \alpha(t) + K\int_0^t\int z(s,y)\,m_{t,x}(s,dy)\,ds,$$

then

$$z(t,x) \leq \alpha(t) + K\int_0^t e^{K(t-s)}\alpha(s)\,ds.$$

We will construct the finite-difference approximations for stochastic process $\xi(t,x)$. Let $0=t_0<t_1<\dots<t_n=T$ be a certain subdivision of interval $[0,T]$. For $t\in[t_k,t_{k+1})$ we assume the following

$$\xi_\delta(t,x) = \xi_\delta(t_k,x) + u(t,x) - u(t_k,x) + \int_{t_k}^{t}\int A(s,y,\xi(t_k,y))\,\mu_{T-t,x}(T-t_k,dy)\,ds +$$
$$+ \int_{t_k}^{t}\int B(s,y,\xi(t_k,y))\,\mu_{T-t,x}(T-t_k,dy)\,dw(s),$$

$$\xi_\delta(0,x) = \varphi(x).$$

If we define a stochastic process $\eta_\delta(t,x)$ and measure $\nu^\delta_{t,x}(A)$ as follows: for $t\in[t_k,t_{k+1})$

$$\eta_\delta(t,x) = \xi_\delta(t_k,x), \quad \nu^\delta_{t,x}(A) = \mu_{T-t,x}(T-t_k,A),$$

then $\xi_\delta(t,x)$ can be represented in the following way

$$\xi_\delta(t,x) = u(t,x) + \int_0^t\int A(s,y,\eta_\delta(s,y))\,\nu^\delta_{t,x}(dy)\,ds + \int_0^t\int B(s,y,\eta_\delta(s,y))\,\nu^\delta_{t,x}(dy)\,dw. \quad (4)$$

Further C will mean different constants.

<u>Theorem 1.</u> If the conditions 1) - 2) are satisfied we have

$$M|\xi(t,x) - \xi_\delta(t,x)|^2 \leq C|\delta|(1+|x|^2),$$

where $|\delta| = \max\limits_k |t_{k+1} - t_k|$.

<u>Proof</u>. According to the conditions 3) and 4), and to the properties of stochastic integrals we obtain

$$M|\xi(t,x)-\xi_\delta(t,x)|^2 \le C\int_0^t\!\!\int M|\xi(s,y)-\xi_\delta(s,y)|^2\mu_{T-t,x}(T-s,dy)\,ds+$$

$$+\int_0^t\!\!\int M|\xi_\delta(s,y)-\eta_\delta(s,y)|^2\mu_{T-t,x}(T-s,dy)\,ds+ C\int_0^t\Big|\int A(s,y,\tag{5}$$

$$\eta_\delta(s,y))\mu_{T-t,x}(T-s,dy)-\int A(s,y,\eta_\delta(s,y))\nu^\delta_{t,x}(dy)\Big|^2ds+$$

$$+C\int_0^t\Big|\int B(s,y,\eta_\delta(s,y))\mu_{T-t,x}(T-s,dy)-\int B(s,y,\eta_\delta(s,y))\nu^\delta_{t,x}(dy)\Big|^2ds$$

Since $u(t,x)=M\varphi(\xi_{T-t,x}(T))$ and $M|\xi_{t,x}(s)|^2\le C(1+|x|^2)$ [2] it can be easily obtained that

$$M|\xi_\delta(s,x)-\eta_\delta(s,x)|^2\le C|\delta|(1+|x|^2),\tag{6}$$

if $s\in[t_k,t_{k+1})$.

Let give the estimation for the last part of the inequality (5). If $t\in[t_k,t_{k+1})$ we have

$$\int_0^t M\Big|\int B(s,y,\eta_\delta(s,y))(\mu_{T-t,x}(T-s,dy)-\nu^\delta_{t,x}(dy))\Big|^2ds\le$$

$$\le\sum_{k=0}^{n-1}M\int_{t_k}^{t_{k+1}}\Big|\int B(s,y,\xi(t_k,y))(\mu_{T-t,x}(T-s,dy)-\mu_{T-t,x}(T-t_k,dy))\Big|^2ds.\tag{7}$$

Define $\tilde\mu_{s,t_k}(A,B)=P\{\xi_{T-t,x}(T-s)\in A,\ \xi_{T-t,x}(T-t_k)\in B\}$, then

$$M\Big|\int B(s,y,\xi(t_k,y))(\mu_{T-t,x}(T-s,dy)-\mu_{T-t,x}(T-t_k,dy))\Big|^2\le$$

$$\le\iint M|B(s,y,\xi(t_k,y))-B(s,z,\xi(t_k,z))|^2\tilde\mu_{s,t_k}(dy,dz)\le$$

$$\leq C \iint |y-z|^2 \tilde{\mu}_{s,tx}(dy,dz) = C\, M\,|\xi_{T-t,x}(T-s) - \xi_{T-t,x}(T-t_\delta)|^2 \leq$$

$$\leq C(1+|x|^2)\,|\delta|.$$

Therefore, the last part of the inequality (5) is not much more than $C(1+|x|^2)|\delta|$. The next to the last part in (5) may be estimated in the same way. The as it follows from (5), (6) and (7)

$$M|\xi(t,x) - \xi_\delta(t,x)|^2 \leq C(1+|x|^2)|\delta| + \int_0^t \int M|\xi(s,y) - \xi_\delta(s,y)|^2 \mu_{T-t,x}(T-s,dy)\,ds.$$

So, the proof of the theorem is completed by the usage of lemma I.

For the proof of the theorem 2 we'll use the following auxiliary result.

Define the function $\rho(x)$: let $\rho(x)$ be an infinitely diferentiable one, $\rho(x) = 1$ for $|x| \leq \frac{1}{2}$ and $\rho(x) = 0$ for $|x| \geq 1$.

Assume

$$A^n(s,x,y) = A(s,x,y)\,\rho\left(\frac{|x|}{n}\right)\rho\left(\frac{|y|}{n}\right),$$

$$B^n(s,x,y) = B(s,x,y)\,\rho\left(\frac{|x|}{n}\right)\rho\left(\frac{|y|}{n}\right),$$

and define the sequence of stochastic processes $\xi_n(t,x)$ as a solution to the following equation

$$\xi_n(t,x) = \varphi(x) + \int_0^t [L_{s,x}\xi_n(s,x) + A^n(s,x,\xi_n(s,x))]\,ds + \int_0^t B^n(s,x,\xi_n(s,x))\,dw.$$

<u>Lemma 2.</u> If the conditions I) and 2) are satisfied the sequaence of stochastic processes $\xi_n(t,x)$ is converted to

stochastic process $\xi(t,x)$ in the mean square for $n\to\infty$.

Proof. It can be easily seen that

$$M|\xi(t,x)-\xi_n(t,x)|^2 \le C g_n(t,x) + C\int_0^t\int M|\xi(s,y)-\xi_n(s,y)|^2 \mu_{T-t,x}(T-s,dy)\,ds,$$

where

$$g_n(t,x) = \int_0^t\int [M|A(s,y,\xi_n(s,y))(1-\rho(\tfrac{|y|}{n})\rho(\tfrac{|\xi_n(s,y)|}{n}))|^2 +$$

$$+ M|B(s,y,\xi_n(s,y))(1-\rho(\tfrac{|y|}{n})\rho(\tfrac{|\xi_n(s,y)|}{n}))|^2]\,\mu_{T-t,x}(T-s,dy)\,ds.$$

Owing to the forth of lemma I it is sufficient to show, that $g_n(t,x)\xrightarrow[n\to\infty]{}0$. We have

$$\int_0^t\int M|A(s,y,\xi_n(s,y))|^2 M(1-\rho(\tfrac{|y|}{n})\rho(\tfrac{|\xi_n(s,y)|}{n}))\mu_{T-t,x}(T-s,dy)ds \le$$

$$\le C\int_0^t\int (1+|y|^2)\, M\chi(|\xi_n(s,y)|<N)(1-\rho(\tfrac{|y|}{n})\rho(\tfrac{|\xi_n(s,y)|}{n}))\mu_{T-t,x}(T-s,dy)ds+$$

$$+ C\int_0^t\int (1+|y|^2)\, P\{|\xi_n(s,y)|\ge N\}\,\mu_{T-t,x}(T-s,dy)ds \le$$

$$\le C\int_0^t\int (1+|y|^2)\, M\chi(|\xi_n(s,y)|<N)(1-\rho(\tfrac{|y|}{n})\rho(\tfrac{|\xi_n(s,y)|}{n}))\mu_{T-t,x}(T-s,dy)ds+$$

$$+ \frac{C}{N^2}\int_0^t\int (1+|y|^2)^2\mu_{T-t,x}(T-s,dy)\,ds.$$

At first we choose N being sufficiently large and then n is directed to ∞, and we can see that the limit of the left-hand part of inequality mentioned above is equal to 0. The part including function B may be estimated analogously. Lemma 2 is proved.

Consider the family of the stochastic processes $\xi_\lambda(t,x)$ which are dependent on parameter λ . Parameter λ belongs to a certain set Λ where λ_0 is a limit point. Let

$$\xi_\lambda(t,x)= \varphi(x)+\int_0^t [L^\lambda_{s,x}\xi_\lambda(s,x)+A_\lambda(s,x,\xi_\lambda(s,x))]ds + \int_0^t B_\lambda(s,x,\xi_\lambda(s,x))\,dw_\lambda(s), \tag{8}$$

where $L^\lambda_{s,x}=\frac{1}{2}\sum_{i,j=1}^{n} a^\lambda_{ij}(s,x)\frac{\partial^2}{\partial x_i\partial x_j}+\sum_{i=1}^{n} b^\lambda_i(s,x)\frac{\partial}{\partial x_i}$.

The function $F_\lambda(s,x,y)$ is integrably continuous on λ for $\lambda=\lambda_0$ if for any (s_1,s_2,x,y)

$$\lim_{\lambda\to\lambda_0}\int_{s_1}^{s_2} F_\lambda(s,x,y)ds=\int_{s_1}^{s_2} F_{\lambda_0}(s,x,y)ds \tag{9}$$

If the equality (9) is satisfied uniformly for s_1, s_2, x, y, $0<s_2-s_1<C$ then the function $F_\lambda(s,x,y)$ is called unoformly integrably continuous.

Let introduce one more condition:

3) the functions a^λ_{ij} and b^λ_i are uniformly bounded and continuous on x being uniformly relative to (t,x,λ) .

Symbol $\Longrightarrow$ means a convergence of finite-dimensional distributions.

<u>Theorem 2.</u> If the conditions 1) - 3) where the constants are independent on λ are satisfied for any λ, and the functions $a^\lambda_{ij}(t,x)$, $b^\lambda_i(t,x)$ are unoformly integrably continuous on λ for $\lambda=\lambda_0$, and the functions $A_\lambda(t,x,y)$ are integrably continuous on λ for $\lambda=\lambda_0$ and for fixed (t,x,y)

$$\int_0^t B_\lambda(s,x,y)\,dw_\lambda \underset{\lambda\to\lambda_0}{\Rightarrow} \int_0^t B_{\lambda_0}(s,x,y)\,dw_{\lambda_0},$$

then $\xi_\lambda(t,x) \underset{\lambda\to\lambda_0}{\Rightarrow} \xi_{\lambda_0}(t,x)$, where

$$\xi_{\lambda_0}(t,x) = \varphi(x) + \int_0^t [L^{\lambda_0}_{s,x}\xi_{\lambda_0}(s,x) + A_{\lambda_0}(s,x,\xi_{\lambda_0}(s,x))]\,ds + \int_0^t B_{\lambda_0}(s,x,\xi_{\lambda_0}(s,x))\,dw_{\lambda_0}.$$

<u>Proof.</u> To prove the theorem we will use the method suggested in [2] .Consider at first the particular case of equation (9). Let

$$A_\lambda(s,x,y) = A_\lambda(s)\,A(x,y),$$

$$B_\lambda(s,x,y) = B_\lambda(s)\,B(x,y).$$

Costruct for $\xi_\lambda(t,x)$ finite-difference approximation $\xi_{\lambda\delta}(t,x)$. Then for any function $f(z_1,\dots,z_r)$ which is continuous and bounded with its partial derivatives $f(\xi_{\lambda\delta}(t_1,x_1),\dots,\xi_{\lambda\delta}(t_r,x_r))$ is continuous and bounded function of

$$u_\lambda(t_i,x_k),\ \int_{t_i}^{t_{i+1}} A_\lambda(s)\,ds,\ \int_{t_i}^{t_{i+1}} B_\lambda(s)\,dw_\lambda(s),\ \int g_\lambda(y)\,\mu^\lambda_{T-t_i}(T-t_j,dy)$$

and $g_\lambda(y,w) \underset{\lambda\to\lambda_0}{\Rightarrow} g_{\lambda_0}(y,w)$.

A convergence $u_\lambda(t_i,x_k)$ to $u_{\lambda_0}(t_i,x_k)$ follows from [3] . Therefore for δ chosen here we have

$$\lim_{\lambda\to\lambda_0} Mf(\xi_{\lambda\delta}(t_1,x_1),\dots,\xi_{\lambda\delta}(t_r,x_r)) = Mf(\xi_{\lambda_0\delta}(t_1,x_1),\dots,\xi_{\lambda_0\delta}(t_r,x_r)).$$

This particular case is proved by the usage of the theorem I.

Let the coefficients $A_\lambda(s,x,y)$ and $B_\lambda(s,x,y)$ be uniformly bounded. For $\varepsilon>0$, $S_N=\{x: |x|\le N\}$, $x_1,\dots,x_{m_N}$ is ε - net in S_N. The functions $g_i(x)$ are chosen to satisfy the following

I. $g_j(x)\ge 0$, $g_j(x)=0$ if $|x-x_j|\ge\varepsilon$.

2. $\sum_{i=1}^{m_N} g_i(x)=1$ if $x\in S_N$

3. $g_j(x)$ have two bounded derivatives.

Assume

$$A_\lambda^\varepsilon(s,x,y)=\sum_{i,k=1}^{m_N} g_i(x)\,g_k(y)\,A_\lambda(s,x_i,x_k),$$

$$B_\lambda^\varepsilon(s,x,y)=\sum_{i,k=1}^{m_N} g_i(x)\,g_k(y)\,B_\lambda(s,x_i,x_k).$$

The theorem is proved for stochastic processes $\xi_\lambda^\varepsilon(t,x)$ which are satisfied (8) with functions A_λ^ε and B_λ^ε. Then

$$M|\xi_\lambda(t,x)-\xi_\lambda^\varepsilon(t,x)|^2\le C(\mathcal{J}_1+\mathcal{J}_2)+C\int_0^t\int M|\xi_\lambda(s,y)-\xi_\lambda^\varepsilon(s,y)|^2\mu_{T-t,x}^\lambda(T-s,dy)\,ds, \quad (\text{IO})$$

where

$$\mathcal{J}_1=M\Big|\int_0^t\int[A_\lambda(s,y,\xi_\lambda^\varepsilon(s,y))-A_\lambda^\varepsilon(s,y,\xi_\lambda^\varepsilon(s,y))]\,\mu_{T-t,x}^\lambda(T-s,dy)\,ds\Big|^2,$$

$$\mathcal{J}_2=M\Big|\int_0^t\int[B_\lambda(s,y,\xi_\lambda^\varepsilon(s,y))-B_\lambda^\varepsilon(s,y,\xi_\lambda^\varepsilon(s,y))]\,\mu_{T-t,x}^\lambda(T-s,dy)\,dw\Big|^2.$$

Giving the estimation for $\mathcal{J}_2$ we obtain

$$\mathcal{J}_2 \le \int_0^t \int_{|y|>N} M \left| B_\lambda(s,y,\xi_\lambda^\varepsilon(s,y)) - B_\lambda^\varepsilon(s,y,\xi_\lambda^\varepsilon(s,y)) \right|^2 \mu^\lambda_{T-t,x}(T-s,dy)\, ds +$$

$$+ \int_0^t \int_{|y|\le N} M\, \chi\left(|\xi_\lambda^\varepsilon(s,y)| < N\right) \left| B_\lambda(s,y,\xi_\lambda^\varepsilon(s,y)) - B_\lambda^\varepsilon(s,y,\xi_\lambda^\varepsilon(s,y)) \right|^2 \mu^\lambda_{T-t,x}(T-s,dy)\, ds +$$

$$+ \int_0^t \int_{|y|\le N} M\, \chi\left(|\xi_\lambda^\varepsilon(s,y)| \ge N\right) \left| B_\lambda(s,y,\xi_\lambda^\varepsilon(s,y)) - B_\lambda^\varepsilon(s,y,\xi_\lambda^\varepsilon(s,y)) \right|^2 \mu^\lambda_{T-t,x}(T-s,dy)\, ds$$

$$= \mathcal{J}_2^1 + \mathcal{J}_2^2 + \mathcal{J}_2^3$$

According to boundary B_λ and B_λ^ε, and the estimations for stochastic integrals [2]

$$\mathcal{J}_2^1 + \mathcal{J}_2^3 \le C \int_0^t P\{ |\xi^\lambda_{T-t,x}(T-s)| > N \}\, ds + C \int_0^t \int P\{ |\xi_\lambda^\varepsilon(s,y)| >$$

$$> N \} \mu^\lambda_{T-t,x}(T-s,dy)\, ds \le \frac{ct(1+|x|^2)}{N^2}$$

and we have $\mathcal{J}_2^2 \le C_N \varepsilon^2 t$, i.e. $\mathcal{J}_2 \le C_N \varepsilon^2 t + \frac{ct(1+|x|^2)}{N^2}$.

Analogues estimations may be obtained for $\mathcal{J}_1$. Thus, using (IC), the estimations mentioned above and lemma I the theorem is proved for this particalar case. Let A_λ and B_λ are satisfied the condition 2) only. The functions A_λ^n and B_λ^n are constructed according to lemma 2. The theorem is proved for any n and the general case follows from lemma 2.

References

I. Баклан В.В. Об одном классе стохастических уравнений в частных производных. Сб. Поведение систем в случайных средах. Киев, 1976, с. 3-7.

2. Гихман И.И., Скороход А.В. Стохастические дифференциальные уравнения. Киев, Наукова думка, 1968.

3. Хасьминский Р.З. О принципе усреднения для параболических и эллиптических дифференциальных уравнений и марковских процессов с малой диффузией. Теория вероятн. и ее примен. 1963, 8, № 1, с. 3-24.

FORMULA FOR CONDITIONAL WIENER INTEGRALS

V.Mackevičius

Let $(C[0,t],B,P)$ be a Wiener measure space: $C[0,t]$ is the space of the real valued continuous functions ω on $[0,t]$ with $\omega(0)=0$ for fixed $t\in(0,\infty)$, B is the σ-algebra of Borel subsets of $C[0,t]$, P is a Wiener measure on B. Here by a conditional Wiener integral we mean a conditional expectation $E(Y|X(t))$ of a P-integrable random variable Y on $C[0,t]$ conditioned by $X(t)$, where $X(s)=X(s,\omega)=\omega(s)$ for $(s,\omega)\in[0,t]\times C[0,t]$. Let us remark that $(X(s),s\in[0,t])$ is a standart Brownian motion on $(C[0,t],B,P)$. It is well known that there exists a Borel measurable function $e:R\to R$ such that $E(Y|X(t))=e(X(t))$ P-a.s. We shall denote this function $e(x)=E^x(Y), x\in R$. J.Yeh ([2],[3]) has derived several Fourier inversion formulas for retrieving $E^x(Y)$ given the function $E(e^{iuX(t)}Y), u\in R$. Here we give a simple formula of another type for evaluating $E^x(Y)$.

<u>Theorem</u>. A version of a conditional Wiener integral is given by

$$E^x(Y)=E(Y^x),\quad Y^x=Y\circ\omega_x,$$

where the map $C[0,t]\ni\omega\mapsto\omega_x=\omega_x(\omega)\in C[0,t]$ is defined by

$$\omega_x(s)=\omega_x(s,\omega)=\frac{x}{t}s+\omega(s)-(t-s)\int_0^s\frac{\omega(u)}{(t-u)^2}du,\ s\in[0,t), \tag{1}$$

$$\omega_x(t)=\lim_{s\uparrow t}\omega_x(s)=x.$$

<u>Proof</u>. Let P^x be a probability measure on $C[0,t]$ for which a coordinate function $X(s)$ on $C[0,t]$ is a reciprocal process derived from Brownian motion $(B(s),s\in[0,t])$ by tying it down at $B(0)=0$ and $B(t)=x$ (see, e.g., B.Jamison [1]). The

measures $P^x(d\omega)$, constitute a regular conditional distribution of P given X(t). Therefore for P-integrable Y we have

$$E^x(Y)=\int_{C[0,t]} Y(\omega)P^x(d\omega) \text{ for a.e. } x\in R.$$

It is well known (see, e.g., theorem 2 of [1]) that P^x coincides with the measure on C[0,t] induced by the process $Z(s), s\in[0,t]$, which satisfies a stochastic differential equation

$$Z(s)=\int_0^s \frac{x-Z(u)}{t-u}du+B(s), s\in[0,t),$$

for some Brownian motion $B(s), s\in[0,t)$. This shows that the process $b(s), s\in[0,t)$, defined by

$$b(s)=b(s,\omega)=\omega(s)-\int_0^s \frac{x-\omega(u)}{t-u}du, s\in[0,t), \tag{2}$$

is the Brownian motion with respect to measure P^x. Solving (2) as the linear differential equation with respect to ω we obtain

$$\omega(s)= \frac{x}{t}s+b(s)-(t-s)\int_0^s \frac{b(u)}{(t-u)^2}du, s\in[0,t), \tag{3}$$

for smooth functions $b\in C[0,t]$. An extension of (3) as the solution of (2) for all $b\in C[0,t]$ is obvious by simple limiting argument.

Comparison of (1) and (3) shows that P^x coincides with the P-distribution of ω_x. That completes the proof.

Remark. Integration by parts of (1) gives

$$\omega_x(s)=\frac{x}{t}s+(t-s)\int_0^s \frac{dX(u)}{t-u} \text{ P-a.s.}, s\in[0,t),$$

where the integral is uderstood as a stochastic one with respect to Brownian motion $X(u), u\in[0,t)$, relative to P.

Example. As an example of application of our formula we shall find two conditional Wiener integrals which were evaluated in [3] in more complicated way:

$$E^x\left(\frac{1}{t}\int_0^t X(s)ds\right)=\frac{1}{t}E\left(\int_0^t \left(\frac{x}{t}s+(t-s)\int_0^s \frac{dX(u)}{t-u}\right)ds\right)=\frac{x}{2};$$

$$E^x\left(\int_0^t X^2(s)ds\right)=E\left(\int_0^t \left(\frac{x}{t}s+(t-s)\int_0^s \frac{dX(u)}{t-u}\right)^2 ds\right)=$$

$$=\frac{x^2}{t^2}\int_0^t s^2 ds+\int_0^t (t-s)^2\int_0^s \frac{du}{(t-u)^2}ds= x^2\cdot\frac{t}{3}+\frac{t^2}{6}.$$

(Here we used well known properties of stochastic integrals).

REFERENCES

1. Jamison B.,TheMarkov processes of Schrödinger,Z. Wahrsch.v.Geb.,1975,32,323-331.
2. Yeh J.,Inversion of conditional expectations,Pacific J.Math.,1974,52,2,631-640.
3. Yeh J.,Inversion of conditional Wiener integrals, Pacific J.Math.,1975,59,2,623-638.

ON THE ASYMPTOTIK BEHAVIOR OF THE SOLUTION OF THE DIMENTIONAL STOCHASTIC DIFFUSION EQUATION.

G. L. KULINIČ

Let exists the solution $\xi(t)$ of one dimentional stochastic diffusion equation

$$d\xi(t) = a(\xi(t))dt + \sigma(\xi(t))dw(t)\,,\ t \geq 0, \qquad (1)$$

where $a(x), \sigma(x) > 0$ are nonrandom real function, $w(t)$ is the Wiener process given on the probability space $(\Omega, \mathfrak{F}, P)$, $\xi(0)$ is the given random value, which is independent of $w(t)$.

Theorem 1. Let $\xi(t)$ be the solution of (1), $M|\xi(0)|^4 < \infty$ and

$$|x a(x) - \bar{\alpha}(x)| + |\sigma(x) - \bar{\sigma}(x)| \leq q(x),$$

where $q(x) \leq C$, $\lim_{|x| \to \infty} \frac{1}{|x|} \int_0^x \frac{q(v)}{\sigma^2(v)} dv = 0,$

$$\bar{\alpha}(x) = \begin{cases} \alpha_1, & x > 0, \\ \alpha_2, & x < 0, \end{cases} \qquad \bar{\sigma}(x) = \begin{cases} \sigma_1, & x > 0, \\ \sigma_2, & x < 0. \end{cases}$$

1. If $2\alpha_1\sigma_1^{-2} > 1, 2\alpha_2\sigma_2^{-2} < 1$, then the stochastic process $T^{-1/2}\xi(tT)$ weakly converges, as $T \to \infty$, to the process $z(t)$, which satisfies the Ito's equation

$$z^2(t) = (2\alpha + \sigma^2)t + 2\sigma \int_0^t z(s)\, d\hat{w}(s)\ , \qquad (2)$$

for $\alpha = \alpha_1$, $\sigma = \sigma_1$.

2. If $2\alpha_1\sigma_1^{-2} < 1$, $2\alpha_2\sigma_2^{-2} > 1$, then stochastic process $-T^{-1/2}\xi(tT)$ weakly converges to the solution $z(t)$ of equation (2), as $T \to \infty$, for $\alpha = \alpha_2$, $\sigma = \sigma_2$.

3. If $\alpha_1=\alpha_2=\alpha$, $\sigma_1=\sigma_2=\sigma$ and $2\alpha+\sigma^2>0$, then the stochastic process $T^{-1/2}|\xi(tT)|$ weakly converges to the solution $\tau(t)$ of equation (2).

Proof of the statement 1. Using the inequalities $2\alpha_1\sigma_1^{-2}>1$, $2\alpha_2\sigma_2^{-2}<1$ we can show that $f(-\infty)=-\infty$, $f(+\infty)\leq C$, where

$$f(x)=\int_0^x \exp\left\{-2\int_0^u \frac{a(v)}{\sigma^2(v)}\,dv\right\}du. \tag{3}$$

Hence, [1] $P\{\lim_{t\to\infty}\xi(t)=+\infty\}=1$. Since by Ito's formula

$$M\int_0^t q(\xi(s))\,ds = M[\Phi(\xi(t))-\Phi(\xi(s))] - M\int_0^t \Phi'(\xi(s))a(\xi(s))\,ds,$$

where

$$\Phi(x)=2\int_0^x\left(\int_0^u \frac{q(v)}{\sigma^2(v)}\,dv\right)du,$$

and $x^{-2}\Phi(x)\to 0$, $\Phi'(x)a(x)\to 0$ as $x\to+\infty$, $Mt^{-1}\xi^2(t)\leq C$, then

$$\lim_{t\to\infty} M\frac{1}{t}\int_0^t q(\xi(s))\,ds=0. \tag{4}$$

Introdduce the parametr $T>0$ and denote, as $0\leq t\leq 1$,

$$z_T(t)=\frac{|\xi(tT)|}{\sqrt{T}},\quad \hat{w}_T(t)=\int_0^t (\operatorname{sign}\xi(sT))\,dw_T(s),$$

$$w_T(t)=\frac{w(tT)}{\sqrt{T}},\quad J_T^{(1)}(t)=2\int_0^t z_T(s)\left[\sigma(\xi(sT))-\bar{\sigma}(\xi(sT))\right]d\hat{w}_T(s),$$

$$J_T^{(2)}(t)=\frac{1}{T}\int_0^{tT}\left[2\xi(s)a(\xi(s))+\sigma^2(\xi(s))-2\bar{\alpha}(\xi(s))-\bar{\sigma}^2(\xi(s))\right]ds.$$

Then
$$z_T^2(t)=z_T^2(0)+\frac{1}{T}\int_0^{tT}\left[2\bar{\alpha}(\xi(s))+\bar{\sigma}^2(\xi(s))\right]ds+$$

$$+2\int_0^t z_T(s)\bar{\sigma}(\xi(sT))d\hat{w}_T(s)+J_T^{(1)}(t)+J_T^{(2)}(t).$$

It is fairly to show that

$$M\sup_{0\le t\le 1} z_T^2(t)\le c,\qquad \sup_{0\le t\le 1}\left|J_T^{(2)}(t)\right|\le\frac{c}{T}\int_0^T q(\xi(s))ds,$$

$$P\left\{\sup_{0\le t\le 1}\left|J_T^{(1)}(t)>\varepsilon\right.\right\}\le P\left\{\sup_{0\le t\le 1} z_T(t)>N\right\}+\frac{c}{\varepsilon^2}N^2\frac{1}{T}M\int_0^T q(\xi(s))ds.$$

For the arbitrary $\varepsilon>0$, $N>0$, therefore taking into acount (4) and convergence of $\xi(t)\to+\infty$ with probability 1, as $t\to\infty$, we obtain

$$z_T^2(t)=z_T^2(0)+(2\alpha_1+\sigma_1^2)t+2\sigma_1\int_0^t z_T(s)d\hat{w}_T(s)+\beta_T(t),\qquad (5)$$

where $M\sup_{0\le t\le 1}|\beta_T(t)|\to 0$, as $T\to\infty$.

Process $(z_T(t), \hat{w}_T(t), \beta_T(t))$ is satisfies the condition of Skorokhod's A.V. theorem on compectness [2] . Therefore we shall suppose that any any subsequence $T_n'\to\infty$ there exists subsequence $T_n\to\infty$ such, that $z_{T_n}(t)\xrightarrow{P}z(t)$, $\hat{w}_{T_n}(t)\xrightarrow{P}$ $\xrightarrow{P}\hat{w}(t)$.

Going to the limit by $T_n \to \infty$ in (5), we obtain, that the process $\tau(t)$ satisfies the equation (2) for $\alpha = \alpha_1$, $\sigma = \sigma_1$. From the uniqueness of solution of (2) and from arbitrariness of subsequence $T_n \to \infty$, it follows the convergence of distributions of the process $\tau_T(t)$, as $T \to \infty$, to those of $\tau(t)$. Moreover it is easily to show, that for any

$$\lim_{h \to 0} \overline{\lim_{T \to \infty}} P\left\{\sup_{|t'-t''|<h} |\tau_T(t') - \tau_T(t'')| > \varepsilon\right\} = 0.$$

Hence the statement 1 is proved.

The proof of statement 2 and 3 of the theorem is analogous, but with some difference. Under the conditions of statement 2 we can show the convergence $\xi(t) \to -\infty$ with probability 1, as $t \to \infty$, and under the condition of statement 3 we can show that

$$\lim_{t \to \infty} \frac{1}{t} \int_0^t P\{|\xi(s)| < c\} ds = 0$$

for any $c > 0$.

<u>Remark 1.</u> Process $\sigma^{-1}\tau(t)$, where $\tau(t)$ is the solution of equation (2), is a Bessel's diffusion process of index $2\alpha\sigma^{-2}+1$ and the transition density may bewriten in obvious form [3].

<u>Remark 2.</u> The results analogous to the statement 3 for n-dimensional case are obtained in [4] under the condition

$$\lim_{|x| \to \infty} \frac{1}{|x|} \int_0^{|x|} q(v) dv = 0.$$

<u>Theorem 2.</u> Let real function $a_T(x), \sigma_T(x) > 0$ be such, that for every $T > 0$ there exists the solution $\xi_T(t)$ of the equation

$$d\xi_T(t) = a_T(\xi_T(t))dt + \sigma_T(\xi_T(t))dw_T(t), \quad 0 \leq t \leq 1, \tag{6}$$

where $w_T(t)$ is the family of Wiener processes given on the probability space $(\Omega, \mathfrak{F}, P)$, and let the family of functions

$$f_T(x) = C_T^{(1)} \int_0^x \exp\left\{-2\int_0^u \frac{a_T(v)}{\sigma_T^2(v)}\,dv\right\} du + C_T^{(2)}$$

be such, that $M f_T^2(\xi_T(0)) \le C$, $0 < \delta \le f_T'(x)\sigma_T(x) \le C$. Then the family of stochastic processes $\hat{\xi}_T(t) = f_T(\xi_T(t))$ is weakly compact and for any subsequence $T_n \to \infty$, for which take place the convergence

$$\lim_{T_n \to \infty} \int_{-N}^{N} \left| G_{T_n}(x) - G(x) \right| dx = 0 \qquad (7)$$

for any $N > 0$, where

$$G_T(x) = \int_0^x \left[f_T'(\varphi_T(u))\sigma_T(\varphi_T(u)) \right]^{-2} du,$$

$\varphi_T(x)$ is the inverse function to $f_T(x)$, and function $G(x)$ is differentiable and $0 < \delta \le G'(x) \le C$, stochastic process $\hat{\xi}_{T_n}(t) = f_{T_n}(\xi_{T_n}(t))$ converges weakly, as $T_n \to \infty$, to the solution $\hat{\xi}(t)$ of Ito's equation

$$d\hat{\xi}(t) = \frac{1}{\sqrt{G'(\hat{\xi}(t))}}\, d\hat{w}(t). \qquad (8)$$

Proof. From condition $0 < \delta \le f_T(x)\sigma_T(x) \le C$ it follows the existence of function $G(x)$, which satisfies the condition of theorem. Without loss of generality we suppose that convergence of (7) take place for $G_T(x)$, as $T \to \infty$, and consider the family of processes $\hat{\xi}_T(t) = f_T(\xi_T(t))$. Using the formula of Ito's [5] we obtain

$$\hat{\xi}_T(t) = \hat{\xi}_T(0) + \int_0^t \left(G'(\hat{\xi}_T(s))\right)^{-1/2} d\eta_T(s), \tag{9}$$

where

$$\eta_T(t) = \int_0^t \sqrt{G'(\hat{\xi}_T(s))}\,\hat{\sigma}_T(\hat{\xi}_T(s))\,dw_T(s),$$

$\hat{\sigma}(x) = f'_T(\varphi_T(x))\sigma_T(\varphi_T(x))$. It is easily to see, that process $(\hat{\xi}_T(t), \eta_T(t), w_T(t))$ satisfies the conditions of the compactness [2]. Therefore we shall suppose that for any subsequence $T'_n \to \infty$ there exists subsequence $T_n \to \infty$, for which $\hat{\xi}_{T_n}(t) \xrightarrow{P} \hat{\xi}(t)$, $\eta_{T_n}(t) \xrightarrow{P} \hat{w}(t)$, $w_{T_n}(t) \xrightarrow{P} w(t)$. Let's show, that the characteristic $\langle \eta_{T_n}(t) \rangle$ of the family of martingales $\eta_{T_n}(t)$ converges to t, as $T_n \to \infty$. For this purpose we introduce the function

$$\Phi_{T_n}(x) = 2\int_0^x [G(u) - G_{T_n}(u)]\,du$$

and by the Ito's formula we obtain

$$\langle \eta_{T_n}(t) \rangle - t = \Phi_{T_n}(\hat{\xi}_{T_n}(t)) - \Phi_{T_n}(\hat{\xi}_{T_n}(0)) - \int_0^t \Phi'_{T_n}(\hat{\xi}_{T_n}(s))\hat{\sigma}_{T_n}(\hat{\xi}_{T_n}(s))\,dw_{T_n}(s).$$

Since for any $N > 0$

$$|\Phi_{T_n}(x)|\chi_{|x| \le N} \le 2\int_{-N}^{N} |G(u) - G_{T_n}(u)|\,du$$

and furthermore on the base of Krylov's N.V. estimate for the integrals of diffusion processes [5],

$$M\int_0^t [\Phi'_{T_n}(\hat{\xi}_{T_n}(s))\hat{\sigma}_{T_n}(\hat{\xi}_{T_n}(s))]^2 \chi_{|\hat{\xi}_{T_n}(s)| \le N}\,ds \le$$

$$\leq C_{\mathcal{N}} \int\limits_{-\mathcal{N}}^{\mathcal{N}} |G(u) - G_{T_n}(u)|^2 du,$$

then $\langle \eta_{T_n}(t) \rangle \xrightarrow{P} t$, as $T_n \to \infty$. Hence the limit process $\hat{w}(t)$ for sequence $\eta_{T_n}(t)$ being the Wiener process. Smoothing the function $G'(x)$ in (9) we can go to the limit for $T_n \to \infty$ [5] . Therefore $\hat{\xi}_{T_n}(t)$ converges to the solution of equation (8). From weak uniqueness [6] of solution (8), and from arbitrariness of subsequence $T_n \to \infty$ and

$$\lim_{h \to \infty} \overline{\lim_{T \to \infty}} P\{\sup_{|t'-t''| \leq h} |\hat{\xi}_T(t') - \hat{\xi}_T(t'')| > \varepsilon\} = 0 \quad (10)$$

for any $\varepsilon > 0$, it follows the proof of theorem 2.

Remark 3. The asymptotic behaviour of the processes $f_T(\xi_T(t))$ in the case when in equation (6) $W_T(t)$ is the family of the continuos martingales with characteristic $\langle W_T(t) \rangle$ converging to t , as $T \to \infty$, is considered in the paper [7] under the more firm conditions on the function $G'(x)$ and some additional conditions on $a_T(t)$ and $\sigma_T(x)$.

As a consequence of theorem 2 we have

Theorem 3. Let $\xi(t)$ be the solution of equation (1) and $0 < \delta \leq f'(x)\sigma(x) \leq C$, where the function $f(x)$ is given by (3), $M f^2(\xi(0)) < \infty$. If

$$\left| \frac{1}{x} \int\limits_0^x [f'(\varphi(u))\sigma(\varphi(u))]^{-2} du - \frac{1}{\bar{\sigma}(x)} \right| \leq q(x),$$

where $\varphi(x)$ is the inverse function to $f(x)$,

$$\bar{\sigma}(x) = \begin{cases} \sigma_1, & x > 0, \\ \sigma_2, & x < 0, \end{cases} \quad \lim_{|x| \to \infty} \frac{1}{|x|} \int\limits_0^x q(v) dv = 0,$$

then the stochastic process $\hat{\xi}_T = T^{-1/2} f(\xi(tT))$ weakly converges to the solution of equation (8), when $G'(x)=\frac{1}{\bar{\sigma}(x)}$. Realy, if we consider the equation for process $\xi_T(t)=T^{-1/2}\xi(tT)$, then the condition of theorem 2 are fulfilled with $C_T^{(1)}=1$, $C_T^{(2)}=0$ in function $f_T(x)$, and for $G_T(x)$ (7) is fulfilled, as $T\to\infty$ with $G(x)=x\frac{1}{\bar{\sigma}(x)}$.

Remark 4. The proof of theorem 3 in the case when $q(x)\to 0$ as $|x|\to\infty$ is given in the paper [8]. Futhermore in theorem 3 for the limit process $\hat{\xi}(t)$ the transition density may be writen in obvious form [9].

Theorem 4. Let the suppositions of theorem 2 be fulfilled with covergence (7) for $G_T(x)$, as $T\to\infty$, and let $g_T(x)$ be the family of real Borel's functions integrated in any finite region for which there exists real Borel's function $g(x)$ square integrated in any finite region, and such, that for any $N>0$

$$\lim_{T\to\infty}\int_{-N}^{N}|Q_T(x)-g(x)|^2 dx = 0,$$

where

$$Q_T(x)=\int_0^x g_T(\varphi_T(u))\left[f'_T(\varphi_T(u))\sigma_T(\varphi_T(u))\right]^{-2} du,$$

the functions $f_T(x)$, $\varphi_T(x)$ are defined in theorem 2. Then the family of stochastic processes $\beta_T(t)=\int_0^t g_T(\xi_T(s))ds$ weakly converges, as $T\to\infty$, to the process

$$\beta(t)=2\left[\int_{\hat{\xi}(0)}^{\hat{\xi}(t)} g(x)dx-\int_0^t g(\hat{\xi}(s))d\hat{\xi}(s)\right],$$

where $\hat{\xi}(t)$ is the solution of equation (8).

Proof. Let us take the subsequence $T_n\to\infty$ for which in theorem 2 $\hat{\xi}_{T_n}(t)\xrightarrow{P}\hat{\xi}(t)$. By Ito's formula

$$\beta_{T_n}(t) = \hat{\Phi}_{T_n}(\hat{\xi}_{T_n}(t)) - \hat{\Phi}_{T_n}(\hat{\xi}_{T_n}(0)) - \int_0^t \hat{\Phi}'_{T_n}(\hat{\xi}_{T_n}(s))\, d\hat{\xi}_{T_n}(s),$$

where
$$\hat{\Phi}_{T_n}(x) = 2\int_0^x Q_{T_n}(u)\,du.$$

Using analogous estimates which have we applied in theorem 2 for the proof of convergence $<\eta_{T_n}(t)> \xrightarrow{P} t$, it is easy to check, that

$$\beta_{T_n}(t) = 2\left[\int_{\hat{\xi}_{T_n}(0)}^{\hat{\xi}_{T_n}(t)} g(x)\,dx - \int_0^t g(\hat{\xi}_{T_n}(s))\,d\hat{\xi}_{T_n}(s)\right] + O(1),$$

where $O(1) \xrightarrow{P} 0$, as $T_n \to \infty$. By the same reason which we use going to the limit in (9), in this case we can go tothe limit under the symbol of integral too. Therefore the distribution of process $\beta_{T_n}(t)$ conveges to those of $\beta(t)$. Further it is not difficult to show that for process $\beta_T(t)$ the relation of the type (10) is fulfilled. From weak convergence of solution (8) it follows the proof of theorem 4.

REFERENCES.

1. Gihman I.I., Skorokhod A.V., Stochastic differential equations "Naukova dumka", 1968 (in Russian).

2. Skorokhod A.V., Investigations on the theory of stochastic processes, Kiev University, 1961, (in Russian).

3. Shiga T., Watanabe S., Bessel diffusions as a one-parameter family of diffusion processes, Z. Wahrscheinlichkeitstheore und Verw. Gebeit, 1973, 27, N 1, 37-46.

4. Kulinič G.L. On asymptotic behaviour of solution of stochastic differential equations of diffusion type with random coefficient, collect. Limit theorems for stochastic processes, Kiev, publish. IM AN UkrSSR, 137-151, (in Russian).

5. Krylov N.V. Controllable processes of diffusion type, publish. "Nauka" , 1977, (in Russian).

6. Veretennikov A.Y. Strong and weak solutions of stochastic equations, autosynopsis of candidate dissertation, Moscow University, 1978, (in Russian).

7. Kulinič G.L. Limit theorems for one dimensional stochastic differential equations under nonregular dependence of coefficients on parameter, Theory of probab. and math. statistic, 1976, 15, 99-114 (in Russian).

8. Kulinič G.L. Asymptotic behaviour of instable solution of stochastic homogenous diffusion equation, Theory of probab. and math. statistic, 1971, 5 , 81-87 (in RUSSIAN).

9. Kulinič G.L. On asymptotic behaviour of distribution of stochastic diffusion equation, Theory of probab. and its applications, 1967, XII, 3, 348-551, (in Russian).

ON A DIRICHLET PROBLEM
WITH RANDOM COEFFICIENTS

V.V.Jurinskii (Novosibirsk)

This communication considers the asymptotic behaviour for small ε of the solution of the Dirichlet problem

$$-div\left(a(x/\varepsilon)\, grad\, u_\varepsilon\right) = F(x),\ x \in \mathcal{G};\quad u_\varepsilon/_{\partial\mathcal{G}} = 0 \tag{1}$$

where $a(y),\ y \in R^k$ is a matrix-valued measurable random field and the bounded domain $\mathcal{G}$ and non-random F are smooth enough (cf., e.g.[7]) to ensure the continuity in $\overline{\mathcal{G}}$ of third-order derivatives of the solution of the averaged Dirichlet problem

$$-div\left(A\, grad\, U\right) = F\ ,\quad U/_{\partial\mathcal{G}} = 0 \tag{2}$$

with a constant positive-definite matrix A.

The measurable random field a satisfies w.p. 1 the ellipticity condition

$$\forall_{b,y \in R^k}\quad a_1 |b|^2 \leqslant (a(y)b,b) \leqslant a_2 |b|^2 \tag{3}$$

with non-random $a_i > 0$. It is homogeneous with respect to integer translations in R^k and satisfies the strong mixing condition

$$|E\xi_1\xi_2 - E\xi_1\, E\xi_2| < \beta(d_{12}),\quad \lim_{d\to\infty} \beta(d) = 0 \tag{4}$$

where $\xi_i, |\xi_i| \leq 1$ are measurable with respect to sigma-fields $\mathcal{M}(\mathcal{G}_i)$ generated by random variables (RV's) $\int a_{j\ell}(y)\phi(y)\,dy$, $\phi \in \dot{C}^{\infty}(\mathcal{G}_i)$ and $d_{1,2}$ is the distance between the bounded domains $\mathcal{G}_i$.

The main result of this communication is

<u>THEOREM 1.</u> If $\varepsilon \to 0$,

$$E\|u_\varepsilon - U\|^2_{L_2(\mathcal{G})} \to 0$$

where $U(x)$ is the solution of (2) with the matrix A defined by the equality

$$\forall_b \quad (Ab, b) = \lim_{N\to\infty} N^{-k} E \min_{\varphi} \int_{K_N} (a(y)\{\nabla\varphi + b\}, \nabla\varphi + b)\,dy . \qquad (5)$$

Here $\nabla = \|\partial/\partial y(j)\|$ denotes the "fast" derivatives with respect to $y = x/\varepsilon$. Minimum is taken over all $\varphi \in \overset{\circ}{W}{}^1_2(K_N)$, $K_N = [0, N]^k$.

The proof of Theorem 1 occupies § 2. The necessary preparations are carried through in § 1.

For equations with non-random coefficients similar results are listed in [1 - 5]. (After submitting the abstract of this communication to the Vilnius symposium the author learned that essentially same result was independently obtained by S.M.Kozlov who used a different technique. A summary of his work appeared in [6]).

§ 1. <u>PRELIMINARIES.</u> Let $\varphi = \varphi_b(y)$ be the solution of the auxiliary "fast" Dirichlet problem

$$-\nabla^T(a(y)\{\nabla\varphi + b\}) = 0, \quad \varphi/\partial K = 0 \qquad (6)$$

where $K = K_N = [0,N]^k$ (in the matrix notation T denotes transposition). It is well known that φ minimizes the quadratic functional of (5). φ_b is linear in b ($b \in R^k$ is constant). It satisfies the energy inequality

$$\int_K (a\nabla\varphi, \nabla\varphi)\, dy \leq \int_K (ab, b)\, dy \tag{7}$$

which, combined with the Friedrichs' inequality, implies

$$\int_K \varphi^2 dy \leq cN^2 |b|^2 \int_K dy . \tag{8}$$

In the sequel c denotes all positive constants defined in terms of a_i, F, g.

LEMMA 1. For $\alpha \in [0, \frac{1}{2}(a_1/a_2)^2]$, $j = 1, \ldots k$

$$\int_{K_N} (y^{(j)}/N)^{-\alpha} (1 - y^{(j)}/N)^{-\alpha} |\nabla\varphi|^2 dy \leq c|b|^2 \int_{K_N} dy .$$

The proof of this lemma can be carried through by a method similar to that of [8].

LEMMA 2. The limit in (5) exists. The limiting matrix A satisfies (3) with the same constants.

PROOF. Let $N = ML + H$ ($M, L, H < M$ - integers), $K(n)$, $n = 1, \ldots L^k$ be the disjoint cubes of size M which form a partition of K_{ML}. Let $\tilde{\varphi} = 0$ in $K_N \setminus K_{ML}$; in $K(n)$ $\tilde{\varphi}$ is the solution of (6) with $K = K(n)$. Since φ minimizes the quadratic functional in $\overset{\circ}{W}{}_2^1(K_N)$

$$\xi_N = N^{-k} \int_{K_N} (a\{\nabla\varphi + b\}, \nabla\varphi + b)\, dy \leq$$
$$\leq \left(\frac{M}{N}\right)^k \sum_n M^{-k} \int_{K(n)} (a\{\nabla\tilde{\varphi} + b\}, \nabla\tilde{\varphi} + b)\, dy + c\,\frac{M}{N} . \tag{9}$$

The random summands in the right-hand side of (9) are distributed as ξ_M because of the homogeneity of a. ξ_N is quadratic in b. It is easily checked that ξ_N satisfies (3). A passage to the limit for $N \to \infty$ and then for $M \to \infty$ shows that

$$\limsup_{N\to\infty} E\xi_N \leq \liminf_{M\to\infty} E\xi_M$$

which implies the statement of the Lemma.

If $N = ML$, $M \to \infty$ and $\varphi, \tilde{\varphi}$ are defined as before

$$E N^{-k} \int_{K_N} (a\{\nabla\varphi - \nabla\tilde{\varphi}\}, \nabla\varphi - \nabla\tilde{\varphi})\, dy = E\xi_M - E\xi_N \to 0 \quad (10)$$

This is proved by a straightforward computation.

LEMMA 3. $\lim\limits_{N\to\infty} E(\xi_N - E\xi_N)^2 \to 0$.

PROOF. It follows from (10) that w.p. 1

$$(\xi_N - E\xi_N)^+ \leq \left(\frac{M}{N}\right)^k \left|\sum_n \eta(n)\right| + c\left(\frac{M}{N} + |E\xi_N - E\xi_M|\right),$$

where the RV's $\eta(n)$ are distributed as $\xi_M - E\xi_M$ and depend on values of a in the disjoint cubes $K(n)$ only. They are also bounded by a non-random constant in view of (7). Hence, the strong mixing condition leads to the estimate

$$E\left|\sum_n \eta(n)\right|^2 \leq c\left\{\left(\frac{N}{M}\right)^k + \beta(M)\left(\frac{N}{M}\right)^{2k}\right\}$$

where $M \uparrow \infty$ can be chosen so that $M = o(N)$. Combined with Lemma 2 it implies that

$$E(\xi_N - E\xi_N)^+ \to 0.$$

Since $E(\xi_N - E\xi_N)^- = E(\xi_N - E\xi_N)^+$, $|\xi_N| < c|b|^2$ this proves the lemma.

§ 2. PROOF OF THEOREM 1. Let disjoint cubes $C(n)$ of size εN form a partition of R^k. $C(n)$ is called inner if it lies inside $\mathcal{G}$ at a distance greater than $\varepsilon k N$ from $\partial\mathcal{G}$. Otherwise, $C(n)$ is referred to as a boundary one. Let $g(x)$ be the random vector field whose components $g^{(i)}(x) = 0$ in the boundary cubes, and are defined by the equality $g^{(i)}(x) = \varepsilon\varphi^{(i)}(x/\varepsilon)$ in the inner cubes $C(n)$, where $\varphi^{(i)}(y)$ is the solution of the "fast" Dirichlet problem (6) for $K = \varepsilon^{-1}C(n)$ and $b = e_i$ (the unit vector of the i-th coordinate axis).

Since by (8) w.p. 1

$$\| g^{(i)} \|_{L_2(\mathcal{G})} \leqslant c\,\varepsilon N \tag{11}$$

to prove Theorem 1 it suffices to obtain for the remainder term w_ε in the representation (cf. [1–4])

$$u_\varepsilon = U + (g, DU) + w_\varepsilon \tag{12}$$

an estimate of the form

$$\| w_\varepsilon \|_{W_2^1(\mathcal{G})} \leqslant \zeta_\varepsilon, \qquad E\zeta_\varepsilon^2 \to 0 \tag{13}$$

(in the sequel $D = \|\partial/\partial x^{(j)}\|$ denotes the column of "slow" derivatives with respect to x).

w_ε satisfies the obvious integral identity

$$\int_{\mathcal{G}} (aDw_\varepsilon, D\phi)\,dx = -\int_{\mathcal{G}} (qDU, D\phi)\,dx + \ldots \tag{14}$$

where the random matrix q is defined by

$$q(x) = a\{Dg^T + I\} - A = \{a(\nabla\varphi^T + I) - A\}\Big|_{y = x/\varepsilon} \quad (15)$$

in the inner cubes, $q(x) = -A$ elsewhere and $\phi \in \dot{C}^\infty(\mathcal{G})$ is arbitrary.

Because of (3) it remains to check that the right-hand side of (14) (RHS) admits the bound

$$|RHS| \leq \zeta_\varepsilon \, |\phi|_1 \quad \text{w.p. } 1, \quad |\phi|_1 = \|\phi\|_{W_2^1(\mathcal{G})} . \quad (16)$$

The more cumbersome part of this checking deals with the first summand in RHS of (14) with integration restricted to the union of inner cubes $\mathcal{G}_\varepsilon$.

Let $\theta = \Theta_{M,N,\varepsilon}(x)$, $0 \leq \theta \leq 1$ be a smooth function vanishing outside $\mathcal{G}_\varepsilon$ and in the $2\varepsilon M$ -vicinity of all hyperplanes dissecting $\mathcal{G}_\varepsilon$ into cubes $C(n)$. $\theta(x) = 1$ in $\mathcal{G}_\varepsilon$ at a distance greater than $3\varepsilon M$ from these hyperplanes; its derivatives of order m are bounded by $c(\varepsilon M)^{-m}$ uniformly in x .

Since U is smooth and q is constructed from the gradients of solutions of (6) in $\varepsilon^{-1}C(n)$, Lemma 1 permits to substitute q by θq . An integration by parts then yields (w.p. 1)

$$\Big| \int_{\mathcal{G}_\varepsilon} (qDU, D\phi)\,dx - \int_{\mathcal{G}_\varepsilon} \phi \cdot tr\{ q\, D(\theta D^T U)\}\,dx \Big| \leq c\left(\frac{M}{N}\right)^{\alpha} |\phi|_1 \quad (17)$$

where $tr(\cdot)$ is the trace of a matrix and α is that of Lemma 1.

Let $\tau * \rho(x)$ be the convolution of ρ with a smooth kernel $\tau(x)$ supported by a ball of size εL ($L = o(M)$ in the sequel) such that $|D\tau| \leq c(\varepsilon L)^{-k-1}$

Using bounds (7) on q and a routine technique of spatial averaging it is easy to show that w.p. 1

$$|\int_{\mathcal{G}_\varepsilon} \phi \cdot tr\,(q D(\theta D^T U))\,dx - \int_{\mathcal{G}_\varepsilon} \phi \cdot tr\,\{(q * \tau) D(\theta D^T U)\}\,dx| \leq \tag{18}$$
$$\leq c|\phi|_1\ \varepsilon L(\varepsilon M)^{-2}.$$

Besides, $D^T(q * \tau) = 0$ in $\mathcal{G}_\varepsilon \cap \{\theta \neq 0\}$.
Hence another integration by parts and a change in the order of integration produces

$$\int_{\mathcal{G}_\varepsilon} (q DU, D\phi)\,dx = \int_{\mathcal{G}_\varepsilon} (\tau * q \cdot \theta DU,\ D\phi)\,dx + R =$$
$$= \int_{\mathcal{G}_\varepsilon} tr\,\{q \cdot \tau * (\theta DU (D\phi)^T)\}\,dx + R, \tag{19}$$

where $R \leq c\{(M/N)^\alpha + L/(\varepsilon M^2)\}\,|\phi|_1,\ \alpha > 0.$

Let now $N = HP$ (H, P - integers). Subdividing each of the cubes $C(n)$ into cubes $C(n,m)$ of size εH and constructing $\tilde{q}$ from solutions of (6) in $C(n,m)$ in the same way as q was "sewn" from $\overset{\circ}{W}{}^1_2(C(n))$ "rags", it is easy to see that (10) implies w.p. 1

$$|\int_{\mathcal{G}_\varepsilon} tr\,(q\,\psi)\,dx - \int_{\mathcal{G}_\varepsilon} tr\,(\tilde{q}\,\psi)\,dx| \leq c\,\zeta_1\,|\phi|_1 \tag{20}$$

where $\psi = \tau * [\theta DU\,(D\phi)^T]$ and

$$E\zeta_1^2 = \int_{\mathcal{G}_\varepsilon} tr\,(q - \tilde{q})^2\,dx \leq f(H),\ \lim_{H\to\infty} f(H) = 0.$$

The smoothness of the kernel τ and (7) ensure estimates

$|D\psi_{\ell m}|_{L_2(\mathcal{G}_\varepsilon)} \leq c(\varepsilon L)^{-1}|\phi|_1$. Hence, if piecewise-constant $\hat{q}(x)$ is defined by

$$\hat{q}(x) = \int_{C(n,m)} \tilde{q}\, dx \Big/ \int_{C(n,m)} dx \,, \qquad x \in C(n,m)$$

in each of the sub-cubes $C(n,m)$ whose union is $\mathcal{G}_\varepsilon$, a routine computation leads to

$$\left|\int_{\mathcal{G}_\varepsilon} tr(\tilde{q}\,\psi)dx - \int_{\mathcal{G}_\varepsilon} tr(\hat{q}\,\psi)\,dx\right| \leq c\varepsilon H(\varepsilon L)^{-1}|\phi|_1 \,. \tag{21}$$

The averages $\hat{q}$ in (21) are precisely those of (5) and Lemma 2 (up to the constants $-A_{\ell m}$). To see this one should use the integral identity equivalent to (6). Hence, uniformly in x, ℓ, m

$$E\,\hat{q}_{\ell m}(x)^2 \leq f_1(H) \,, \lim_{H\to\infty} f_1(H) = 0$$

and

$$\left|\int_{\mathcal{G}_\varepsilon} tr(\hat{q}\,\psi)dx\right| \leq \zeta_2 \left|tr(\psi\psi^T)\right|^{1/2}_{L_1(\mathcal{G}_\varepsilon)} \leq c\zeta_2|\phi|_1 \tag{22}$$

with

$$\zeta_2^2 = \int_{\mathcal{G}_\varepsilon} tr(\hat{q}\hat{q}^T)\,dx \,, \quad E\zeta_2^2 \leq c\,f_1(H) \to 0$$

A combination of (17 - 22) yields the final estimate

$$\left|\int_{\mathcal{G}_\varepsilon} (qDU, D\phi)\,dx\right| \leq \zeta_3|\phi|_1 \,, \quad E\zeta_3^2 \to 0 \tag{23}$$

provided integers $N, M, \ldots$ satisfy relations $\varepsilon N \to 0$, $L/(\varepsilon M^2) \to 0$, $M/N \to 0$, $H/L \to 0$ (these restrictions hold, e.g. for $N \sim \varepsilon^{-1+\delta}$, $M \sim \varepsilon^{-1+2\delta}$, $L \sim \varepsilon^{-1+5\delta}$, $H \sim \varepsilon^{-1+6\delta}$ with $0 < \delta < 1/6$).

Evaluations of the remaining part of RHS in (14) use (11) and the fact that

$$\int_{\mathcal{G} \setminus \mathcal{G}_\varepsilon} dx \leq C\varepsilon N.$$

They are omitted because of their routine character.

REFERENCES

1. Bahvalov N.S. Average Characteristics of Bodies with Periodic Structure, Doklady Akad. Nauk SSSR, (1974), 218, 5, 1046-1048.
2. Berdicevskii V.L. Spatial Averaging of Periodic Structures, Doklady Akad. Nauk SSSR (1975), 222, 3, 565-567.
3. Bensoussan A., Lions J.-L., Papanicolaou G. Sur quelques phénomènes asymptotiques stationnaires, CR Acad.Sci.Paris, Ser. A. (1975), 281, № 2-3, 89-94.
4. Freidlin M.I. On a diffusion process with a small parameter, Proc. of the Third USSR-Japan Symposium on Probability Theory, (1975), Tashkent, "Fan", 179-181.
5. Kozlov S.M. Averaging of Differential Operators with Almost-Periodical Highly Oscillating Coefficients, Doklady Akad. Nauk SSSR (1977), 236, 5, 1068-1071.
6. Kozlov S.M. Averaging of Random Structures, Doklady Akad.Nauk SSSR (1978), 241, 5, 1016-1019.
7. Ladyzhenskaya O.A., Uraltseva N.N., Linear and Quasilinear Equations of Elliptic Type, (1973), Moscow, "Nauka".
8. Mikhailov V.P. On a Dirichlet Problem for Elliptic Equation of Second Order, Diff. Uravnenia (1976), 12, 10, 1877-1891.

STOCHASTIC SPECTRAL EQUATIONS

V.L.Girko

The necessity of the study of linear stochastic operator spectrum has arisen for the first time while solving of some many-dimensional statistic analysis problems (1,2), in the study of the heavy nuclei energy levels (7 - 10) and in the investigation of some disorder structures (6, 8). The heavy nuclui and crystals are of very complicated structure therefore the statistical approach seems to be expedient to the solution of some nuclear physics and solid state physics problems just as in the solving if many statistical physics problems (5 - 11). The creation of random matrix spectral theory would allow to solve certain fundamental physics problems (7). The stochastic linear operators are now widely used not only in physics and many-dimensional stochastic analysis but also in the theory of stability of stochastic differential eguation solutions, in the theory of control of linear stochastic systems, in the numerical analysis, in circular accelerator and waneguide theories and so on.

§ 1. Distribution of the proper values and proper vectors of Hermitian random matrices

Let $H_n = (\xi_{ij})$ be the Hermitian matrix of the n -th order which elements are the complex random variables and X_n the nonrandom Hermitian n -th order matrix. We suppose the real and imaginary parts of the matrix H_n elements which are on the main diagonal and above it have the common distribution density which will be denoted by $p(X_n)$ (the function $p(X_n)$ depends on the real and imaginary parts of the maxtix X_n elements). The pfoper values λ_i and vectors $\vec{\theta_i}$ of matrix H_n are the solutions of the equation's system with random coefficients $(H_n - \lambda_i \bar{I})\vec{\theta_i} = 0$.

. We order the proper values of matrix H_n in the nondecreasing sequence $\lambda_1 \geqslant \lambda_2 \geqslant \cdots \geqslant \lambda_n$ and choose the proper vectors $\vec{\theta_i}$, $i=\overline{1,n}$ so that their first components are equal to certain constants c_i, $0 \leqslant c_i \leqslant 2\pi$. It is obvious that proper values and vectors such chosen will be single-valued determined with probability 1 . Denote by Θ_n the matrix which columns are vectors $\vec{\theta_i}$, $i=\overline{1,n}$.

Let Γ be the group pf the n -th order unitary matrices, ν - the normalized Haar's measure on it, B - σ - algebra of Borel sets of the group Γ.

Theorem 1. If the density $p(X_n)$ of the random Hermitian matrix H_n exists then for any subset $E \subset B$ and real numbers α_i, β_i, $i=\overline{1,n}$

$$P\{\Theta_n \in E, \alpha_i < \lambda_i < \beta_i, i=\overline{1,n}\} = c_n \int p(U_n Y_n U_n^*) \prod_{i=1}^{n} \delta(\arg u_{1i} - c_i) \times \prod_{i>j} (y_i - y_j)^2 \nu(dU_n)\, dY_n,$$

where the integration is over the region $y_1 > y_2 > \dots > y_n$, $\alpha_i < y_i < \beta_i$, $i=\overline{1,n}$, $U_n \in E$, $\delta(x)$ - delta-function,

$$c_n = \left[\pi^{-n^2+n/2} \prod_{j=0}^{n-1} j! \int \prod_{i=1}^{n} \delta(\arg u_{1i} - c_i)\, \nu(dU_n) \right]^{-1}.$$

Proof. When $y_1 > y_2 > \ldots > y_n$ and $\arg u_{1i} = c_i$, the transformation $X_n = U_n Y_n U_n^*$ is mutually single-values since one can easily verify that the numbers of the independent parameters on the left and right parts of the equality $X_n = U_n Y_n U_n^*$ are equal. The Eulerian angles of the matrix U_n are chosen as its parameters. It follows from [9] that the Iacobian of transformation $X_n = U_n Y_n U_n^*$ is equal to $J(U_n, Y_n) = \prod_{p>l} (y_p - y_l)^2 \varphi(U_n)$, where $\varphi(U_n)$ is some Borel function of matrix U_n parameters. For any bounded and continuous function $f(\Theta_n, \Lambda_n)$ of matrices Θ_n and Λ_n elements one has $M f(\Theta_n, \Lambda_n) = \int f(U_n, Y_n) p(X_n) dX_n$ where intergation is over the region $\{\arg u_{1i} = c_i, i = \overline{1,n}, y_1 > \ldots > y_n\}$ and matrices Y_n and U_n satisfy to the equation $X_n = U_n Y_n U_n^*$. Since the Iacobian of transformation $X_n = H_n Z_n H_n^*$, where $H_n \in \Gamma$, Z_n - Hermitian matrix, is equal to 1, one has

$$M f(\Theta_n, \Lambda_n) = \int f(H_n U_n, Y_n) p(H_n X_n H_n^*) dX_n, \qquad (1.1)$$

where the region integration of is $\{\arg \sum_{k=1}^{n} h_{1k} u_{ki} = c_i, \quad i = \overline{1,n}, y_1 > \ldots > y_n\}$. After the replacement of variables $X_n = U_n Y_n U_n^*$ if follows from (1.1)

$$M f(\Theta_n, \Lambda_n) = \int f(H_n U_n, Y_n) p(H_n U_n Y_n U_n^* H_n^*) \prod_{i=1}^{n} \delta(\arg \sum_{k=1}^{n} h_{1k} u_{ki} - c_i) \prod_{p>l} (y_p - y_l)^2 \varphi(U_n) \, dY_n \prod_{i=1}^{l} du_i, \qquad (1.2)$$

where the region of integration is $y_1 > \ldots > y_n$ the matrix U_n parameters chahge in some ragion

The expression on the right part of (1.2) does not depend on matrix H_n. Therefore we can integrate both parts of

(1.2) over Haar's measure ν determined on the group of matrices Γ. Then

$$M f(\Theta_n, \Lambda_n) = c_n \int f(H_n, Y_n) p(H_n Y_n H_n^*) \prod_{i=1}^{n} \delta(\arg h_{1i} - c_i) \times$$
$$\times \prod_{p>l} (y_p - y_l)^2 dY_n \nu(dH_n).$$

Determine the constant c_n. To do this we put in the last fotmula $f \equiv 1$, $p(X_n) = 2^{-n/2} \pi^{-n^2} \exp(-\frac{1}{2} Sp X_n X_n^*)$. Then [9]

$$1 = c_n \int \prod_{i=1}^{n} \delta(\arg h_{1i} - c_i) \nu(dH) \pi^{-n^2+n/2} (n!)^{-1} [\Gamma(2)]^{-n} \prod_{j=1}^{n} \Gamma(1+j).$$

Theorem 1 is proved.

Similar assertions can be obtained for the symmetrical, antisymmetrical, complex, nonsymmetrical, orthogonal and unitary random matrices.

Theorem 1 has namy applications in the theory of random matrices but when n is large the calculations with it become cumbersome. In the next section we shall consider another appoach to the determination of random matrices proper values and vectors distribution.

§ 2. Distribution of proper values and vectors of matrix additive random pfocesses

Let $\Xi_n(t) = (\xi_{ij}(t))$ be the random process with the independent additive increments, the matrix $\Xi_n(t)$ is Hermitian for any t. The increments $\Xi_n(t) - \Xi_n(s)$ are normally multi-dimensionaly distributed with zero vector of mean values and covariation matrix $M(\xi_{ij}(t) - \xi_{ij}(s))(\xi_{pl}(t) - \xi_{pl}(s)) = r_{ijpl} |t-s|$, $A = (\delta_{ij} a_i)$ - diagonal matrix of the n -th order, $a_i \neq a_j$, $i \neq j$.

The proper numbers $\lambda_i(t)$ and proper vectors $\vec{\varphi}_i(t)$ of matrix $A+\Xi(t)$ are functions of t. But we can not choose them here in the same way as the previous section because we have to do the ordering of the proper numbers for any t and therefore there on no equation ofr $\lambda_i(t)$. We determine the proper values $\lambda_i(t)$ and vectors $\vec{\varphi}_i(t)$ of matrix $A+\Xi(t)$ using the formulas of matrix perturbation

$$\lambda_j(t)=\sum_{m=0}^{\infty}\lambda_j^{(m)}(s)\,(t-s)^{m/2}, \quad \vec{\varphi}_j(t)=\sum_{m=0}^{\infty}\vec{\varphi}_j^{(m)}(s)\,(t-s)^{m/2}, \quad \lambda_j^{(0)}(0)=a_i,$$
$$\vec{\varphi}_j^{(0)}(0)=\vec{e}_j, \quad t\geqslant s, \tag{2.1}$$

where $\vec{e}_j$ - the unit vectors of the n -th order (see [12]). Note that choosing $t-s$ sufficiently small one can secure the convergence of there series with probability 1. Obvisously proper values and vector such choosen will be random numbers and random vectors respectively. Using the perturbation formulas (2.1) and also the next inequality (see [12]),

$$|\lambda_j^{(m)}(s)|\leqslant \frac{1}{m}\,\alpha^m(\Delta t)\,(2/\varepsilon)^{m-1}, \tag{2.2}$$

with the condition $|\lambda_i(s)-\lambda_j(s)|\geqslant\varepsilon,\ i\neq j$, where $\alpha(\Delta t)=\|\Xi(t)-\Xi(s)\|$. It follows from the Theorem 1 that $M|\lambda_i(t)-\lambda_i(s)|^p\leqslant L\,|t-s|^{1+\tau}$, $p>0$, $\tau>0$, $L>0$ - the certain constants. Thus the next assertion is true.

<u>Lemma 1</u>. The process $\{\lambda_i(t),\ i=\overline{1,n}\}$ is continuous with the probability 1 on any finite segment. $[0,T]$

By the inequalities for martingales, formulas (2.1) and (2.2) we obtain with certain calculations the next assertion.

Lemma 2. The continuous random process $\{ \lambda_i(t),\ i=\overline{1,n} \}$ has for any real $T>0$ the following property

$$P\{ \inf_{t\in[0,T]} |\lambda_i(t)-\lambda_j(t)|>0,\ i\neq j,\ i,j=\overline{1,n} \} = 1 . \tag{2.3}$$

For the first two terms of the perturbation formulas (2.2) we have the followihg expression [12] $\lambda_K^{(1)}(s) = (B(t,s)\vec{\varphi}_K(s), \vec{\varphi}_K(s))$, $\lambda_K^{(2)}(s) = \sum_{m\neq K} (B(t,s)\vec{\varphi}_K(s), \vec{\varphi}_m(s))^2 (\lambda_K(s)-\lambda_m(s))^{-1}$, $\varphi_K^{(1)}(s) = \sum_{m\neq K} \vec{\varphi}_m(s) \times$ $\times (B(t,s)\vec{\varphi}_K(s), \vec{\varphi}_m(s))(\lambda_K(s)-\lambda_m(s))^{-1}$, $\varphi_K^{(2)}(s) = \sum_{m,p\neq K} \frac{(B(t,s)\vec{\varphi}_m(s), \vec{\varphi}_p(s))}{\lambda_K(s)-\lambda_p(s)} \times$ $\times \frac{(B(t,s)\vec{\varphi}_p(s), \vec{\varphi}_K(s))}{\lambda_K(s)-\lambda_m(s)} \vec{\varphi}_m(s) - \sum_{m\neq K} \frac{(B(t,s)\vec{\varphi}_K(s), \vec{\varphi}_K(s))(B(t,s))\vec{\varphi}_m(s), \vec{\varphi}_K(s))}{(\lambda_K(s)-\lambda_m(s))^2} \vec{\varphi}_m(s) -$ $- \frac{1}{2}\vec{\varphi}_K(s) \sum_{m\neq K} \frac{(B(t,s)\vec{\varphi}_K(s), \vec{\varphi}_m(s))}{(\lambda_K(s)-\lambda_m(s))^2}$, $t \geqslant s$, where $B(t,s) = \frac{\Xi(t)-\Xi(s)}{\sqrt{t-s}}$.

Using these formulas and also Lemmas 1 and 2 we obtain the following assertion.

<u>Theorem 2.</u> The proper values $\lambda_K(t)$ and functions $\vec{\varphi}_K(t)$ defined by formulas (2.1) satisfy the system of stochactic spectral differential equations with the drift vector

$$\{ \alpha_{\lambda_K(t)},\ K=\overline{1,n},\ \alpha_{\varphi_{ij}(t)},\ i,j=\overline{1,n} \} = \Big\{ \sum_{m\neq K} \frac{B(\varphi_K,\varphi_m,\varphi_K,\varphi_m)}{\lambda_K(t)-\lambda_m(t)},\ K=\overline{1,n},$$

$$\sum_{m,p\neq i} \frac{B(\varphi_m,\varphi_p,\varphi_p,\varphi_i)}{(\lambda_i(s)-\lambda_p(s))(\lambda_i(s)-\lambda_m(s))} \varphi_{mj}(s) - \sum_{m\neq i} \frac{B(\varphi_i,\varphi_i,\varphi_m,\varphi_i)}{(\lambda_i(s)-\lambda_m(s))^2} \varphi_{mj}(s) -$$

$$- \frac{1}{2}\varphi_{ij}(s) \sum_{m\neq i} B(\varphi_i,\varphi_m,\varphi_i,\varphi_m)(\lambda_i(s)-\lambda_m(s))^{-2},\ i,j=\overline{1,n} \Big\}.$$

and the diffusion matrix

$$b_{\lambda_K,\lambda_m} = B(\varphi_K,\varphi_K,\varphi_m,\varphi_m),\quad K,m=\overline{1,n},$$

$$b_{\lambda_K,\varphi_{ei}} = \sum_{m\neq K} \frac{B(\varphi_K,\varphi_K,\varphi_e,\varphi_m)}{\lambda_e(s)-\lambda_m(s)} \varphi_{mi},\quad K,e,i=\overline{1,n},$$

$$b_{\varphi_{Ki},\varphi_{ej}} = \sum_{m_1 \neq K,\ m_2 \neq e} \frac{B(\varphi_K, \varphi_{m_1}, \varphi_e, \varphi_{m_2})}{(\lambda_K(s) - \lambda_{m_1}(s))(\lambda_e(s) - \lambda_{m_2}(s))} \varphi_{m_1 i} \varphi_{m_2 j},$$

$K, e, i, j = \overline{1,n},$

where $B(\varphi_K, \varphi_p, \varphi_s, \varphi_q) = \sum_{i,j,m,e} \tau_{ijme} \varphi_{iK} \varphi_{jp} \varphi_{ms} \varphi_{eq}.$

Simular spectral stochastic equations can be found for other kinds of additive and multiplicative random processes with the independent increments.

§ 3. Canonical spectral equation

We consider in this section the limit theorems for the normalised spectral functions of symmetrical random matrices when their ofder increaces to the infinity.

Theorem 3. Let for any value n the random elements $\xi_{ij}^{(n)}, i \geq j, i,j = \overline{1,n}$ of symmetrical matrix $\Xi_n = (\xi_{ij}^{(n)})$ are independent, infinitesimal and all given on one probability space, $M\xi_{ij}^{(n)} = 0$, $K_n(u,v,z) \Rightarrow K(u,v,z)$, where

$$K_n(u,v,z) = n \int_{-\infty}^{z} y^2 \, dP\{\xi_{ij}^{(n)} < y\}, \quad \frac{i}{n} \leq u < \frac{i+1}{n}, \quad \frac{j}{n} \leq v < \frac{j+1}{n},$$

the function $K(u,v,z)$ is nondecreasing has limited variation over z and is continuous in u and v in the region $0 \leq u \leq 1$, $0 \leq v \leq 1$.

Then with the probability 1 $\lim_{n \to \infty} \mu_n(x) = \mu(x)$. at any point of continuuety of the nonrandom spectral function $\mu(x)$

with the Stiltjes transformation

$$\int (1+itx)^{-1} d\mu(x) = \int_0^1 \int_0^1 x \, d G(x,z,t) dz, \quad \mu_n(x) = \frac{1}{n} \sum_{\lambda_i < x} 1,$$

$G(x,z,t)$ is the distribution function over x $(0 \le x \le 1, 0 \le z \le 1, -\infty < t < \infty)$, which satisfibs at the continuuty points to the canonical spectral equation

$$G(x,z,t) = P\{[1 + t^2 \xi(G(\cdot,\cdot,t),z)]^{-1} < x\}, \tag{3.1}$$

$\xi(G(\cdot,\cdot,t),z)$ - random functional which has the next Laplace transformation of one-dimensional distributions

$$M \exp\{-s\xi(G(\cdot,\cdot,t),z)\} = \exp\{ \int_0^1 \int_0^1 [\int_0^\infty (e^{-syx^2} - 1) x^{-2} \times$$

$$\times d K(v,z,x) d G(y,v,t) dv\}, \quad s \geq 0.$$

The solution of equation (3.1) exists and is unique in the class of function $G(x,z,t)$ which are the distribution function in x $(0 \le x \le 1)$ at any fixed $0 \le z \le 1$, $-\infty < t < \infty$ such that the functions $\int_0^1 x^k d G(x,z,t)$, $k = 1, 2, \ldots$ are analitical in t.

Pfoof. Denote $R_t = (I + it \Xi_n)^{-1} = (r_{pe}(t))$. According to [3] one with the pfobability 1

$$\lim_{n\to\infty} \frac{1}{n} [Sp R_t - M Sp R_t] = 0,$$

$$r_{kk} = [1 + it\xi_{kk}^{(n)} + t^2 (R_t^k \vec{\xi}_k, \vec{\xi}_k)]^{-1},$$

where $\vec{\xi}_k = (\xi_{1k}, \ldots, \xi_{k-1k}, 0, \xi_{k+1k}, \ldots, \xi_{nk})$, $R_t^k = (I + it\Xi_n^k)^{-1} = (r_{ij}^k)$, the matrix Ξ_n^K is obtained from the matirx Ξ_n by the changing of elements ξ_{pK}, $p=\overline{1,n}$ by zeros. Thus for proving the Theorem one needs find the limit

$$\lim_{n\to\infty} \frac{1}{n} M\, Sp R_t .$$

It is obvious

$$r_{KK} = \left[1 + t^2 \sum_{p=1}^{n} r_{pp}^K \xi_{pK}^2\right]^{-1} + \varepsilon_n, \quad \underset{n\to\infty}{plim}\, \varepsilon_n = 0. \qquad (3.2)$$

At the condition that random values r_{pp}^K are fixed we obtain by using limit theorems for the sums of independent infinitesimal random variables

$$M\exp\left\{-s \sum_{p=1}^{n} r_{pp}^K \xi_{pK}^2\right\} = M\exp\left\{\sum_{p=1}^{n} \alpha_p^{(n)}\right\} + o(1),$$

where

$$\alpha_p^{(n)} = M\left[\exp\left(-s\, r_{pp}^K \xi_{pK}^2\right) - 1 / r_{pp}^K\right].$$

Just as in [3] we obtain

$$\lim_{n\to\infty} \left[M\exp\left\{\sum_{p=1}^{n} \alpha_p^{(n)}\right\} - \exp\left\{M \sum_{p=1}^{n} \alpha_p^{(n)}\right\}\right] = 0.$$

It follows from here taking into account (3.2) that (3.1) is valid. The proof of uniqueness of its solution is obvious. The Theorem is pfoved.

L I T E R A T U R E

I. Андерсон Т., Введение в многомерный статистический анализ, М., Физматгиз, I963.

2. Уилкс С., Математическая статистика, М., Наука, I967.

3. Гирко В.Л., Случайные матрицы, Киев, "Вища школа", I975.

4. Гирко В.Л., Предельные теоремы общего вида для спектральных функций случайных матриц, Теория вероятностей и ее применение, I977, т.22, вып.I, I60 - I63.

5. Марадудун А., Монтролл Э., Вейсс Д., Динамическая теория кристаллической решетки в гармоническом приближении, М., Мир, I965.

6. Лифшиц И.М., О структуре энергетического спектра и квантовых состояниях неупорядоченных конденсированных систем, Успехи физических наук, I964, 83, 6I7 - 655.

7. Wigner E.P., Random matrices in physics, SIAM Rev., 1967, 9, N° 1, 1 - 23.

8. Dyson F.J. The dynamics of a disozdered linear chain, Phys. Rev. 1953, 6, 92, 1331 - 1338.

9. Mehta M.J. Random matrices and the statistical theory of energy levels. New York, Acad. Press, 1968.

10. Porter C.E., Statistical theories of spectra: Fluctuations. New York - London, Acad. Press, 1965.

11. Dyson F.J. A Brownian - motion model for the eigenvalues of a random matrix, J;Math.Phys, 1962, 3, 6.

12. Kato T., Perturbation theory for lineaf operators, Springer - Verlag Berlin. Heidelberg - New York, 1966.

Lecture Notes in Control and Information Sciences

Edited by A. V. Balakrishnan and M. Thoma

Vol. 22: Optimization Techniques
Proceedings of the 9th IFIP Conference on
Optimization Techniques,
Warsaw, September 4–8, 1979
Part 1
Edited by K. Iracki, K. Malanowski, S. Walukiewicz
XVI, 569 pages. 1980

Vol. 23: Optimization Techniques
Proceedings of the 9th IFIP Conference on
Optimization Techniques,
Warsaw, September 4-8, 1979
Part 2
Edited by K. Iracki, K. Malanowski, S. Walukiewicz
XV, 621 pages. 1980

Vol. 24: Methods and Applications
in Adaptive Control
Proceedings of an International Symposium
Bochum, 1980
Edited by H. Unbehauen
VI, 309 pages. 1980

Vol. 25: Stochastic Differential Systems –
Filtering and Control
Proceedings of the IFIP-WG7/1 Working Conference
Vilnius, Lithuania, USSR, Aug. 28 – Sept. 2, 1978
Edited by B. Grigelionis
X, 362 pages. 1980

Vol. 26: D. L. Iglehart, G. S. Shedler
Regenerative Simulation of Response
Times in Networks of Queues
XII, 204 pages. 1980

Vol. 27: D. H. Jacobson, D. H. Martin, M. Pachter, T. Geveci
Extensions of Linear-Quadratic Control Theory
XI, 288 pages. 1980

Vol. 28: Analysis and Optimization of Systems
Proceedings of the Fourth International
Conference on Analysis and Optimization of Systems
Versailles, December 16–19, 1980
Edited by A. Bensoussan and J. L. Lions
XIV, 999 pages. 1980